J Elphick

Optimal Control Theory

International Series in
Management Science/Operations Research

Suresh P. Sethi
University of Toronto

Gerald L. Thompson
Carnegie-Mellon University

Optimal Control Theory

Applications to Management Science

Martinus Nijhoff Publishing
Boston/The Hague/London

DISTRIBUTORS FOR NORTH AMERICA:
Martinus Nijhoff Publishing
Kluwer Boston, Inc.
190 Old Derby Street
Hingham, Massachusetts 02043, U.S.A.

DISTRIBUTORS OUTSIDE NORTH AMERICA:
Kluwer Academic Publishers Group
Distribution Centre
P.O. Box 322
3300 AH Dordrecht, The Netherlands

Library of Congress Cataloging in Publication Data

Sethi, Suresh P.
Optimal control theory.

(International series in management science/
operations research; v. 1)
Bibliography: p.
Includes index.
1. Management–Mathematical models.
2. Control theory. 3. Operations research.
I. Thompson, Gerald Luther, 1923- joint author.
II. Title. III. Title: Management science.
IV. Series.
HD30.25.S47 658.4'033 80-28980

ISBN 0-89838-061-8

Printed in the United States of America

To the memory of my mother and to Billie
–Suresh P. Sethi

To my father
–Gerald L. Thompson

CONTENTS

PREFACE

The purpose of this book is to exposit, as simply as possible, some recent results obtained by a number of researchers in the application of optimal control theory to management science. We believe that these results are very important and deserve to be widely known by management scientists, mathematicians, engineers, economists, and others. Because the mathematical background required to use this book is two or three semesters of calculus plus some differential equations and linear algebra, the book can easily be used to teach a course in the junior or senior undergraduate years or in the early years of graduate work. For this purpose, we have included numerous worked-out examples in the text, as well as a fairly large number of exercises at the end of each chapter. Answers to selected exercises are included in the back of the book. A solutions manual containing completely worked-out solutions to all of the 205 exercises is also available to instructors.

The emphasis of the book is not on mathematical rigor, but on modeling realistic situations faced in business and management. For that reason, we have given in Chapters 2 and 7 only heuristic proofs of the continuous and discrete maximum principles, respectively. In Chapter 3 we have summarized, as succinctly as we can, the most important model types and terminal conditions that have been used to model management problems. We found it convenient to put a summary of almost all the important management science models on two pages; see Tables 3.1 and 3.3.

One of the fascinating features of optimal control theory is the extraordinarily wide range of its possible applications. We have tried to cover a wide variety of applications as follows: Chapter 4 covers finance; Chapter 5 considers production and inventory; Chapter 6 covers marketing; Chapter 8 treats machine maintenance and replacement; Chapter 9 deals with problems of optimal consumption of natural resources (renewable or exhaustible); and Chapter 10 discusses several economic applications.

In Chapter 11 we treat some computational algorithms for solving optimal control problems. This is a very large and important area that needs more development.

Chapter 12 treats several more advanced topics of optimal control: differential games, distributed parameter systems, optimal filtering, stochastic optimal control, and impulsive control. We believe that some of these models are capable of wider applications and further theoretical development.

Finally, four appendixes cover either elementary material, such as differential equations, or advanced material, whose inclusion in the main text would spoil its continuity. Also at the end of the book is a bibliography of works actually cited in the text. While it is extensive, it is by no means an exhaustive bibliography of management science applications of optimal control theory. Several surveys of such applications, which contain many other important references, are cited.

We have benefited greatly during the writing of this book by having discussions with and obtaining suggestions from various colleagues and students. Our special thanks to go Gustav Feichtinger for his careful reading and suggestions for improvement of the entire book. Carl Norstrom contributed two examples to Chapters 4 and 5 and made many suggestions for improvement. Jim Bookbinder used the manuscript for a course at the University of Toronto, and Tom Morton suggested some improvements for Chapter 5. The book has also benefited greatly from various coauthors with whom we have done research over the years. Both of us also have received numerous suggestions for improvements from the students in our applied control theory courses taught during the past several years. We would like to express our gratitude to all these people for their help.

The book has gone through several drafts, and we are greatly indebted to Eleanor Balocik and Rosilita Jones for their patience and careful typing.

Although the applications of optimal control theory to management science are recent and many fascinating applications have already been made, we believe that much remains to be done. We hope that this book will contribute to the popularity of the area and will enhance future developments.

1 INTRODUCTION

Many management science applications involve the control of dynamic systems, i.e., systems that evolve over time. They are called continuous time systems or discrete time systems depending on whether the time variable t varies continuously or discretely. We shall deal with both kinds of systems in this book, although the main emphasis will be on continuous time systems.

Optimal control theory is a relatively new branch of mathematics developed to find the optimal way to control a dynamic system. The purpose of this book is to give an elementary introduction to the mathematical theory and then to apply it to a wide variety of different situations arising in management science applications. We have deliberately kept the level of mathematics as elementary as possible in order to make the book accessible to a large audience. The only mathematical requirements for this book are calculus, including partial differentiation, some knowledge of vectors and matrices, and elementary differential equations.

The principle management science applications discussed in this book are to the following areas: finance; economics; production and inventory; marketing; maintenance and replacement; and consumption of natural resources. Other applications are also taken up in Chapter 10 and in various exercises. In each major area we have exposited one or more simple models followed by a more complicated model. The reader may wish at first reading to cover only the simpler models in each area to get an overall idea of what it is possible to do with optimal control theory. Later the reader may wish to go in more depth into one or more of the applied areas.

1.1 What is Optimal Control Theory?

We shall use the word system as a primitive term in this book. The only property that we require of a system is that it is capable of existing in various states. Let the (real) variable $x(t)$ be the state variable of the system at time t. For example, $x(t)$ could measure the inventory level at time t, the amount of advertising goodwill at time t, the amount of unconsumed wealth or natural resources at time t, etc.

We assume that there is a way of controlling the state of the system. Let the (real) variable $u(t)$ be the control variable of the system at time t. For example, $u(t)$ could be the production rate at time t, the advertising rate at time t, the consumption rate at time t, etc.

Given the values of the state variable $x(t)$ and the control variable $u(t)$, the state equation

$$\dot{x} = f(x,u,t), \quad x(0) = x_o \tag{1.1}$$

determines the instantaneous rate of change in the state variable, where f is a given function of x, u, and t, and x_o is the initial value of x. If we know the initial value x_o and the control trajectory, that is, the values of $u(t)$ over the whole time interval $0 \leq t \leq T$, then we can integrate (1.1) to get the state trajectory, that is, the values of $x(t)$ over the same time interval. We want to choose the control trajectory so that the state and control trajectories maximize the objective function

$$J = \int_0^T F(x,u,t)dt + S[x(T)]. \tag{1.2}$$

In (1.2), F is a given function of x, u, and t which could measure the negative of the cost of production, the benefits minus the costs of advertising, the utility of consumption, etc. Also in (1.2) the function S gives the "salvage value" of the ending state $x(T)$. The salvage value is needed so that the solution will make "good sense" at the end of the problem.

Usually the control variable $u(t)$ will be constrained. We indicate this as

$$u(t) \in \Omega(t) \tag{1.3}$$

where $\Omega(t)$ is the set of possible values of the control variable at time t.

Finally, we note that (1.1) and (1.3), when taken together limit the set of possible terminal values $x(T)$. We denote this by saying

$$x(T) \in X(T) \tag{1.4}$$

where $X(T)$ is called the <u>reachable set</u> of the state variable. Here $X(T)$ is the set of all possible terminal values that can be reached when $x(t)$ obeys (1.1) and the control variable obeys (1.3).

Although the above description of the control problem may seem complicated, you will find that in each specific application the various variables and parameters will have reasonable interpretations which make them easy to understand and remember. The examples to be discussed next will illustrate this point.

1.2 Formulation of Simple Control Models

We now formulate three simple models chosen from the areas of production, advertising, and economics. Our only objective here is to identify and interpret for these models each of the variables and functions described in the previous section. The solutions for these models will be given in detail in later chapters.

Example 1.1. A Production-Inventory Model. The various quantities that define this model are summarized in Table 1.1 for easy comparison with the later models.

We consider the production and inventory storage of a given good, say of steel, in order to meet an exogeneous demand. The state variable $I(t)$

State Variable	$I(t)$ = Inventory level
Control Variable	$P(t)$ = Production rate
State Equation	$\dot{I}(t) = P(t) - S(t), \quad I(0) = I_o$
Objective Function	Maximize $\{J = \int_0^T - [h(I) + c(P)]dt\}$
State Constraint	$I(t) \geq 0$
Control Constraints	$0 \leq \underline{P} \leq P(t) \leq \overline{P}$
Terminal Constraint	$I(T) \geq I_{min}$
Exogeneous Functions	$S(t)$ = Demand Rate $h(I)$ = Inventory Holding Cost $c(P)$ = Production Cost
Parameters	T = Terminal Time I_{min} = Minimum Ending Inventory $\underline{P}$ = Lower Production Rate $\overline{P}$ = Upper Production Rate I_0 = Initial Inventory Level

Table 1.1 Production-Inventory Example

measures the number of tons of steel we have on hand at time t. At time t we have exogeneous demand rate, $S(t)$, in tons of steel per day; and must choose the production rate, $P(t)$, also measured in tons of steel per day. Given initial inventory of I_o tons of steel on hand at $t = 0$, the state equation $\dot{I}(t) = P(t) - S(t)$ describes how the steel variable changes. Since $h(I)$ is the cost of holding inventory I in dollars per day, and $c(P)$ is the cost of producing at rate P, also in dollars per day, the objective function is clearly to maximize the negative of the sum of the total holding plus production costs over the time interval $[0,T]$. Of course, maximizing a negative sum is the same as minimizing the sum of holding plus production costs. The state variable constraint, $I(t) \geq 0$, is imposed so that the demand is satisfied for all t, in other words, the "backlogging" of demand is not permitted. (An alternative formulation would be to make $h(I)$ become very large when I becomes negative, that is, to impose a "stockout" penalty cost.) The control constraints keep the production rate $P(t)$ between the lower rate $\underline{P}$ and an upper rate $\overline{P}$. Finally, the terminal constraint $I(T) \geq I_{min}$ is imposed so that the terminal inventory is at least I_{min}. If we did not impose this constraint, the optimum policy would certainly drive $I(T)$ to zero, which would be equivalent to closing down the factory at time T.

The statement of the problem is lengthy, because of the number of variables, functions, and parameters. However, with the production-inventory interpretations as given, it is not difficult to see the reasons for each condition. In Chapter 5 various versions of this model will be solved in detail.

Example 1.2. An Advertising Model. The various quantities that define this model are summarized in Table 1.2.

We consider a special case of the Nerlove-Arrow advertising model which will be discussed in great detail in Chapter 6. The problem is to determine the rate at which to advertise a product at each time t. Here the state variable is advertising "goodwill," $G(t)$, which measures how well the product is known. We assume there is a "forgetting" coefficient δ which measures the rate at which customers tend to forget the product. To counteract forgetting, advertising is carried out at a rate measured by the control variable $u(t)$. Hence the state equation is $\dot{G} = u - \delta G$

State Variable	$G(t)$ = Advertising goodwill
Control Variable	$u(t)$ = Advertising rate
State Equation	$\dot{G}(t) = u(t) - \delta G(t), \quad G(0) = G_o$
Objective Function	Maximize $\{J = \int_0^\infty [\pi(G) - u]e^{-\rho t}dt\}$
State Constraint	———————
Control Constraint	$0 \leq u \leq Q$
Terminal Condition	———————
Exogeneous Function	$\pi(G)$ = Profit rate
Parameters	δ = Goodwill Decay Constant
	ρ = Discount Rate
	Q = Upper Bound on Advertising Rate
	G_o = Initial Goodwill Level

Table 1.2 Advertising Example

with $G(0) = G_o$ giving the initial goodwill for the product.

The objective function J requires special discussion. Note that the integral defining J is from time $t = 0$ to $t = \infty$; we will later call a problem with upper time limit of ∞ an <u>infinite horizon problem</u>. Because of this upper limit the integrand of the objective function includes the discount factor $e^{-\rho t}$ where ρ is the (constant) discount rate. Without this discount factor the integral would (in most cases) diverge to infinity; hence we will see that such discount factors are an essential part of infinite horizon models. The rest of the integrand in the objective function consists of the profit rate $\pi(G)$ which results from goodwill level G less (the cost of) the advertising level which is proportional to u (proportionality factor = 1); thus $\pi(G) - u$ is the net profit rate at time t. Also $[\pi(G) - u]e^{-\rho t}$ is the net discounted or present value profit rate. Hence J can be interpreted as the total value of discounted future profits, and is the quantity we are trying to maximize.

Note that there is no state constrant. (It can be seen from the state equation that G in fact never becomes negative.) However there is the control constraint $0 \leq u \leq Q$ where Q is the upper bound on the advertising level.

You will find it beneficial to compare this model with the previous one and note similarities and differences.

<u>Example 1.3</u>. <u>A Consumption Example</u>. Rich Rentier plans to retire at age 65 with a lump sum pension of W_o dollars. Rich estimates his remaining life span to be T years and he wants to consume his wealth during these T years in a way that will maximize his total utility of consumption.

Since he does not want to take investment risks, he plans to put his money into an insured savings account that pays interest at a continuously compounded rate of r. If we let the state variable $W(t)$ be the principal at time t and the control variable $C(t)$ the rate of consumption, it is easy to see that the state equation is $\dot{W}(t) = rW(t) - C(t)$, with initial condition $W(0) = W_o$. Letting $U(C)$ be the utility of consuming an amount C, and $B(W)$ be the bequest function we see that the problem can be stated as an optimal control problem with the various variables, equations, and constraints shown in Table 1.3.

Note that the objective function has two parts: first the integral of the discounted utility of consumption from 0 to T, where ρ is the discount rate; and second the bequest function $B(W)$ which measures Rich's utility of

State Variable	$W(t)$ = Wealth
Control Variable	$C(t)$ = Consumption rate
State Equation	$\dot{W}(t) = rW(t) - C(t), \quad W(0) = W_o$
Objective Function	Maximize $\{J = \int_0^T U(C)e^{-\rho t}dt + B[W(T)]\}$
State Constraint	$W(t) \geq 0$
Control Constraint	$C(t) \geq 0$
Terminal Condition	————
Exogeneous Functions	$U(C)$ = Utility of Consumption
	$B(W)$ = Bequest (Salvage) Function
Parameters	T = Terminal Time
	W_o = Intial Wealth
	ρ = Discount Rate
	r = Interest Rate

Table 1.3 Consumption Example

leaving an estate $W(T)$ to his heirs. If he has no heirs and does not care about about charity then $B(W) \equiv 0$. But if he has heirs or a favorite charity to whom he wishes to leave money, then $B(W)$ measures the strength of his desire to leave an estate. The nonnegativity constraints on state and control variables are obviously natural requirements that must be imposed.

1.3 History of Optimal Control Theory

Optimal control theory is an extension of the calculus of variations problems (See Appendix A2), so we discuss the history of the latter first.

The creation of the calculus of variations by Newton, Leibnitz and the Bernoulli brothers occurred almost immediately after the invention (or formalization) of calculus by the first two authors. An important problem in calculus is to find an argument of a function at which the function takes on its maximum or minimum. The extension of this problem posed in the calculus of variations is to find a function which maximizes or minimizes the value of an integral (or functional) of that function. As might be expected, the calculus of variations extremum problem is much harder than the calculus extremum problem. The Bernoulli brothers, Euler, and Lagrange are generally considered the founders of the variational calculus.

The problem first considered in the calculus of variations was the "path of least time" or the brachistochrone problem. The problem is illustrated in Figure 1.1. We want to find a curve Γ connecting the two points A and B with the property that a particle sliding along the curve under the influence of gravity will move from A to B in the shortest possible time. In Appendix A2 we solve the Brachistochrone problem and show that the form of the solution curve is a cycloid.

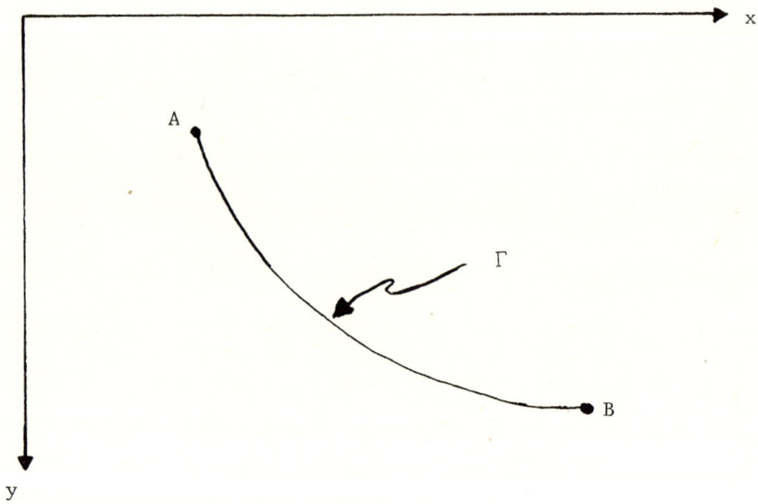

Figure 1.1 The Brachistochrone Problem

In the nineteenth and early twentieth centuries, many authors contributed to the theory of the solution of calculus of variations problems. The names of Weierstrass, Bliss, and Bolza and others are important in this history.

To change calculus of variations problems into control theory problems requires one more conceptual extension—the addition of control variables to the state equations. R. Isaacs made such an extension to two-person pursuit games in the period 1948-1955 [Isaacs, 1965]. R. Bellman made a similar extension in the ideas of dynamic programming, published in [Bellman, 1957].

However, it is safe to say that the firm starting date of modern control theory was the publication in Russian in 1958 (1962 in English) of the book, The Mathematical Theory of Optimal Processes, by Pontryagin, Boltyanski, Gamkrelidze and Mischenko [Pontryagin, et al., 1962]. (Other important American mathematicians associated with the maximum principle are Valentine, McShane, and Hestenes.) The importance of the book by Pontryagin et al. lies not only in a rigorous formulation of a calculus of variations problem with (constrained) control variables, but also the proof of the maximum principle for optimal control problems. The maximum principle permits the decoupling of the dynamic problem over time into a series of problems holding at each instant. The optimal solutions of the instantaneous problems can be shown to give the optimum solution to the overall problem.

In this book we will be principally concerned with the application of the maximum principle in various forms to the solution of a wide variety of applied problems in management science and economics. It is hoped that the reader, after reading some of these problems and solutions, will appreciate, as we do, the importance of the maximum principle.

Some important books and surveys of the application of the maximum principle to management science are [Bensoussan, Hurst and Naslund, 1974], [Tapiero, 1977], [Connors and Teichroew, 1967], [Clark, 1976], [Stoeppler, 1975], [Sethi, 1977, 1978a], [Wickwire, 1977], [Bookbinder and Sethi, 1980], [Kleindorfer, 1978], and [Pekelman, 1978]. Because of the existence of these ample surveys, we have included in our bibliography only the works actually cited in the text.

1.4 Notation Used

In order to make the book readable we shall adopt the following standard notation which will hold throughout the book.

Let y be an h-component column vector and z be a k-component row vector; i.e.,

$$y = (y_1,\ldots,y_h)' \quad \text{and} \quad z = (z_1,\ldots,z_k).$$

We denote the transpose of a vector by a prime ($'$). If y and z are functions of time then the time derivatives $\dot{y}$ and $\dot{z}$ are defined as

$$\dot{y} = (\dot{y}_i,\ldots,\dot{y}_h)' \quad \text{and} \quad \dot{z} = (\dot{z}_1,\ldots,\dot{z}_k) \tag{1.5}$$

where $\dot{y}_i$ and $\dot{z}_j$ denote the time derivatives dy_i/dt and dz_j/dt.

When $h = k$ we can define the inner product

$$zy = \sum_{i=1}^{k} z_i y_i \tag{1.6}$$

More generally if $A = \{a_{ij}\}$ is an m×k matrix and $B = \{b_{ij}\}$ is a k×n matrix we define the matrix product $C = AB$, which is an m×n matrix with components

$$c_{ij} = \sum_{\ell=1}^{k} a_{i\ell} b_{\ell j}. \tag{1.7}$$

If $F(y,z)$ is a scalar function, i.e., $F\colon\ E^h \times E^k \to E^1$, where E^s denotes the s-dimensional Euclidean space, F_y and F_z are defined as

$$F_y = (F_{y_1},\ldots,F_{y_h}) \quad \text{(a row vector)} \tag{1.8}$$

$$F_z = (F_{z_1},\ldots,F_{z_k}) \quad \text{(a row vector)} \tag{1.9}$$

where F_{y_i} and F_{z_j} denote the partial derivatives with respect to the subscripted variables.

If f is an n-dimensional vector function, either row or column, i.e.,

$$f = (f_1,\ldots,f_n) \quad \text{or} \quad f = (f_1,\ldots,f_n)'$$

where each component $f_i = f_i(y,z)$ depends on y and z; then f_y will denote the nxh matrix

$$f_y = \begin{pmatrix} \partial f_1/\partial y_1 & \cdots & \partial f_1/\partial y_h \\ \cdot & \cdots & \cdot \\ \partial f_n/\partial y_1 & \cdots & \partial f_n/\partial y_h \end{pmatrix} . \tag{1.10}$$

Applying this rule to (1.8) we can define $F_{yz} = (F_y)_z$ to be

$$F_{yz} = \begin{pmatrix} F_{y_1 z_1} & \cdots & F_{y_1 z_k} \\ \cdot & \cdots & \cdot \\ F_{y_h z_1} & \cdots & F_{y_h z_k} \end{pmatrix} \tag{1.11}$$

That is, F_{yz} is the hxk matrix whose (i,j)th element is the second partial derivative

$$F_{y_i z_j} = \partial^2 F/\partial y_i \partial z_j . \tag{1.12}$$

Let g be an n-component row vector function and f be an n-component column vector function of an m-component vector x. Then in Exercise 1.8 you are asked to verify the following identity:

$$(gf)_x = gf_x + f'g_x . \tag{1.13}$$

In Exercise 1.9 you are asked to show further that if $m = n$ and $g_x = g_x'$ then

$$(gf)_x = gf_x + (g_x f)' \quad . \tag{1.14}$$

The latter result will be used in Chapter 2.

Most mathematical expressions in this book will be vector equations or inequalities. In most cases it will be possible to read them as one variable expressions with the correct meaning. In fact, it may be a good idea to do this on the first reading. Then in the second and further readings the extension to vector form will be easy.

The norm of a vector z is defined to be

$$\| z \| = \sqrt{z_1^2 + \ldots + z_k^2} \; . \tag{1.15}$$

The norm of a vector is commonly used to define a neighborhood N_{z_o} of a point; e.g.,

$$N_{z_o} = \{z \mid 0 < \| z - z_o \| < \varepsilon\}. \tag{1.16}$$

We shall occasionally make use of the so-called "little-o" notation, $o(z)$. What this means is that a function $f(z)$ is said to be of the order $o(z)$, if

$$\lim_{\| z \| \to 0} \frac{f(z)}{\| z \|} = 0 \quad .$$

The most common use of this notation will be to collect higher order terms in a series expansion.

In the theoretical continuous time sections of this book we will use x to denote the the state vector, u the control vector, and λ the adjoint vector. When talking about optimal state, control, or adjoint vectors we put a asterisk "*" as a superscript; for instance as x^*, u^*, λ^*.

The norm of a k-dimensional vector function $z(t)$, $t \in [0,T]$ is defined to be

$$\| z \| = \sqrt{\sum_{j=1}^{k} \int_0^T z^2(\tau) d\tau} \qquad (1.17)$$

We use the symbol "$\triangleq$" to mean "defined to be equal to" and the symbol "$\approx$" to mean "approximately equal to." The double arrow "$\Rightarrow$" means "implies" and the reverse epsilon "$\in$" means "is a member of".

In the discrete time models we use x^k, u^k, λ^k to denote state, control, and adjoint variables at time k. We also denote the difference operator by $\Delta x^k \triangleq x^{k+1} - x^k$. As in the continuous case the optimal variables have an asterisk as a superscript; thus x^{*k}, u^{*k}, and λ^{*k} are optimal quantities.

In order to specify the optimal control for linear control problems we will define a special notation

$$u = \text{bang}[b_1, b_2; W(x,\lambda,t)] \ . \qquad (1.18)$$

The definition of this function is:

$$\text{bang}[b_1, b_2; W(x,\lambda,t)] = \begin{cases} b_1 & \text{if } W(x,\lambda,t) < 0 \\ \text{undefined} & \text{if } W(x,\lambda,t) = 0 \\ b_2 & \text{if } W(x,\lambda,t) > 0 \end{cases} \qquad (1.19)$$

In the case when $W(x,\lambda,t) = 0$ the value of the optimal control requires additional analysis which is dealt with under the topic of singular control for each of the models that requires it.

Another useful notation for specifying the optimal control for linear quadratic problems we define another special function

$$u = \text{sat}[y_1, y_2;\ W(x,\lambda,t)] \tag{1.20}$$

which is defined by

$$\text{sat}[y_1,y_2;W(x,\lambda,t)] = \begin{cases} y_1 & \text{if } W(x,\lambda,t) < y_1 \\ W(x,\lambda,t) & \text{if } y_1 \leq W(x,\lambda,t) \leq y_2 \\ y_2 & \text{if } W(x,\lambda,t) > y_2 \end{cases} . \tag{1.21}$$

The word "sat" is short for the word "saturation." The latter name comes from an electrical engineering application to saturated amplifiers.

In several applications to be discussed we will need the concept of impulse control, which is sometimes needed in cases when an unbounded control can be applied for a very short time. An example is the advertising model in Table 1.2 when $Q = \infty$. We apply unbounded countrol for a short time in order to cause a jump discontinuity in the state variable. For the example in Table 1.2 this might mean an intense advertising campaign (media blitz) in order to increase advertising goodwill in a very short time. The impulse function defined below is required to evaluate the integral in the objective function which measures the cost of the advertising campaign.

Suppose we want to apply an impulse control at time t to change the state variable from $x(t) = x_1$ to the value x_2 immediately after t, that is, action $x(t+) = x_2$. To compute the effect on the objective function (1.2) of this we use the following procedure: Given $\varepsilon > 0$ and a constant control $u(\varepsilon)$, integrate (1.1) from t to $t + \varepsilon$ with $x(t) = x_1$ and choose $u(\varepsilon)$ so that $x(t+\varepsilon) = x_2$; this gives the trajectory $x(\tau,\varepsilon,\ u(\varepsilon))$ for $\tau \in [t,\ t+\varepsilon]$. We can now compute

$$\mathrm{Imp}(x_1,x_2;t) = \lim_{0} \int_t^{t+\varepsilon} F(x,u,\tau)d\tau \quad . \tag{1.23}$$

If the impulse is applied only at time t, then we can calculate (1.2) as

$$J = \int_0^t F(x,u,\tau)d\tau + \mathrm{Imp}(x_1,x_2;t) + \int_t^T F(x,u,\tau)d\tau + S[x(T)]. \tag{1.24}$$

If there are several instants at which impulses are applied, then this procedure is easily extended. Examples of the use of (1.24) occur in Chapters 4 and 6. We frequently omit t in (1.24) when the impulse function is independent of t.

Exercises for Chapter 1

1.1 For Example 1.1 let the functions and parameters of the production-inventory model be given by:

$h(I) = 10I$, $c(P) = 20P$, $T = 10$, $I_0 = 1{,}000$

$\underline{P} = 600$, $\overline{P} = 1200$, $I_{min} = 800$, $S(t) = 900 + 10t$

(a) Set $P(t) = 1000$ for $0 \le t \le 10$. Determine whether this control is feasible; if it is feasible, compute the value J of the objective function.

(b) If $P(t) = 800$, show that the terminal constraint is violated and hence the control is infeasible.

(c) If $P(t) = \underline{P}$ for $0 \leq t \leq 6$ and $P(t) = \overline{P}$ for $6 < t \leq 10$, show that the control is infeasible because the state constraint is violated.

1.2 For the advertising model in Example 1.2 let $\pi(G) = 2\sqrt{G}$, $\delta = .05$, $\rho = .2$, $Q = 2$, $G_0 = 16$. Set $u(t) = .8$ for $t \geq 0$, and show that $G(t)$ is constant for all t; compute the value J of the objective function.

1.3 Rich Rentier in Example 1.3 has initial wealth $W_0 = \$1{,}000{,}000$. Assume $B = 0$, $\rho = .1$, $r = .15$, and Rich expects to live for exactly 20 years.

(a) What is the maximum constant consumption level which Rich can afford during his remaining life?

(b) If Rich's utility function is $U(C) = \ln C$, what is the present value of the total utility in part (a)?

(c) Suppose Rich sets aside \$100,000 to start the Rentier Foundation What is the maximum constant grant level which the foundation can support if it is to last forever?

1.4 Suppose Rich in Exercise 1.3 takes on a part-time job which yields an income of $y(t)$ at time t. Assume $y(t) = 10{,}000e^{-.05t}$ and he has a bequest function $B(W_T) = .5e^{-\rho T}\ln W(T)$.

(a) Reformulate this new optimal control problem.

(b) If Rich (no longer a <u>rentier</u>) consumes at the constant rate found in Exercise 1.3(a), find his terminal wealth, and his new total utility

1.5 In Example 1.1, suppose there is a cost associated with changing the rate of production. One way to formulate this problem is to let the control variable $u(t)$ denote the rate of change of the production rate $P(t)$, having a cost βu^2 associated with such changes. Formulate the new problem. [Hint: Let $P(t)$ be an additional state variable.]

1.6 In Example 1.2, suppose G measures the number of people who know about the product; hence if A is the total population, then $A - G$ is the number of people who do not know about the product. If $u(t)$ measures the advertising rate at time t, assume that $u(A-G)$ is the corresponding rate of increase of G due to this advertising. Formulate the new model.

1.7 Consider the following educational policy question. Let $S(t)$ denote the total number of scientists at time t, and let δ be the retirement rate of scientists. Let $E(t)$ be the number of teaching scientists, and $R(t)$ be the number of research scientists, so that $S(t) = E(t) + R(t)$. Assume $\gamma E(t)$ is the number of newly graduated government scientists at time t, of which the policy allocates $u\gamma E$ to the pool of teachers, where $0 \leq u \leq 1$. The remaining graduates are added to the pool of researchers. The government has a target of maximizing the function $\alpha E(T) + \beta R(T)$ at a future time T. Formulate the optimal control problem for the government.

1.8 Let x be an m-component vector; let g be an n-component row vector function of x, and let f be an n-component column vector function of x. Use the ordinary chain rule of calculus to derive the vector chain rule

$$(gf)_x = gf_x + f'g_x.$$

1.9 If $m = n$, and $g_x = (g_x)'$ show that the chain rule in 1.8 can be written

$$(gf)_x = gf_x + (g_xf)' \quad .$$

1.10 Sketch the optimal control

$$u^*(t) = \text{bang}[-1,1;\ W(t)]$$

for $0 \leq t \leq 5$ when

(a) $W(t) = t - 2$

(b) $W(t) = t^2 - 4t + 3$

(c) $W(t) = \sin \pi t$.

1.11 Sketch the optimal control

$$u^*(t) = \text{sat}[2,3;\ W(t)]$$

for $0 \leq t \leq 5$, when

(a) $W(t) = 4 - t$

(b) $W(t) = 2 + t^2$

(c) $W(t) = 4 - 4e^{-t}$

1.12 Compute $\text{Imp}(G_1,G_2;t)$ where $G_2 > G_1$ for the advertising model of Table 1.2 when $Q = \infty$.

2 THE MAXIMUM PRINCIPLE: CONTINUOUS TIME

The main purpose of this chapter is to introduce the maximum principle as a necessary condition for optimal control. The statement of the basic problem of the optimal control of a dynamic system appears in Section 2.1. Although vector notation is used, the reader can think of the problem as having only a single variable on the first reading. In Section 2.2 the method of dynamic programming [Bellman, 1957; Bellman and Kalaba, 1965] is used to derive the maximum principle. We use this method because of the simplicity of the dynamic programming concept and its familiarity. Although restricted in scope, the derivation yields significant economic interpretations. (In Appendix A3, the maximum principle is also derived by using a more general method similar to that of [Pontryagin et. al., 1962], but with certain simplifications.) In Section 2.4, the maximum principle is shown to be sufficient for optimal control under an appropriate concavity condition, which holds for many management science problems.

2.1 Statement of the Problem

Optimal Control Theory deals with the problem of optimization of dynamic systems. The problem must be well-posed before any solution can be attempted. This requires a clear mathematical description of the system to be optimized, the constraints imposed on the system, and the objective function to be maximized (or minimized).

2.1.1 The Mathematical Model

An important part of any control problem is the process of modeling the dynamic system, physical, business, or otherwise, under consideration. The aim is to arrive at a mathematical description which is simple enough to deal with, and realistic enough to be able to predict the response of the system to any given inputs. The model of the system for our purpose is restricted to systems that can be characterized by ordinary differential equations. (For the discrete time case dealt with in Chapter 7, the appropriate model involves ordinary difference equations.) Thus, given the initial state x_o of the system and control history $u(t)$, $t \in [0,T]$, of the process, the evolution of the system may be described by the first-order differential equation, known also as the <u>state equation</u>,

$$\dot{x}(t) = f(x(t),u(t),t), \quad x(0) = x_o, \tag{2.1}$$

where the vector of <u>state variables</u>, $x(t) \in E^n$, the vector of <u>control variables</u>, $u(t) \in E^m$, and $f\colon E^n \times E^m \times E^1 \to E^n$. Furthermore, the function f is assumed to be continuously differentiable. The path $x(t)$, $t \in [0,T]$ is called a <u>state trajectory</u> and $u(t)$, $t \in [0,T]$ is called a <u>control trajectory</u> or simply, a <u>control</u>. The terms, vector of state variables, state vector, and state will be used interchangeably; similarly for the terms, vector of control variables, control vector, and control. When no confusion arises, we will usually suppress the time notation (t); thus, e.g., $x(t)$ will be written simply as x. Furthermore, whether x denotes the state at time t or the entire state trajectory should be inferred from the context. A similar statement holds for u.

2.1.2 Constraints

In this chapter, we are concerned with problems with no constraints (other than the state equation (2.1)) on the state variables. (State variable constraints are considered in Chapter 3.) We do impose constraints on the control variables. We define an <u>admissible control</u> to be a control trajectory $u(t)$, $t \in [0,T]$, which is piecewise-continuous and satisfies, in addition

$$u(t) \in \Omega(t) \subset E^m \tag{2.2}$$

for each $t \in [0,T]$. Note that the set $\Omega(t)$ represents physical constraints on the values of the control variables at time t.

2.1.3 The Objective Function

The objective function is a quantitative measure of the performance of the system over time. An optimal control is defined to be an admissible control which maximizes (or minimizes) the objective function. In business problems, a typical objective function gives some appropriate measure of profits, sales, or costs. Mathematically, we let

$$J = \int_0^T F(x,u,t)dt + S[x(T)] \tag{2.3}$$

denote the objective function, where $F: E^n \times E^m \times E^1 \to E^1$ and the salvage value function $S: E^n \to E^1$. For instance, in a typical business application $F(x,u,t)$ could be the instantaneous profit rate and, $S[x(T)]$ could be the salvage value of the firm at the terminal time, T. We assume F and S to be continuously differentiable.

2.1.4 The Optimal Control Problem

Given the preceding definitions we can state the optimal control problem with which we will be concerned for the rest of this chapter. The problem is to find an admissible control u^* which maximizes the objective function (2.3) subject to the state equation (2.1) and control constraints (2.2). We now restate the optimal control problem as:

$$\boxed{\begin{aligned} &\max_{u(t) \in \Omega(t)} \left\{J = \int_0^T F(x,u,t)dt + S[x(T)]\right\}, \\ &\text{subject to} \\ &\dot{x} = f(x,u,t), \quad x(0) = x_0. \end{aligned}} \tag{2.4}$$

The control u^* is called an <u>optimal control</u> and x^*, determined by means of the state equation, is called an <u>optimal trajectory</u> or an <u>optimal path</u>.

The optimal control problem (2.4) specified above is said to be in <u>Bolza form</u> because of the form of the objective function. It is said to be in <u>Lagrange form</u> when $S[x(T)] \equiv 0$ and it is said to be in <u>Mayer form</u> when $F \equiv 0$. We say the problem is in <u>linear Mayer form</u> when S is linear, i.e.,

$$\boxed{\begin{array}{l} \max\limits_{u(t)\in\Omega(t)} \{J = c\,x(T)\} \\ \text{subject to} \\ \dot{x} = f(x,u,t), \quad x(0) = x_0, \end{array}} \tag{2.5}$$

where $c = (c_1, c_2, \ldots, c_n)$ is an n-dimensional row vector of given constants. In the next paragraph and in Exercise 2.3 it will be demonstrated that all of these forms can be converted into the linear Mayer form.

To show that the Bolza form can be reduced to the linear Mayer form we define a new state vector y, having $n+1$ components defined as follows: $y_i = x_i$ for $i = 1,\ldots,n$, and

$$\dot{y}_{n+1} = \bar{F}(y,u,t) + \dot{\bar{S}}(y), \quad y_{n+1}(0) = \bar{S}(y_0) \tag{2.6}$$

where $\bar{F}[(y_1,\ldots,y_n,y_{n+1}),u,t] = F[(y_1,\ldots,y_n),u,t]$ and $\bar{S}[(y_1,\ldots,y_n,y_{n+1})] = S[(y_1,\ldots,y_n)]$. We also put $c = (0,\ldots,0,1)$ where c has $n+1$ components, so that the objective function is $J = cy(T) = y_{n+1}(T)$. If we now integrate (2.6) from 0 to T we have

$$\begin{aligned} y_{n+1}(T) &= \int_0^T \bar{F}(y,u,t)dt + \bar{S}[y(T)] \\ &= \int_0^T F(x,u,t)dt + S[x(T)] \end{aligned} \tag{2.7}$$

which is the same as the objective function J in (2.4). Of course, the price paid for going from Bolza to linear Mayer form is the addition of one state variable and its associated differential equation (2.6).

Exercise 2.3 poses the question of showing in a similar way that the Lagrange and Mayer forms can also be reduced to the linear Mayer form.

In Section 2.2 we derive necessary conditions for optimal control in the form of the maximum principle, and in Section 2.4 we derive sufficient conditions. In any particular application the existence of a solution will be demonstrated by actually finding a solution that satisfies both the necessary and the sufficient conditions for optimality. We thus avoid the necessity of having to prove general existence theorems, which require advanced and difficult mathematics.

2.2 Dynamic Programming and the Maximum Principle

We shall now derive the maximum principle by using a dynamic programming approach. The proof is intuitive in nature and is not intended to be mathematically regorous. For more rigorous derivations we refer the reader to Appendix A3, [Pontryagin, 1962], [Berkovitz, 1961], [Halkin, 1967], [Boltyanskii, 1971], [Bryant and Mayne, 1974], and [Clarke, 1976].

2.2.1 The Hamilton-Jacobi-Bellman Equation

Suppose $V(x,t)\colon E^n \times E^1 \to E^1$ is a function whose value is the maximum value of the objective function of the control problem given that we start it at time t at state x. We initially assume that the value function $V(x,t)$ exists for all x and t in the relevant ranges; later we will make additional assumptions about V.

Bellman in his book on dynamic programming [Bellman, 1957] states the principle of optimality as follows:

> "An optimal policy has the property that, whatever the initial state and initial decision are, the remaining decisions must constitute an optimal policy with regard to the outcome resulting from the first decision."

Intuitively this principle is obvious, for if we were to start in state x at time t and did not follow an optimal path from thereon, then there would exist (by assumption) a better path from t to T, hence we could improve by following the better path from time t on. We will use the principle of optimality to derive conditions on the value of function $V(x,t)$.

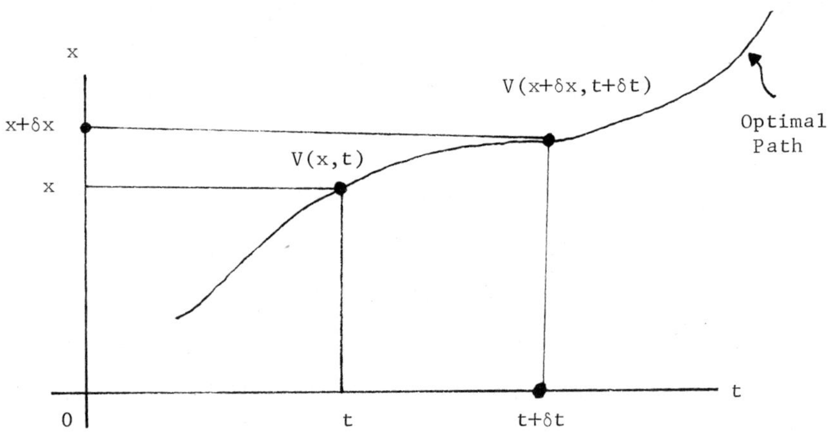

Figure 2.1 Optimal Path in State-Time Space

Figure 2.1 is a schematic picture of the optimal path in state-time space, and two nearby points (x,t) and $(x+\delta x,\ t+\delta t)$ where δt is a small increment of time. The value function changes from $V(x,t)$ to $V(x+\delta t,\ t+\delta t)$ between these two points. By the principle of optimality, the change in the objective function is made up of two parts: first, the incremental change in J from t to $t+\delta t$ which is given by the integral of $F(x,u,t)$ from t to $t+\delta t$; second, the value function $V(x+\delta x,\ t+\delta t)$ at time $t+\delta t$.

The current control action should be chosen within constraints Ω to maximize the sum of these two terms. In equation form this is

$$V(x,t) = \max_{\substack{u(\tau)\in\Omega(\tau)\\ \tau\in[t,t+\delta t]}} \left\{\int_t^{t+\delta t} F[x(\tau),u(\tau),\tau]\,d\tau + V[x(t+\delta t),t+\delta t]\right\} \tag{2.8}$$

where δt represents a small increment in t.

Since F is continuous the integral in (2.8) is approximately $F(x,u,t)\delta t$ so that we can rewrite (2.8) as

$$V(x,t) = \max_{u\in\Omega(t)} \{F(x,u,t)\delta t + V[x(t+\delta t),t+\delta t]\} + o(\delta t) \tag{2.9}$$

where $o(\delta t)$ denotes a collection of higher order terms in δt. (By definition $o(\delta t)$ is a function such that $\lim_{\delta t\to 0} \frac{o(\delta t)}{\delta t} = 0$.)

We now make an assumption of which we will talk more later. We assume that the return function or payoff function V is a continuously differentiable function of its arguments. This allows us to use the Taylor's series expansion of V with respect to δt to obtain

$$V[x(t+\delta t),t+\delta t] = V(x,t) + [V_x(x,t)\dot{x} + V_t(x,t)]\delta t + o(\delta t). \tag{2.10}$$

where V_x and V_t are partial derivatives of $V(x,t)$ with respect to x and t, respectively.

Substituting for $\dot{x}$ from (2.1) in the above equation and then using it in (2.9), we obtain

$$V(x,t) = \max_{u\in\Omega(t)} [F(x,u,t)\delta t + V(x,t) + V_x(x,t)f(x,u,t)\delta t + V_t(x,t)\delta t] + o(\delta t). \tag{2.11}$$

Cancelling $V(x,t)$ on both sides and then dividing by δt we get

$$0 = \max_{u \in \Omega(t)} [F(x,u,t) + V_x(x,t)f(x,u,t) + V_t(x,t)] + \frac{o(\delta t)}{\delta t}. \quad (2.12)$$

Now we let $\delta t \to 0$ and obtain the following equation

$$0 = \max_{u \in \Omega(t)} [F(x,u,t) + V_x(x,t)f(x,u,t) + V_t(x,t)] \quad (2.13)$$

with the boundary condition

$$V(x(T),T) = S(x(T)). \quad (2.14)$$

This boundary condition is obvious from the fact that if we are at $t = T$ then the value function is simply the salvage value function.

Note that the components of the vector $V_x(x,t)$ can be interpreted as the marginal contributions of the state variables x to the objective function being maximized. We denote this marginal return vector by an adjoint (row) vector $\lambda(t) \in E^n$, i.e.,

$$\lambda(t) \triangleq V_x(x,t). \quad (2.15)$$

From the preceding remark, we can also interpret λ to be the per unit change in the objective function for small changes in x. (See section 2.2.4.) Next we introduce the so-called <u>Hamiltonian</u>

$$H[x,u,V_x,t] \triangleq F(x,u,t) + V_x(x,t)f(x,u,t) \quad (2.16)$$

or, simply

$$H(x,u,\lambda,t) = F + \lambda f. \quad (2.17)$$

we can rewrite equation (2.13) as the following equation, called the <u>Hamilton-Jacobi-Bellman equation</u>

$$0 = \max_{u \in \Omega(t)} [H + V_t] . \quad (2.18)$$

Note that it is possible to take V_t out of the maximizing operation since it does not depend on u.

The Hamiltonian maximizing condition of the maximum principle can be obtained from (2.18) by observing that if $x^*(t)$ and $\lambda^*(t)$ are optimal values of state and adjoint variables at time t, then the optimal control $u^*(t)$ must satisfy (2.18), i.e., for all $u(t)\epsilon\Omega(t)$,

$$H[x^*(t),u^*(t),\lambda^*(t),t] + V_t(x^*(t),t) \geq H[x^*(t),u(t),\lambda^*(t),t] + V_t(x^*(t),t). \quad (2.19)$$

Cancelling the term V_t on both sides, we obtain

$$H[x^*(t),u^*(t),\lambda^*(t),t] \geq H[x^*(t),u(t),\lambda^*(t),t] \quad (2.20)$$

for all $u(t)\epsilon\Omega(t)$.

In order to complete the statement of the maximum principle we must still obtain the adjoint equations.

2.2.2 Derivation of the Adjoint Equations

The derivation of the adjoint equation proceeds from the Hamilton-Jacobi-Bellman equation (2.18). (This derivation appears in [Feldbaum, 1965] and [Kirk, 1970]. Note that, given the optimal path x^*, the optimal control u^* maximizes the right side of (2.18), and its maximum value is zero. We now consider small perturbations of the values of the state variables in a neighborhood of the optimal path x^*. Thus, let

$$x(t) = x^*(t) + \delta x(t) , \quad (2.21)$$

where $\|\delta x(t)\| < \epsilon$, be such a perturbation.

We now consider a 'fixed' time instant t. We can then write (2.18) as

$$H[x^*(t),u^*(t),V_x(x^*(t),t),t] + V_t(x^*(t),t) \geq H[x(t),u^*(t),$$

$$V_x(x(t),t),t] + V_t(x(t),t). \qquad (2.22)$$

To explain, we note that the left hand side of (2.22) equals zero. The right hand side can attain the value zero only if $u^*(t)$ is also an optimal control for $x(t)$. In general for $x(t) \neq x^*(t)$, i.e., $\delta x(t) \neq 0$, this will not be so. From this observation, it follows that the expression on the right side of (2.22) attains its maximum (of zero) at $x(t) = x^*(t)$. Furthermore, $x(t)$ is not explicitly constrained. In other words $x^*(t)$ is an unconstrained local maximum of the right hand side of (2.22), so that the derivative of this expression with respect to x must vanish at $x^*(t)$; i.e.,

$$H_x[x(t),u^*(t),V_x(x(t),t),t] + V_{tx}(x(t),t) = 0 \qquad (2.23)$$

In order to take the derivative as we did in (2.23) we must further assume that V is a twice differentiable function of its arguments. Using the definition of the Hamiltonian in (2.16) and the identity (1.14) we obtain

$$F_x + V_x f_x + (V_{xx} f)' + V_{tx} = 0 \qquad (2.24)$$

where the prime denotes the transpose of a vector. (The reason for the appearance of the transpose in (2.24) is explained in Exercises 1.8 and 1.9.)

The derivation of the necessary condition (2.24) is crux of the reasoning in the derivation of the adjoint equation. It is easy to obtain the so called adjoint equation from it. We begin by taking the time derivative of $V_x(x,t)$; thus

$$\frac{dV_x}{dt} = (V_{xx}\,\dot{x})' + V_{xt} = (V_{xx}\,f)' + V_{tx} \qquad (2.25)$$

Since the terms on the right of (2.25) are the same as the last two terms in (2.24), we see that (2.25) becomes

$$\frac{dV_x}{dt} = - F_x - V_x f_x \tag{2.26}$$

Because λ was defined in (2.15) to be V_x, we can rewrite (2.26) as

$$\dot{\lambda} = - F_x - \lambda f_x .$$

To see that the right hand side of this equation can be written simply as $-H_x$, we need to go back to the definition of H in (2.17) and recognize that when taking the partial derivative of H with respect to x, the adjoint variables λ are considered to be independent of x. We note further that along the optimal path, λ is a function of t only. Thus,

$$\dot{\lambda} = - H_x . \tag{2.27}$$

Also, from the definition of λ in (2.15) and the boundary condition (2.14), we have the <u>terminal boundary condition</u>, which is also called the <u>transversality condition</u>:

$$\lambda(T) = \left. \frac{\partial S(x)}{\partial x} \right|_{x=x(T)} = S_x[x(T)] . \tag{2.28}$$

The <u>adjoint equation</u> (2.27) together with its boundary condition (2.28) determine the adjoint variables.

From the definition of the Hamiltonian in (2.17), it is also obvious that the state equation can be written as

$$\dot{x} = f = H_\lambda \tag{2.29}$$

using an argument similar to the one used in deriving (2.27). The system of equations (2.27) and (2.29) together with their respective boundary conditions

$$\boxed{\begin{aligned} \dot{x} &= H_{\lambda}, \quad x(0) = x_0 \\ \dot{\lambda} &= -H_x, \quad \lambda(T) = S_x[x(T)] \end{aligned}} \tag{2.30}$$

is called a canonical system of equations or canonical adjoints. Hence the name adjoint vector for λ.

This completes our derivation of the maximum principle using dynamic programming. We can now summarize the main results in the following section.

2.2.3 The Maximum Principle

The necessary conditions for u^* to be an optimal control are:

$$\boxed{\begin{aligned} &\dot{x}^* = f(x^*,u^*,t), \quad x^*(0) = x_0, \\ &\dot{\lambda}^* = -H_x[x^*,u^*,\lambda^*t], \quad \lambda^*(T) = S_x[x^*(T)], \\ &H[x^*(t),u^*(t),\lambda^*(t),t] \geq H[x^*(t),u(t),\lambda^*(t),t] \\ &\text{for all admissible controls, i.e., piecewise-} \\ &\text{continuous controls such that} \\ &\qquad u(t) \in \Omega(t), \quad t \in [0,T]. \end{aligned}} \tag{2.31}$$

It should be emphasized that the state and the adjoint arguments of the Hamiltonian are $x^*(t)$ and $\lambda^*(t)$, respectively, on both sides of the Hamiltonian-maximizing condition in (2.31). Furthermore, $u^*(t)$ must provide a global maximum of the Hamiltonian $H[x^*(t),u(t),\lambda^*(t),t]$. For this reason the necessary conditions in (2.31) are called the maximum principle.

Note that to apply the maximum principle we must simultaneously solve two sets of differential equations with u obtained from the Hamiltonian maximizing condition in (2.31). With the control variable u so obtained the state equations for x depends on the initial value x_o, and the adjoint equations for λ depends on the terminal value $\lambda(T)$. Such a system of equations is called a _two-point boundary value problem_. The general solution of such problems can be very difficult, see [Roberts and Shipman, 1972]. However, there are certain special cases which are easy. One such is the case in which the adjoint equation is independent of the state and the control variables; here we can solve the adjoint equation first, then get the optimal control u^*, and then solve for x. In subsequent chapters we will solve many such two-point boundary value problems of varying degrees of difficulty. Chapter 11 contains a summary of the various methods currently being used to solve such problems.

Note also that if we can solve the Hamiltonian maximizing condition for an optimal control function in closed form as

$$u^*(t) = u[x^*(t),\lambda^*(t),t],$$

then we can substitute these into the state and adjoint equations to get a set of reduced differential equations which is the two-point boundary value problem. In Exercise 2.17 you are asked to derive such a two-point boundary value problem.

One final remark should be made. Because an integral is unaffected by values of the integrand at a finite set of points, some of the arguments made in this chapter may not hold at a finite set of points. This does not affect the validity of the results.

In the next section we give economic interpretations of the maximum principle, and in Section 2.3 we solve five simple examples by using the maximum principle.

2.2.4 Economic Interpretations of the Maximum Principle

Recall from Section 2.1.3 that the objective function (2.3) is

$$J = \int_0^T F(x,u,t)dt + S[x(T)]$$

where F will be considered to be the instantaneous profit rate measured in dollars per unit of time, and $S[x(T)]$ is the salvage value in dollars of the firm at time T when the terminal state is $x(T)$. For purposes of discussion it will be convenient to consider the state $x(t)$ as the stock of capital at time t.

In (2.15) we interpreted $\lambda(t)$ to be the per unit change in the value function $V(x,t)$ for small changes in capital stock x. In other words $\lambda(t)$ is the marginal value per unit capital at time t, and it is also referred to as the "price" or "shadow price" of a unit of capital. In particular, the value of $\lambda(0)$ is the marginal rate of change of the maximum value of J, (the objective function) with respect to the change in the initial capital stock, x_0.

The interpretation of the Hamiltonian function in (2.17) can now be derived. Multiplying (2.17) by dt gives

$$\begin{aligned} Hdt &= Fdt + \lambda f dt \\ &= Fdt + \lambda \dot{x} dt \\ &= Fdt + \lambda dx \end{aligned}$$

where we made use of the state equation (2.1). The first term $F(x,u,t)dt$ represents the <u>direct contribution</u> to J in dollars from time t to $t + dt$ if we are in state x and we apply control u. The differential $dx = f(x,u,t)dt$ represents the change in capital stock from t to $t + dt$, when we are in state x and control u is applied. Therefore, the second term λdx represents the value in dollars of the incremental capital stock, dx, and hence can be considered as the <u>indirect contribution</u> to J in dollars. Thus Hdt can be interpreted as the <u>total contribution to J</u> from the interval t to $t + dt$ when $x(t) = x$ and $u(t) = u$.

With this interpretation of the Hamiltonian it is easy to see why the Hamiltonian must be maximized at each time t. If we were just to maximize F at each time t, we would not be maximizing J, because we would ignore the effect of control in changing the capital stock, which give rise to indirect contributions to J. The maximimum principle derives prices, the adjoint variables λ, in such a way that $\lambda(t)dx$ is the correct valuation for the indirect contribution from the interval t to $t + dt$. As a consequence the Hamiltonian maximizing problem can be treated as a static problem at each instant t. In other words the maximum principle "decouples" the dynamic maximization problem (2.4) in the interval $[0,T]$ into a series of static maximization problems at each instant t in $[0,T]$. Therefore the Hamiltonian can be interpreted as a surrogate profit rate to be maximized at each time t.

The value of λ to be used in the maximum principle is given by (2.27) and (2.28), i.e.

$$\dot{\lambda} = -\frac{\partial H}{\partial x} = -\frac{\partial F}{\partial x} - \lambda\frac{\partial f}{\partial x}\,; \quad \lambda(T) = S_x[x(T)].$$

Rewriting the equation as

$$-d\lambda = H_x dt = (F_x dt) + \lambda(f_x dt)$$

we can observe that along the optimal path the decrease in the price of capital from t to $t + dt$, which can be considered as the <u>marginal cost of holding</u>

that capital, equals the marginal revenue of investing the capital as is evident from the interpretation of the Hamiltonian given above. The marginal revenue, $H_x dt$, consists of the sum of direct marginal contribution, $F_x dt$, and the indirect marginal contribution, $\lambda f_x dt$. Thus the adjoint equations become the familiar economic equilibrium relation--marginal cost equals marginal revenue.

Further insight can be obtained by integrating the above adjoint equation from t to T as follows:

$$\lambda(t) = \lambda(T) + \int_t^T H_x(x(\tau),u(\tau),\tau)d\tau$$

$$= S_x[x(T)] + \int_t^T H_x d\tau$$

Note that the price $\lambda(T)$ of a unit of capital at time T is its marginal salvage value, $S_x[x(T)]$. The price $\lambda(t)$ of a unit of capital at time t is the sum of its terminal price, $\lambda(T)$, plus the integral of the marginal surrogate profit rate, H_x, from t to T.

The above interpretations show that the adjoint variables behave in much the same way as dual variables in linear (and nonlinear) programming. (Other interpretations are given in [Dorfman, 1969] and [Peterson, 1973].) The differences being that here the adjoint variables are time dependent and satisfy derived differential equations. These connections will become clearer in Chapter 7 which covers the discrete maximum principle.

2.3 Elementary Examples

In order to absorb the maximum principle the reader should study very carefully the examples in this section, all of which are problems having only one state and one control variable. Some or all of the exercises at the end of the chapter should also be worked.

Example 2.1. Consider the problem

$$\text{Maximize} \quad \{J = \int_0^1 (-x)dt\} \tag{2.32}$$

subject to the state equation

$$\dot{x} = u, \quad x(0) = 1 \tag{2.33}$$

and the control constraint

$$u \in \Omega = [-1,1]. \tag{2.34}$$

Note that $T = 1$, $F = -x$, $S = 0$, and $f = u$. Because of the minus sign in F we can interpret the problem as being that of minimizing the (signed) area under the curve $x(t)$ for $0 \leq t \leq 1$.

Solution. First we form the Hamiltonian

$$H = -x + \lambda u \tag{2.35}$$

and note that, because the Hamiltonian is linear in u, the form of the optimal control is

$$u^*(t) = \begin{cases} 1 & \text{if } \lambda > 0 \\ \text{undefined} & \text{if } \lambda = 0 \\ -1 & \text{if } \lambda < 0 \end{cases} \tag{2.36}$$

or, referring to the notation in Section 1.4

$$u^*(t) = \text{bang}[-1,1;\lambda] \tag{2.37}$$

To find λ we write the adjoint equation

$$\dot{\lambda} = -\frac{\partial H}{\partial x} = 1, \quad \lambda(1) = S_x[x(T)] = 0 \tag{2.38}$$

Because this equation does not involve x and u we can easily solve it as

$$\lambda(t) = t-1. \tag{2.39}$$

It follows that $\lambda(t) = t-1 \leq 0$ for all $t\epsilon[0,1]$ and since we can set $u^*(1) = -1$ (which defines u at the single point $t=1$), we have the optimal control

$$u^*(t) = -1 \quad \text{for} \quad t\epsilon[0,1].$$

Substituting this into the state equation (2.33) we have

$$\dot{x} = -1, \qquad x(0) = 1 \tag{2.40}$$

whose solution is

$$x^*(t) = 1-t \quad \text{for} \quad t\epsilon[0,1] . \tag{2.41}$$

The graphs of the optimal state and adjoint trajectories appear in Figure 2.2 Note that the optimum value of the objective function is $J^* = -\frac{1}{2}$.

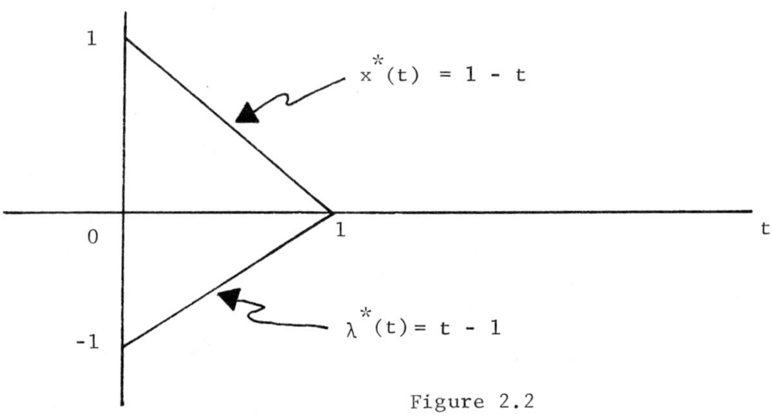

Figure 2.2

<u>Example 2.2.</u> Let us solve the same problem as in Example 2.1 over the interval $[0,2]$ so that the objective function is

$$\text{Maximize } \{J = \int_0^2 - x dt\} \tag{2.42}$$

The constraints are (2.33) and (2.34) as before. Here we want to minimize the signed area between the horizontal axis and the trajectory of $x(t)$ for $0 \leq t \leq 2$.

Solution. The Hamiltonian is (2.35) and the optimum control is (2.37) as before. The adjoint equation is

$$\dot{\lambda} = 1, \quad \lambda(2) = 0 \tag{2.43}$$

is same as (2.38) except $T = 2$ instead of $T = 1$. The solution of (2.43) is easily found to be

$$\lambda(t) = t-2 \text{ for } t\epsilon[0,2].$$

Hence the state equation (2.40) and its solution (2.41) are exactly the same. The graphs of the optimal state and adjoint trajectories appear in Figure 2.3. Note that the optimal value of the objective function here is $J^* = 0$.

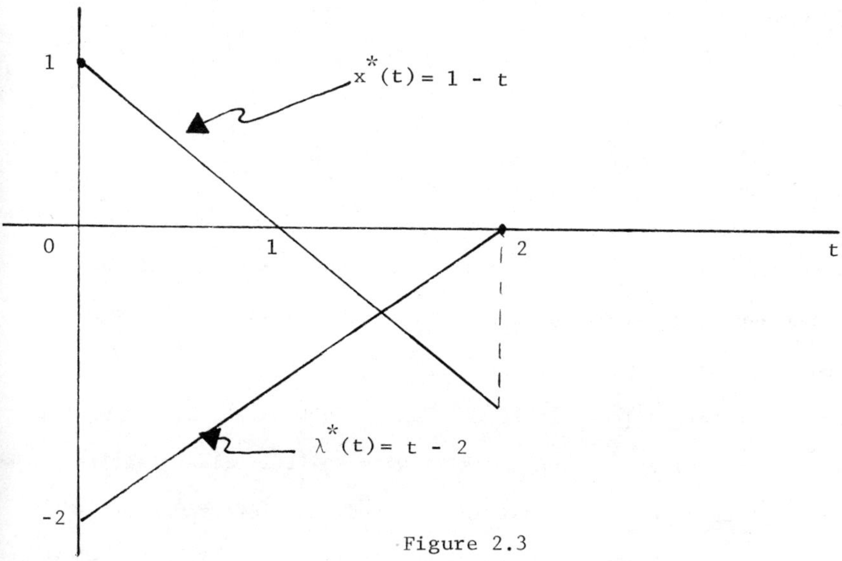

Figure 2.3

Example 2.3. The next example has objective function

$$\text{Maximize } \{J = \int_0^1 (-\frac{1}{2}x^2)dt\} \tag{2.45}$$

subject to the same constraints as in Example 2.1, namely

$$\dot{x} = u, \quad x(0) = 1 \tag{2.46}$$

$$u \in \Omega = [-1,1]$$

Here $F = -(1/2)x^2$ so that the interpretation of the objective function (2.45) is that we are trying to find the trajectory $x(t)$ so that the area under the curve $(1/2)x^2$ is minimized.

Solution. The Hamiltonian is

$$H = -\frac{1}{2}x^2 + \lambda u \tag{2.47}$$

which is linear in u so that the optimal policy is

$$u^*(t) = \text{bang}[-1,1;\lambda] \tag{2.48}$$

The adjoint equation is

$$\dot{\lambda} = -\frac{\partial H}{\partial x} = x, \qquad \lambda(1) = 0 \tag{2.49}$$

Here the adjoint equation involves x so that we cannot solve it initially. Because the state equation (2.45) involves u, which depends on λ, we can't integrate it either!

The way out of this dilemma is to use some intuition. Since we want to minimize the area under $(1/2)x^2$, and since $x(0) = 1$, it is clear that we want to decrease x as quickly as possible. Let us therefore temporarily assume that λ is non-positive in the interval $[0,1]$ so that from (2.48) we have $u = -1$ through out the interval. (In Exercise 2.5 you will be asked to show that this assumption is correct.) With this assumption we can solve (2.46) as

$$x(t) = 1-t \tag{2.50}$$

Substituting this into (2.49) gives

$$\dot{\lambda} = 1-t$$

Integrating both sides of this equation from t to 1 gives

$$\int_t^1 \dot{\lambda}(\tau)d\tau = \int_t^1 (1-\tau)d\tau$$

$$\lambda(1) - \lambda(t) = (\tau - \frac{1}{2}\tau^2)\Big|_t^1$$

which, using $\lambda(1) = 0$, yields

$$\lambda(t) = -\frac{1}{2}t^2 + t - \frac{1}{2} \tag{2.51}$$

The reader may now verify that $\lambda(t)$ is non-positive in the interval [0,1], verifying our original assumption. Hence (2.50) and (2.51) satisfy the necessary conditions. In Exercise 2.6 you will be asked to show that they satisfy sufficient conditions so that they are, in fact, optimal.

Figure 2.4 shows the graphs of the optimal trajectories.

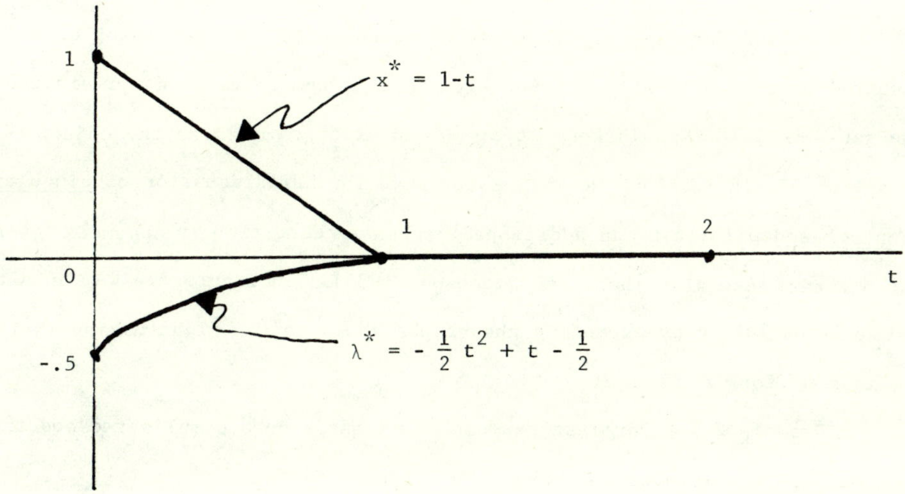

Figure 2.4

Example 2.4. Let us rework Example 2.3 with $T = 2$, that is, with objective function

$$\text{Maximize} \quad \{J = \int_0^2 (-\frac{1}{2} x^2)dt\} \tag{2.52}$$

with constraints (2.46).

Solution. The Hamiltonian is still (2.47) and the form of the optimal policy remains (2.48). The adjoint equation is

$$\dot{\lambda} = x, \qquad \lambda(2) = 0$$

which is the same as (2.49) except $T = 2$ instead of 1. Let us try to extend the solution of the previous example. Note from (2.51) that $\lambda(1) = 0$. If we recall from the definition of the bang function that bang $[-1,1;0]$ is not defined, so that we can choose u in (2.48) arbitrarily when $\lambda = 0$. This is an instance of singular control. Let us see if we can maintain singular control. To do this we choose $u = 0$ when $\lambda = 0$. Since $\lambda(1) = 0$ we set $u(1) = 0$ so that from (2.46) $\dot{x}(1) = 0$. Now note that if we set $u(t) = 0$ for $t > 1$ then by integrating these equations forward from 1 to 2 we see that $x(t) = 0$ and $\lambda(t) = 0$ for $1 \le t \le 2$; in other words $u(t) = 0$ maintains singular control in the interval. Intuitively, this is the correct answer since once we get $x = 0$ we should keep it at 0 in order to maximize the objective function J in (2.52). We will later give further discussion of singular control and will state an additional necessary condition in Appendix A4 for such cases. See also [Bell and Jacobson, 1975]. In Figure 2.4 we can get the singular solution by extending the graphs shown to the right making $x^*(t) = 0$, $u^*(t) = 0$ for $1 \le t \le 2$.

Example 2.5. Our last example is slightly more complicated and the optimal control is not bang-bang. The problem is

$$\text{Maximize} \quad \{J = \int_0^2 (2x - 3u - u^2)dt \tag{2.53}$$

subject to

$$\dot{x} = x + u, \quad x(0) = 5 \tag{2.54}$$

and the control constraint

$$u \in \Omega = [0,2] \tag{2.55}$$

Solution. Here $T = 2$, $F = 2x - 3u - u^2$, $S = 0$ and $f = x + u$. The Hamiltonian is

$$\begin{aligned} H &= (2x - 3u - u^2) + \lambda(x + u) \\ &= (2 + \lambda)x - (u^2 + 3u - \lambda u) \end{aligned} \tag{2.56}$$

Let us find the optimal control policy by differentiating (2.56) with respect to u;

$$\frac{\partial H}{\partial u} = -2u - 3 + \lambda = 0$$

so that the form of the optimal control is

$$u(t) = \frac{\lambda(t) - 3}{2} \tag{2.57}$$

provided this expression stays with in the limits of $\Omega = [0,2]$. (Note that the second derivative of H with respect to u is $\partial^2 H/\partial u^2 = -2 < 0$ so that (2.57) satisfies the second order condition for the maximum of a function.

We next derive the adjoint equation as

$$\dot{\lambda} = -\frac{\partial H}{\partial x} = -2 - \lambda, \quad \lambda(2) = 0 \tag{2.58}$$

which can be rewritten as

$$\dot{\lambda} + \lambda = -2, \quad \lambda(2) = 0.$$

This equation can be solved by the techniques explained in Appendix (A1). Its solution is

$$\lambda(t) = 2\left(e^{2-t} - 1\right) . \tag{2.59}$$

If we substitute this into (2.57) and impose the Ω constraints, we see that the optimal control is

$$u^*(t) = \begin{cases} 2 & \text{if } e^{2-t} - 2.5 > 2 \\ e^{2-t} - 2.5 & \text{if } 0 \le e^{2-t} - 2.5 \le 2 \\ 0 & \text{if } e^{2-t} - 2.5 < 0 \end{cases}$$

or referring to the notation in Section 1.4

$$u^*(t) = \text{sat}[0,2;e^{2-t} - 2.5] .$$

The graph of u^* appears in Figure 2.5.

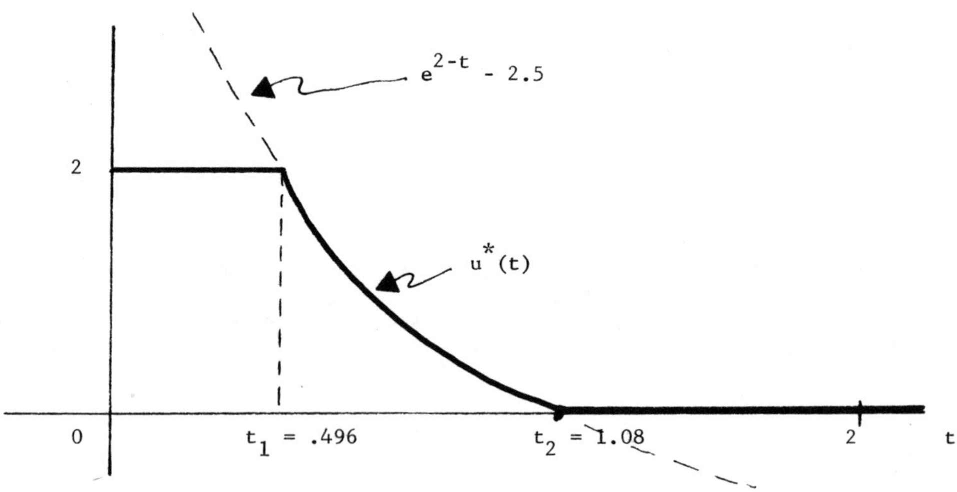

Figure 2.5

In the figure t_1 is the solution of $e^{2-t} - 2.5 = 2$ which is $t_1 = .496$ while t_2 solves $e^{2-t} - 2.5 = 0$ which is $t_2 = 1.08$.

In Exercise 2.4 you will be asked to compute the optimal trajectory $x^*(t)$ by piecing together the solutions of three different differential equations.

2.4 Sufficient Conditions

We next prove a theorem that gives conditions under which the maximum principle conditions are sufficient (as well as necessary). This theorem is important from our point of view since the models derived from many management science applications will satisfy conditions required for the sufficiency result. As remarked earlier, our technique for proving existence will be to display for any given model a solution that satisfies both necessary and sufficient conditions. A good reference for sufficiency conditions is [Seierstad and Sydsaeter, 1977].

We first define a function $H^0: E^n \times E^n \times E^1 \to E^1$ called the <u>derived Hamiltonian</u> as follows:

$$H^0(x,\lambda,t) = \underset{u\in\Omega(t)}{\text{Maximum}}\ H(x,u,\lambda,t) \tag{2.60}$$

We assume that from this equation a function $u = u^0(x,\lambda,t)$ is implicitly and uniquely defined. Given these assumptions we have by definition,

$$H^0(x,\lambda,t) = H(x,u^0,\lambda,t). \tag{2.61}$$

It is also possible to show that

$$H^0_x(x,\lambda,t) = H_x(x,u^0,\lambda,t) \tag{2.62}$$

for all $u\in\Omega(t)$. To see this for the case of differentiable u^0, let us differentiate (2.61) with respect to x:

$$H_x^0(x,\lambda,t) = H_x(x,u^0,\lambda,t) + H_u(x,u^0,\lambda,t)\frac{\partial u^0}{\partial x} \tag{2.63}$$

Let us look at the second term on the right hand side of (2.63). We must show

$$H_u(x,u^0,\lambda,t)\frac{\partial u^0}{\partial x} = 0 \tag{2.64}$$

for all x. There are two cases to consider: (i) Case 1; the unconstrained global maximum of H occurs in the interior or on the boundary of $\Omega(t)$. Here $H_u(x,u^0,\lambda,t)=0$. (ii) Case 2; the unconstrained global maximum of H occurs outside of $\Omega(t)$. Here $\frac{\partial u^0}{\partial x} = 0$ because changing x does not influence the optimum value of u. Thus (2.64) and therefore (2.62) hold. Exercise 2.14 gives a specific instance of this proof.

Remark. We have shown the result in (2.62) for cases where u^0 is a differential function of x. It holds more generally provided $\Omega(t)$ is appropriately qualified, see [Lee and Markus, 1968] and [Sethi, 1974].

THEOREM 2.1. (Sufficient Conditions.) Let $u^*(t)$ and the corresponding $x^*(t)$, $\lambda^*(t)$ satisfy the maximum principle necessary condition (2.31) for all $t\epsilon[0,T]$; then u^* is an optimal control if $H^0(x,\lambda^*,t)$ is concave in x for each t and $S(x)$ is concave in x.

Proof. The proof is a minor extension of the arguments in [Arrow and Kurz, 1970] and [Mangasarian, 1966]. By definition

$$H[x(t),u(t),\lambda^*(t),t] \le H^0[x(t),\lambda^*(t),t] \tag{2.65}$$

From concavity of H^0,

$$\begin{aligned} H^0[x(t),\lambda^*(t),t] &\le H^0[x^*(t),\lambda^*(t),t] \\ &+ H_x^0[x^*(t),\lambda^*(t),t][x(t) - x^*(t)]; \end{aligned} \tag{2.66}$$

thus from (2.61), (2.65) and (2.66)

$$H[x(t),u(t),\lambda^*(t),t] \le H[x^*(t),u^*(t),\lambda^*(t),t] + H_x^0[x^*(t),\lambda^*(t),t][x(t) - x^*(t)]. \tag{2.67}$$

By definition of H in (2.17) and the adjoint equations of (2.31)

$$F[x(t),u(t),t] + \lambda^*(t)f[x(t),u(t),t] \le F[x^*(t),u^*(t),t] + \lambda^*(t)f[x^*(t),u^*(t),t] - \dot{\lambda}^*(t)[x(t)-x^*(t)]. \tag{2.68}$$

Using the state equations of (2.31), transposing and regrouping,

$$F[x^*(t),u^*(t),t] - F[x(t),u(t),t] \ge \dot{\lambda}^*(t)[x(t)-x^*(t)] + \lambda^*(t)[\dot{x}(t) - \dot{x}^*(t)]. \tag{2.69}$$

Furthermore, since S[x(T)] is a concave function, we have

$$S[x(T)] \le S[x^*(T)] + S_x[x^*(T)]\ [x(T) - x^*(T)], \tag{2.70}$$

or

$$S[x^*(T)] - S[x(T)] + S_x[x^*(T)][x(T) - x^*(T)] \ge 0. \tag{2.71}$$

Integrating both sides of (2.69) from 0 to T and adding (2.71), we have

$$J(u^*) - J(u) + S_x[x^*(T)]\ [x(T) - x^*(T)] \ge \lambda^*(T)[x(T) - x^*(T)] - \lambda^*(0)[x(0)-x^*(0)]. \tag{2.72}$$

where J(u) is the value of the objective function associated with a control u.

Since $x^*(0) = x(0) = x_o$ (the initial condition) and since $\lambda^*(T) = S_x[x^*(T)]$ from the terminal adjoint condition in (2.31), we have

$$J(u^*) \geq J(u), \tag{2.73}$$

thus, u^* is an optimal control. This completes the proof of the theorem.

Note that if the problem is given in the Lagrange form (i.e., $S(x) \equiv 0$), then all we need is the concavity of $H^0(x,\lambda^*,t)$ in x for each t. Since λ^* is not known a priori, it is usual to test H^0 for a stronger assumption, i.e., check for the concavity of $H^0(x,\lambda,t)$ in x for each λ and each t. Sometimes the stronger condition given Exercise 2.19 can be used.

<u>Example 2.6.</u> Let us show that the problem in Examples 2.1 and 2.2 satisfy the sufficient conditions. We have from (2.35) and (2.60)

$$H^0 = -x + \lambda u^0$$

where u^0 is given by (2.36). Since u^0 is a function of λ only, H^0 is certainly concave in x for each t and λ (and in particular for $\lambda = \lambda^*$). Since $S(x) = 0$ the sufficient conditions hold.

Exercises for Chapter 2

2.1 (a) In Example 2.1 show $J^* = -\frac{1}{2}$.

(b) In Example 2.2 show $J^* = 0$.

(c) In Example 2.3 show $J^* = -\frac{1}{6}$.

(d) In Example 2.4 show $J^* = -\frac{1}{6}$.

2.2 Rework Example 2.5 with $F = 2x - 3u$.

2.3 Show that both the Lagrange and Mayer forms of the optimal control problem can be reduced to the linear Mayer form (2.5).

2.4 Complete Example 2.5 by writing the optimal $x^*(t)$ in the form of integrals over the three intervals $(0,t_1)$, (t_1,t_2) and $(t_2,2)$ shown in Figure 2.5.

[Hint. It is not necessary to actually carry out the numerical evaluation of these integrals, unless you are ambitious.]

2.5 In Example 2.3 show that λ cannot be positive over any finite interval in $[0,1)$.

2.6 Show that the derived Hamiltonian H^0 found in Examples 2.4 and 2.5 satisfy the concavity conditions required for the sufficiency result in Section 2.4.

2.7 Show that the optimal control obtained from the application of the maximum principle satisfies the principle of optimality: if $u^*(t)$ is an optimal control and $x^*(t)$ is the corresponding optimal path for $0 \le t \le T$ with $x(0) = x_0$, then verify the above proposition by showing that $u^*(t)$ for $\tau \le t \le T$ satisfies the maximum principle for the problem beginning at time τ with the initial condition $x(\tau) = x^*(\tau)$

2.8 Use the maximum principle to solve the following problem given in Meyer form:

$$\text{Maximize } 8x_1(18) + 4x_2(18)$$

$$\dot{x}_1 = x_1 + x_2 + u, \quad x_1(0) = 15$$

$$\dot{x}_2 = 2x_1 - u, \qquad x_2(0) = 20,$$

and the control constraint

$$0 \le u \le 1.$$

[Hint. Use the method in Appendix A1 to solve the simultaneous differential equations.]

2.9 A simple controlled dynamical system is modeled by the scalar equation

$$\dot{x} = x + u.$$

The optimal control problem consists in steering $x(t)$ from an initial state $x(0) = x_0$ to the target $x(1) = 0$ such that

$$J(u) = \frac{1}{4}\int_0^1 u^4 dt$$

is minimized. Use the maximum principle to show that the optimal control is given by

$$u^*(t) = \frac{4x_0}{3}(e^{-4/3}-1)^{-1}e^{-t/3}.$$

2.10 <u>Regional Allocation of Investment</u>: Let K_i, $i = 1,2$, denote the capital stock in region i. Let b_i be the productivity of capital and s_i be the marginal propensity to save in region i. Since the investment funds for the two regions come from the savings in the whole economy, we have

$$\dot{K}_1 + \dot{K}_2 = b_1 s_1 K_1 + b_2 s_2 K_2 = g_1 K_1 + g_2 K_2$$

where $g_i = b_i s_i$. Let u denote the control variable representing the fraction of investment allocated to region 1 with the remainder going to region 2. Clearly,

$$0 \leq u \leq 1, \tag{1}$$

and

$$\begin{aligned} \dot{K}_1 &= u(g_1 K_1 + g_2 K_2), \quad K_1(0) = a_1 > 0 \\ \dot{K}_2 &= (1-u)(g_1 K_1 + g_2 K_2), \quad K_2(0) = a_2 > 0. \end{aligned} \tag{2}$$

The optimal control problem is to maximize the productivity of the whole economy at time T. Thus the objective is to

$$\text{Maximize } \{J = b_1 K_1(T) + b_2 K_2(T)\}$$

subject to (1) and (2).

(a) Use the maximum principle to derive the form of the optimal policy.

(b) Assume $b_2 > b_1$. Show that $u^*(t) = 0$ for $t \in [\hat{t}, T]$ where $\hat{t}$ is a switching point and $0 \leq \hat{t} < T$.

(c)* If you are ambitious, find the $\hat{t}$ of part (b).

2.11* System (2.4) is termed autonomous if F, f and Ω are not explicit functions of time t. In this case, show that the Hamiltonian is constant along the optimal path, i.e., show that

$$\frac{dH}{dt} = 0 .$$

2.12 A water reservoir being used for the purpose of fire-fighting is leaky and its water height $x(t)$ is governed by

$$\dot{x} = -0.1x + u, \qquad x(0) = 10$$

where $u(t)$ denotes the net inflow at time t; $0 \leq u \leq 3$.

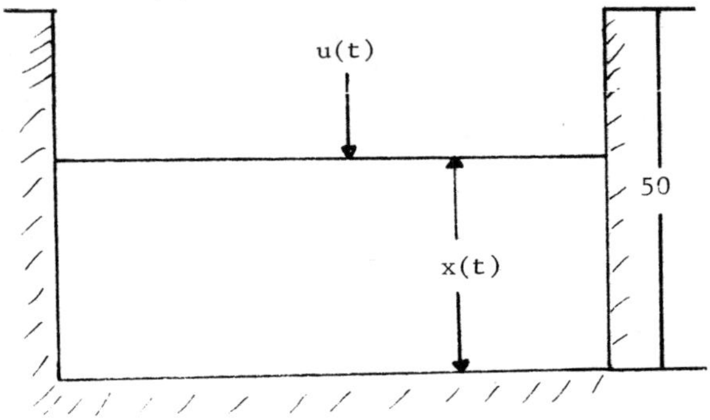

Note that $x(t)$ also represents the water pressure in some appropriate units. Also note that high water pressure is useful for fire-fighting.

a) Find the optimal control which maximizes

$$J = \int_0^{100} x\,dt.$$

Find the maximum level reached.

b) Replace the objective function in (a) by

$$J = 5x(100)$$

and resolve the problem.

c) Redo the problem with $J = \int_0^{100} (x - 5u)\,dt$.

2.13 For the fixed end point problem the transversality condition on the adjoint variable is given by $\lambda(T) = K$ where K is some constant to be determined, instead of the condition in (2.31) for the free end point problem. Using (2.72) show that the sufficiency theorem for the fixed point problem goes through with the new transversality condition.

2.14 (a) Solve the optimal consumption problem of Example 1.3 with $U(C) = \ell n C$ and $B = 0$; also replace the statement constraint $W(t) \geq 0$ by the terminal condition $W(T) = 0$.

(b) Find the rate of change of optimal consumption over time and conclude that consumption remains constant when $r = \rho$, increases when $r > \rho$, and decreases when $r < \rho$.

2.15 Suppose $H(x,u,\lambda,t) = \lambda ux - \frac{1}{2} u^2$ and $\Omega(t) = [0,1]$ for all t.

(a) Show that the optimal control u^* is given by

$$u^*(x) = \begin{cases} \lambda x & \text{if } 0 \leq \lambda x \leq 1 \\ 1 & \text{if } \lambda x > 1 \\ 0 & \text{if } \lambda x < 0 \end{cases}$$

$$= \text{sat}\,[0,1;\lambda x]$$

(b) Verify that (2.62) holds for all values of x.

2.16* Using the equation (2.62), give an alternative treatment of Section 2.2.2. (Hint: Restate (2.18) as $-V_t = H^0$ and differentiate it with respect to x.)

2.17 Derive the two-point boundary value problem whose solution gives the optimal control, $u^*(t)$, and trajectory $x^*(t)$, for the problem:

$$\text{Maximize } \{J = \int_0^1 -\frac{1}{2}(x^2 + u^2)dt\}$$

subject to

$$\dot{x} = -x^3 + u, \quad x(0) = 1.$$

The numerical solution of this TPBVP is asked for in Exercise 11.4.

2.18 Consider the following problem:

$$\max_u \left\{ J = -\int_0^T (g(x) + c^2u^2)dt \right\},$$

subject to

$$\dot{x} = f(x) + b(x)u, \quad x(0) = x_0, \quad x(T) = 0,$$

where functions $g \geq 0$, f, and b are assumed to be continuously differentiable. Derive the two point boundary value problem satisfied by the optimal state and control trajectories.

2.19 If F and f are concave and differentiable in x and u and if $\lambda(t) \geq 0$ and then the derived Hamiltonian H^0 is concave in x so that the sufficiency conditions of Theorem 2.1 holds. (These conditions are easier to check than concavity of H^0.)

3 A SURVEY OF MODEL TYPES

Optimal control theory has been used to solve problems occurring in engineering, economics, management science, and other fields. In each field of application certain general kinds of models which we call model types are likely to occur, and each such model requires a specialized form of a maximum principle. In Chapter 2 we derived in considerable detail a simple form of the continuous maximum principle. However, to continue to provide such details for each different version of the maximum principle we will need in later chapters of this book, would be both repetitive and lengthy.

The purpose of the present chapter is to avoid the latter by listing all the different management science model types that we will use in later chapters. For each model type, we will give a brief description of the corresponding objective function, state equations, terminal conditions, adjoint equations, and form of optimal control policy. We will also indicate where each of these model types is applied in later chapters.

The reader may wish to skim this chapter on first reading to get an idea of what it contains, work a few of the exercises, and go on to the various functional areas discussed in later chapters. Then when specific model types are encountered the reader may then return to read in more detail the relevant parts of the present chapter.

3.1 A General Maximum Principle

First, we state a maximum principle general enough to cover the model types described in this chapter. We will state the maximum principle without proving it and without being rigorous. For further details see [Arrow and Kurz, 1970], [Hadley and Kemp, 1971], [Hestenes, 1966], and [Bensoussan, Hurst, and Näslund, 1974].

Let the system under consideration be described by the following vector differential equation

$$\dot{x} = f(x,u,t), \quad x(0) = x_0 \tag{3.1}$$

given the initial conditions x_0 and control trajectory $u(t)$, $t \in [0,T]$. Note that in the above equation, $x(t) \in E^n$, $u(t) \in E^m$ and $f\colon E^n \times E^m \times E^1 \to E^n$ is assumed to be continuously differentiable.

Let the objective function be

$$\max\{J = \int_0^T F(x,u,t)dt + S[x(T),T]\}, \tag{3.2}$$

where $F\colon E^n \times E^m \times E^1 \to E^1$ and $S\colon E^n \times E^1 \to E^1$ are continuously differentiable and where T denotes the terminal time.

Furthermore, we define a control $u(t)$, $t \in [0,T]$ admissible, if it is piecewise-continuous and satisfies

$$g(x,u,t) \geq 0, \tag{3.3}$$

where $g\colon E^n \times E^m \times 1 \to E^q$ is differentiable in x and u and must contain terms in u. In the present chapter we shall assume that constraints of the form $u(t) \in \Omega(t)$ which we dealt with in Chapter 2 can be expressed as inequality constraints of the form $k(u,t) \geq 0$ so that they can be subsumed in the constraints (3.3); for the rest of this chapter we shall only use constraints of type (3.3). Finally, the terminal state is constraint to be

$$x(T) \in Y \subset X , \tag{3.4}$$

where X is the set of all feasible terminal states, also called the reachable set : i.e.,

$$X = \{x(T) | x(T) \text{ obtained by an admissible control } u \text{ and } (3.1)\}.$$

To state the maximum principle we define the Hamiltonian function $H: E^n \times E^m \times E^n \times 1 \rightarrow E^1$ as

$$H[x,u,\lambda,t] \triangleq F(x,u,t) + \lambda f(x,u,t), \tag{3.5}$$

where $\lambda \epsilon E^n$ (a row vector) and the Lagrangian function $L: E^n \times E^m \times E^n \times E^q \times E^1 \rightarrow 1$ as

$$L[x,y,\lambda,\mu,t] \triangleq H[x,u,\lambda,t] + \mu g(x,u,t), \tag{3.6}$$

where $\mu \epsilon E^q$ (a row vector) whose components are called Lagrange multipliers. The adjoint vector satisfies the differential equation

$$\dot{\lambda} = - L_x[x,u,\lambda,\mu,t] \tag{3.7}$$

with boundary conditions given in (3.8).

The maximum principle states that the necessary conditions for u^* to be an optimal control are that there exist λ^* and μ^* such that (3.8) holds.

Remark. It should be pointed out that if the set Y in (3.8) consists of a single point $Y = \{b\}$, which we call a fixed end point problem, then the transversality condition reduces to simply $\lambda^*(T)$ equals an arbitrary constant, since $x^*(T) = b$. In this case we must have $S = 0$. In case $Y = X$ the terminal condition in (3.8) reduce to (2.28). Further discussion of the terminal conditions is given in Section 3.3, and a summary is given in Table 3.2.

$\dot{x}^* = f(x^*, u^*, t), \quad x^*(0), \; x^*(T) \in Y \subset X$

$\dot{\lambda}^* = -L_x[x^*, u^*, \lambda^*, \mu^*, t]$,

with the terminal conditions

$\lambda^*(T) \geq S_x[x^*(T)]$

and $[\lambda^*(T) - S_x][y - x^*(T)] \geq 0$ for all $y \in Y$,

the Hamiltonian maximizing condition

$$H[x^*(t), u^*(t), \lambda^*(t), t] \geq H[x^*(t), u(t), \lambda^*(t), t] \tag{3.8}$$

for all admissible controls, i.e., piecewise-continuous controls such that for $t \in [0,T]$,

$g[x^*(t), u(t), t] \geq 0$,

and the Lagrange multipliers μ^* are such that

$\left.\frac{\partial L}{\partial u}\right|_{u=u^*} = 0$,

and the complementary slackness conditions

$\mu^* \geq 0$ and $\mu^* g(x^*, u^*, t) = 0$ hold.

<u>Example 3.1</u>. Consider the problem

$$\text{Maximize} \quad \{J = \int_0^1 u\,dt\}$$

subject to

$$\dot{x} = u, \quad x(0) = 1 \tag{3.9}$$

$$u \geq 0, \quad x - u \geq 0 \tag{3.10}$$

Note that constraints (3.10) are of type (3.3). They can also be rewritten, $0 \leq u \leq x$.

Solution. The Hamiltonian is

$$H = u + \lambda u = (1+\lambda)u$$

so that the optimal control has the form

$$u = \text{bang}\ [0,x;\ 1+\lambda] \quad . \tag{3.11}$$

To get adjoint equation and the multipliers associated with the constraints we form the Lagrangian:

$$L = H + \mu_1 u + \mu_2(x-u) \ = \mu_2 x + (1 + \lambda + \mu_1 + \mu_2)u.$$

From the second form of the Lagrangian, it follows that the optimal control is a bang function.

From this we get the adjoint equation

$$\dot{\lambda} = -\frac{\partial L}{\partial x} = -\mu_2 \ , \quad \lambda(1) = 0 \tag{3.12}$$

Also note that the optimal control must satisfy

$$\frac{\partial L}{\partial u} = 1 + \lambda + \mu_1 - \mu_2 = 0 \tag{3.13}$$

and μ_1 and μ_2 must also satisfy the complementary slackness conditions

$$\mu_1 \geq 0, \quad \mu_1 u = 0 \tag{3.14}$$

$$\mu_2 \geq 0 \ , \ \mu_2(x-u) = 0 \quad . \tag{3.15}$$

It is obvious for this simple problem that $u^*(t) = x(t)$ should be the optimal control for all t. We now show that this control satisfies all the conditions of the Lagrangian form of the maximum principle.

Since $x(0) = 1$, the control $u^* = x$ gives $x = e^t$ as the solution of (3.9). Because $x = e^t > 0$, it follows that $u^* = x > 0$; thus $\mu_1 = 0$ from (3.14).

From (3.13) we then have

$$\mu_2 = 1 + \lambda$$

Substituting this and $x = e^t$ into (3.12) and solving gives

$$1 + \lambda(t) = e^{1-t} \tag{3.16}$$

Since the right hand side of (3.16) is always positive, $u^* = x$ satisfies (3.11). Notice that $\mu_2 = e^{1-t} \geq 0$ and $x - u^* = 0$ so (3.15) holds. We have therefore shown that all the necessary conditions (3.8) hold. In Exercise 3.3 you are asked to check sufficiency conditions for this problem.

If the terminal time is unspecified, there is an additional necessary transversality condition for T^* to be optimal (see Exercise 3.5), namely,

$$H[x^*(T^*),u^*(T^*),\lambda^*(T^*),T^*] + S_T[x^*(T^*),T^*] = 0. \tag{3.17}$$

The use of this condition will be illustrated in Example 3.6.

3.1.1 Pure State Variable Inequality Constraints

In many management science and economics problems, it is common to have a non-negativity state space constraint of the form

$$x(t) \geq 0 \quad \text{for} \quad t \in [0,T]. \tag{3.18}$$

Since the control variables do not enter directly into (3.18), these are called pure state variable inequality constraints, see [Arrow and Kurtz, 1970]. To deal with such constraints, we proceed as follows.

At any point where $x_i(t) > 0$, the corresponding constraint (3.18) is not binding and can be ignored. In any interval where $x_i(t) = 0$, we must have $\dot{x}_i(t) \geq 0$ so that x_i does not become negative. Hence the control must be constrained to satisfy $\dot{x}_i = f_i \geq 0$, making $f_i \geq 0$ as constraint of type (3.3) over the interval. We can add the constraint

$$f_i(x,u,t) \geq 0, \quad \text{whenever} \quad x_i(t) = 0, \tag{3.19}$$

to the original set (3.3). We associate multipliers η_i with (3.19) whenever (3.19) must be imposed, i.e., whenever $x_i(t) = 0$. A convenient way to do this is to impose an "either or" condition, $\eta_i x_i = 0$. This will make $\eta_i = 0$ whenever $x_i > 0$. We can now form the Lagrangian,

$$L = H + \mu g + \eta f \tag{3.20}$$

and apply the maximum principle in (3.8) with the additional necessary conditions:

$$\eta^* \geq 0, \ \eta^* x^* = 0, \ \eta^* f(x^*, u^*, t) = 0. \tag{3.21}$$

Example 3.2. Consider the problem

$$\text{Maximize} \quad \{J = \int_0^2 (-x)dt\}$$

Subject to

$$\dot{x} = u, \quad x(0) = 1 \tag{3.22}$$

$$u + 1 \geq 0, \quad 1 - u \geq 0 \tag{3.23}$$

$$x \geq 0 \tag{3.24}$$

Note that (3.23) is a restatement of $-1 \leq u \leq 1$. Note further that this problem is the same as Example 2.2, except for (3.24).

Solution. The Hamiltonian is

$$H = -x + \lambda u$$

which implies the optimal control is

$$u^* = \text{bang}\,[-1,1;\lambda] \text{ whenever } x > 0. \tag{3.25}$$

Whenever $x = 0$ we must impose $\dot{x} = u \geq 0$ in order to insure that (3.24) holds. Therefore

$$u^* = \text{bang}[0,1;\lambda] \text{ whenever } x = 0\ . \tag{3.26}$$

Now we form the Lagrangian

$$L = H + \mu_1(u + 1) + \mu_2(1 - u) + \eta u$$

where μ_1, μ_2, and η satisfy the complementary slackness conditions

$$\mu_1 \geq 0, \quad \mu_1(u + 1) = 0 \tag{3.27}$$

$$\mu_2 \geq 0, \quad \mu_2(1 - u) = 0 \tag{3.28}$$

$$\eta \geq 0, \quad \eta x = 0, \quad \eta u = 0 \tag{3.29}$$

Note that (3.29) is obtained from (3.21). Furthermore the optimal trajectory must satisfy

$$\frac{\partial L}{\partial u} = \lambda + \mu_1 - \mu_2 + \eta = 0\ . \tag{3.30}$$

From the Lagrangian we also get

$$\dot{\lambda} = -\frac{\partial L}{\partial x} = 1, \quad \lambda(2) = 0\ .$$

Here the solution is the same as in Example 2.2, namely,

$$\lambda(t) = t - 2 \tag{3.3}$$

Since $\lambda(t) \leq 0$ in [0,2] and $x(0) = 1 > 0$ the initial optimal control is given by (3.25) as $u^*(t) = 1$. Substituting this into (3.22) we get $x(t) = 1 - t$ which remains positive until $t = 1$. Thus

$$u^*(t) = -1 \quad \text{for} \quad 0 \leq t < 1 \quad .$$

In this interval by (3.28) $\mu_2 = 0$ since $u^* < 1$ and by (3.29) $\eta = 0$ because $x > 0$. Therefore $\mu_1(t) = -\lambda = 2 - t > 0$ for $0 \leq t < 1$, and this with $u = -1$ satisfies (3.27).

At $t = 1$ we have $x(1) = 0$ so the optimal control is given by (3.26), which is $u^*(1) = 0$. Now assume that we continue the control $u^*(t) = 0$ for $1 \leq t \leq 2$. With this control we can solve (3.22) with boundary condition $x(1) = 0$, and observe that $x(t) = 0$ for $1 \leq t \leq 2$. Since $\lambda(t) \leq 0$ in the same interval we see that $u^*(t) = 0$ satisfies (3.26) through this interval.

To complete the problem we calculate the Lagrange multipliers. Since $u^* = 0$ we have $\mu_1 = \mu_2 = 0$ throughout $1 \leq t \leq 2$. Then from (3.30) we obtain $\eta = -\lambda = 2 - t \geq 0$ which, with $x = 0$ and $u^* = 0$ satisfies (3.29). This completes the solution. The graphs of $x(t)$ and $\lambda(t)$ are shown in Figure 3.1.

In Exercise 3.6 you are asked to solve a similar problem with $F = -u$.

A generalization of (3.18) would include pure state inequality constraints of the type $h(x,t) \geq 0$. It is noted that the constraint h_i is called a constraint of rth order if the rth time derivative of h_i is the first time a term in control u appears in the expression by putting $f(x,u,t)$ for $\dot{x}$ after each differentiation. It is through this expression that the control acts to satisfy the constraint h_i. A method for adjoining these types of constraints to form the Lagrangian is given in [Bryson and Ho, 1969]. See also Exercise 3.7.

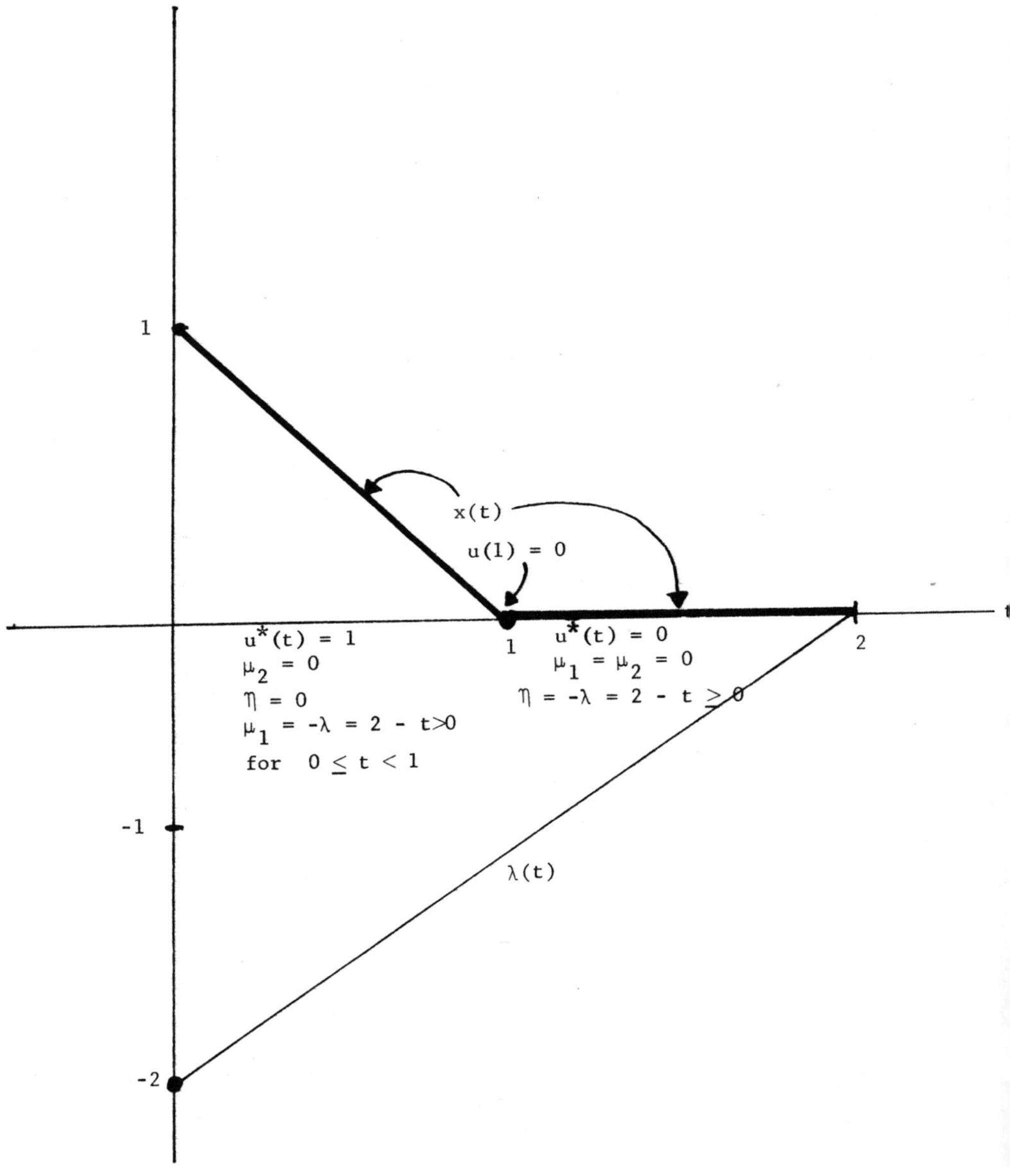

Figure 3.1

Sometimes the optimal trajectory of a state space constrained problem "hits the boundary," that is, satisfies $h(x(t),t) = 0$ for $t_1 \leq t \leq t_2$. When this happens, the adjoint trajectory may contain a <u>jump discontinuity</u> at the entry point t_1. As in Exercise 3.7, the Lagrangian, assuming h is a first order constraint, is

$$L = H + \mu g + \eta b$$

where $\boxed{b = h'(x)\dot{x} = h'(x)f(x,u,t).}$ As in (3.21) η must satisfy

$$\eta \geq 0, \quad \eta h = 0, \quad \eta b = 0$$

$\dot{\eta} \leq 0$ whenever $\dot{\eta}$ is defined (see [McIntyre and Paierwansky, 1967]) and μ and λ satisfy the conditions stated in (3.8). We now state a <u>jump condition</u> for the adjoint variable which must hold under certain conditions; see [Pontryagin, et al., 1962], and [McIntyre and Paierwonsky, 1967].

$$\lambda(t_1-) = \lambda(t_1+) + \eta(t_1+)h_x \ . \tag{3.32}$$

This jump condition usually is equivalent to the requirement that the Hamiltonian should be continuous at t_1. The continuity of the Hamiltonian makes intuitive sense when considered in the light of its interpretation given in Section 2.2.4.

Example 4.1 is an instance where this jump condition holds. Similar jump conditions may hold when the constraint $g(x,u,t) \geq 0$ becomes tight.

This brief discussion of jump conditions is far from complete, and a detailed discussion is beyond the scope of this book. An interested reader should consult the references given in the previous paragraph, and also [Berkovitz and Dryfus, 1965], [Rockafeller, 1972], [Jacobson, Lele, and Speyer, 1971], and [Russak, 1976]. For other references see [Sethi, Drews, and Seger, 1981].

3.1.2 Sufficient Conditions

In this section we will state without proof a sufficiency result due to [Seierstad and Sydsaeter, 1977] for the problem stated in (3.1) - (3.4).

If the Hamiltonian (3.5) is differentiable and concave in (x,u) and if each component of g in (3.3) is differentiable and quasi-concave (see [Mangasarian, 1969]) in (x,u), and if $\lambda(t)$ is continuous, then the necessary conditions in (3.8) are also sufficient for (x^*,u^*) to be optimal.

Another theorem in [Seierstad and Sydsaeter, 1977] replaces the concavity requirement on the Hamiltonian by a concavity requirement on H^0 and the quasi-concavity requirement on g by a constraint qualification on the state constraints (3.3).

3.2 Current-Value Formulation

In most management science and economic problems the objective function is usually formulated in money or utility terms. The future streams of money or utility are usually discounted.

For this section, let us assume a constant continuous discount rate $\rho \geq 0$. The discounted objective function can now be written as a special case of (3.2); thus we assume the time dependence of the relevant functions comes only through the discount factor, thus:

$$F(x,u,t) = \Phi(x,u)e^{-\rho t}$$
$$S[x(T),T] = \Sigma[x(T)]e^{-\rho T}. \tag{3.33}$$

Now, the objective function becomes

$$\max \{J = \int_0^T \Phi(x,u)e^{-\rho t}dt + \Sigma[x(T)]e^{-\rho T}\} \tag{3.34}$$

subject to (3.1) and (3.3).

For this problem, the standard Hamiltonian is

$$H^S \triangleq e^{-\rho t}\Phi(x,u) + \lambda^S f(x,u,t) \tag{3.35}$$

and the standard Lagrangian is

$$L^S \triangleq H^S + \mu^S g(x,u,t), \tag{3.36}$$

with the standard adjoint variables λ^S satisfying the differential equations

$$\dot{\lambda}^S = -L_x^S, \quad \lambda^S(T) = S_x[x(T),T] = e^{-\rho T}\Sigma_x[x(T)] \ . \tag{3.37}$$

We use superscript S in this section to distinguish these from the current-value functions defined in the following. Elsewhere, we do not need to make the distinction explicitly since we will either be using the standard definitions or the current-value definition of these functions. The reader will always be able to tell from the context which is meant.

We now define the current-value Hamiltonian, see [Arrow and Kurz, 1970]

$$H[x,u,\lambda,t] \triangleq \Phi(x,u) + \lambda f(x,u,t) \tag{3.38}$$

and the current-value Lagrangian

$$L[x,u,\lambda,\mu_1 t] \triangleq H + \mu g(x,u,t). \tag{3.39}$$

To see why we can do this, we note that if we define

$$\lambda \triangleq e^{\rho t} \lambda^S \quad \text{and} \quad \mu \triangleq e^{\rho t} \mu^S \tag{3.40}$$

we can rewrite (3.35) and (3.36) as

$$H = e^{\rho t} H^S \quad \text{and} \quad L = e^{\rho t} L^S . \tag{3.41}$$

Since $e^{\rho t} > 0$, maximizing H^S with respect to u at time t is equivalent to maximizing the current-value Hamiltonian H with respect to u at time t. Furthermore, from (3.40),

$$\dot{\lambda} = \rho e^{\rho t} \lambda^S + e^{\rho t} \dot{\lambda}^S . \tag{3.42}$$

The first term on the right hand side of (3.42) is simply $\rho\lambda$ using the definition in (3.40). To simplify the second term we use the differential equation (3.37) for λ^S and the fact that $L_x = e^{\rho t} L^S$ from (3.39). Thus

$$\dot{\lambda} = \rho\lambda - L_x, \quad \lambda(T) = \Sigma_x[x(T)], \tag{3.43}$$

where the terminal condition $\lambda(T)$ follows immediately from the terminal condition $\lambda^S(T)$ in (3.37) and the definition (3.40).

Finally, the complimentary slackness condition satisfied by the standard Lagrange multiplier μ^S, i.e.,

$$\mu^S \geq 0 \quad \text{and} \quad \mu^S g = 0$$

are equivalent to the similar condition satisfied by the current-value Lagrange multipliers μ, i.e.,

$$\mu \geq 0 \quad \text{and} \quad \mu g = 0$$

on account of (3.40).

To summarize, the current-value version of (3.17), i.e., the necessary transversality condition for T^* to be optimal is

$$\boxed{H[x^*(T^*),u^*(T^*),\lambda^*(T^*),T^*] - \rho\Sigma[x^*(T^*)] = 0.} \tag{3.44}$$

You are asked to prove this result in Exercise 3.9.

We will now state the maximum principle in terms of the current-value functions. It states that the necessary conditions for u^* to be an optimal control are that there exist λ^* and μ^* such that conditions (3.45) hold:

$$\dot{x}^* = f(x^*,u^*,t), \quad x^*(0) = x_0, \quad x^*(T) \in Y \subset X,$$

$$\dot{\lambda}^* = \rho\lambda^* - L_x[x^*,u^*,\lambda^*,\mu^*,t]$$

with the terminal condition

$$[\lambda^*(T) - \Sigma_x(x(T))][x - x^*(T)] \geq 0, \quad \forall x \in Y,$$

and the Hamiltonian maximizing condition

$$H[x^*(t),u^*(t),\lambda^*(t),t] \geq H[x^*(t),u(t),\lambda^*(t),t] \tag{3.45}$$

for all admissible controls, i.e., piecewise-continuous control such that for $t \in [0,T]$,

$$g[x^*(t),u^*(t),t] \geq 0,$$

and the Lagrange multipliers μ^* are such that

$$\left.\frac{\partial L}{\partial u}\right|_{u=u^*} = 0,$$ and the complementary slackness

conditions $\mu^* \geq 0$ and $\mu^* g(x^*,u^*,t^*) = 0$

hold.

Note that the remark following (3.8) applies here as well for the fixed end point problem.

Example 3.3. In Exercise 2.14(a) you were asked to solve the following consumption problem (for $\rho = r$):

$$\text{Maximize} \quad \{J = \int_0^T e^{-\rho t} \ell n\, [\, C(t)]\ dt\}$$

$$\text{Subject to} \quad \dot{W} = rW - C, \quad W(0) = W_0, \quad W(T) = 0 \tag{3.46}$$

where $W_0 > 0$. Note that the condition $W(T) = 0$ is sufficient to make $W(t) \geq 0$ for all t. We can interpret $\ell n[\, C(t)]$ as the utility of consuming $C(t)$ at time t.

In Exercise 2.14(a) you used the standard Hamiltonian Formulation to solve the problem.

We now demonstrate the use of the current value Hamiltonian formulation:

$$H = \ell n\, C + \lambda(rW - C) \tag{3.47}$$

where the adjoint equation is

$$\dot{\lambda} = \rho\lambda - \frac{\partial H}{\partial W} = \rho\lambda - r\lambda = 0, \quad \lambda(T) = K \tag{3.48}$$

since we assume $\rho = r$, and where K is some constant to be determined (see Exercise 2.13 and the remark following (3.8)) because of the fixed end point condition $W(T) = 0$. The solution of (3.48) is simply $\lambda(t) = K$ for $0 \leq t \leq T$.

To find the optimal control we maximize H by differentiating (3.47) with respect to C and setting the result to zero:

$$\frac{\partial H}{\partial C} = \frac{1}{C} - \lambda = 0$$

which implies $C = \frac{1}{\lambda} = \frac{1}{K}$. Using this consumption level in (3.46) gives

$$\dot{W} = rW - \frac{1}{K}, \quad W(T) = 0$$

which can be solved as

$$W(t) = W_0 e^{rt} - \frac{1}{Kr}(e^{rt}-1) \quad .$$

Setting $W(T) = 0$ gives

$$K = \frac{1-e^{-rT}}{rW_0} \quad .$$

Therefore the optimal consumption

$$C^*(t) = \frac{1}{K} = \frac{rW_0}{1-e^{-rT}} = \frac{\rho W_0}{1-e^{-\rho T}}$$

since $\rho = r$. Note that the present value of this stream of consumption discounted at rate ρ equals the initial wealth W_0.

In Exercise 3.10 you are asked to solve this problem for $\rho \neq r$ using the current value formulation.

The interpretation of the current-value function is that these functions reflect the values at time t in terms of the current (or, time-t) dollars. The standard functions, on the other hand, reflect the values in terms of time-zero dollars. For example, the standard adjoint variable $\lambda^S(t)$ can be interpreted as the marginal value per unit change in the state at time t in the same units as that of the objective function (3.34), i.e., in terms of time-zero dollars (see Section 2.2.4). On the other hand, $\lambda(t) = e^{\rho t}\lambda^S(t)$ is, obviously, the same value given in current (or, time-t) dollars.

For the consumption problem of Example 3.3 note that the current value adjoint function $\lambda(t) = (1-e^{-rT})/rW_0$ for all t. This gives the marginal value in time t dollars of a unit increase in wealth at time t. In Exercise (2.13) the standard adjoint variable was $\lambda^S(t) = e^{-\rho t}(1-e^{-rT})/rW_0$. Note that $\lambda^S(t)$ expresses the same marginal value in time-zero dollars.

3.3 Terminal Conditions

Terminal conditions on the adjoint variables, also known as transversality conditions, are extremely important in optimal control theory. Because the salvage value function $\Sigma[x(T)]$ is known, we know the marginal value per unit change in the state at the terminal time T. Since $\lambda(T)$ must be equal to this marginal value, it provides us with the boundary conditions for the differential equations for the adjoint variables. We shall now derive the terminal or transversality conditions for the current-value adjoint variables for important special cases of the general problem treated in Section 3.2.

(1) The easiest case is that of the free end point. In this case we do not put any constraint on the terminal state $x(T)$, thus,

$$x(T) \in X.$$

From the terminal conditions in (3.8), it is obvious that for the free-end point problem, i.e., when $Y = X$,

$$\lambda^*(T) = \Sigma_x[x^*(T)]. \tag{3.49}$$

This includes the special case $\lambda^*(T) = 0$ for $\Sigma \equiv 0$; see Example 3.1, specifically (3.12).

The physical interpretation of $\lambda^*(T)$ is as follows: Σ_x gives the marginal value of a unit charge in the terminal state and so must $\lambda^*(T)$.

(2) The case on the other extreme is that of the fixed-end point. In this case

$$x(T) = b \in X,$$

and the transversality condition in (3.8) does not provide any information for $\lambda^*(T)$. However, as remarked after (3.45), $\lambda^*(T)$ will be some constant which will be determined by solving the boundary value problem where the differential equation system consists of the state equations with both initial and terminal conditions and the adjoint equations with no boundary conditions. See Example 3.3, specially (3.48).

(3) The third case is that of one sided constraints. Here, we restrict

$$x(T) \in X \cap [b,\infty)$$

where $b \in X$. In this case it is possible to show that

$$\lambda^*(T) \geq \Sigma_x[x^*(T)] \tag{3.50}$$

and

$$\{\lambda^*(T) - \Sigma_x[x^*(T)]\}\{b - x^*(T)\} = 0. \tag{3.51}$$

For $\Sigma \equiv 0$, these terminal conditions can be written as

$$\lambda^*(T) \geq 0 \quad \text{and} \quad \lambda^*(T)[b - x^*(T)] = 0. \tag{3.52}$$

(4) The last case is the general case

$$x(T) \in Y \subset X$$

already stated in (3.8).

<u>Example 3.4.</u> Consider the problem

$$\text{Maximize} \quad \{J = \int_0^2 (-x)dt\}$$

Subject to

$$\dot{x} = u, \; x(0) = 1 \tag{3.53}$$

$$x(2) \geq 0 \tag{3.54}$$

$$-1 \leq u \leq 1 \quad .$$

<u>Solution</u>. The Hamiltonian is

$$H = -x + \lambda u.$$

Clearly the optimal control is

$$u = \text{bang}[-1,1;\lambda] \tag{3.55}$$

The adjoint equation is

$$\dot{\lambda} = 1$$

with transversality conditions obtained from (3.52)

$$\lambda(2) \geq 0 \quad \text{and} \quad \lambda(2)x(2) = 0 . \tag{3.56}$$

Since λ is monotonically increasing, the control (3.55) can switch at most once, and it can only switch from $u = -1$ to $u = 1$. Let the switching time be $t^* \leq 2$. The optimal control is then of the form

$$u^*(t) = \begin{cases} -1 & \text{for } 0 \leq t \leq t^* \\ +1 & \text{for } t^* < t \leq 2 \end{cases} . \tag{3.57}$$

Since the control switches at t^*, $\lambda(t^*)$ must be 0. Solving $\dot{\lambda} = 1$ we get

$$\lambda(t) = t - t^*.$$

There are two cases $t^* < 2$ and $t^* = 2$. We analyze the first case first. In this case $\lambda(2) > 0$; therefore from (3.56), $x(2) = 0$. Solving for x with u^* given in (3.57) we obtain

$$x(t) = \begin{cases} 1 - t & \text{for } 0 \leq t \leq t^* \\ (t-t^*) + x(t^*) = (t-t^*) + (1-t^*) & \text{for } t^* < t \leq 2 \\ \quad = t + 1 = 2t^* & \end{cases} .$$

Therefore, setting $x(2) = 0$ gives

$$x(2) = (2-t^*) + (1-t^*) = 0$$

which makes $t^* = 3/2$. Since this satisfies $t^* < 2$ we do not have to deal with the case $t^* = 2$. Figure 3.2 shows the optimal state and adjoint trajectories. You should compare Figure 3.2 with Figure 3.1.

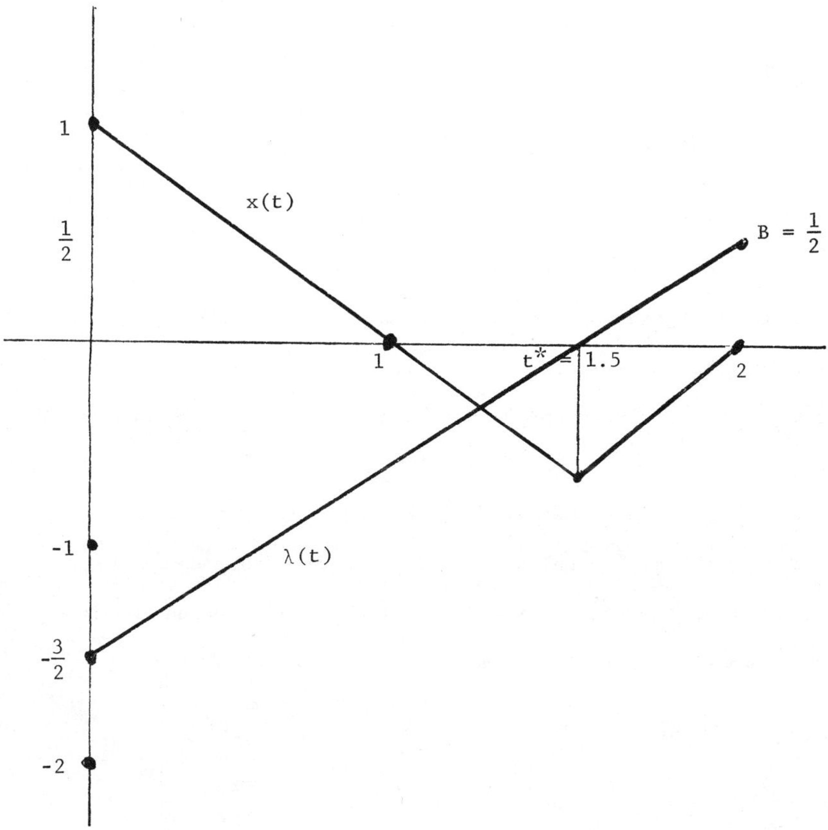

Figure 3.2

In Exercise 3.12 you are asked to rework this problem when $T = \frac{1}{2}$.

An important situation which gives rise to a one sided constraint occurs when there is an isoperimetric or budget constraint of the form

$$\int_0^T \ell(x,u,t)dt \leq K.$$

where $\ell\colon E^n \times E^m \times E^1 \to E^1$ is assumed nonnegative bounded, and continuously differentiable and $K \geq 0$ is a nonnegative constant. To see how this constraint can be converted into a one sided constraint we define an additional state variable x_{n+1} by the state equation

$$\dot{x}_{n+1} = - \ell(x,u,t), \quad x_{n+1}(0) = K, \quad x_{n+1}(T) \geq 0.$$

We employ the index $(n+1)$ simply because we already have n state variables $x \in E^n$. Also the above equation becomes an additional equation which is added to the original system.

In Exercise 3.13 you will be asked to rework the leaky reservoir of Exercise 2.12 with an additional isoperimetric constraint on the total amount of water available.

In Table 3.1, we have summarized all the terminal or transversality conditions discussed above. In the next section we discuss model types. We will see that, given the initial state conditions x_0, we can completely specify a control model by selecting a model type and a transversality condition.

3.3.1 Examples

In the first example we will solve a variation of the consumption problem in Example 3.3. It illustrates the use of one sided transversality conditions in the current value formulation.

Example 3.5. Let us modify the objective function of the consumption problem Example 3.3 to take into account the salvage (bequest) value of terminal wealth. This is the utility to the consumer of leaving an estate to his heirs upon death.

Solution. The objective function becomes:

	Constraint on $x(T)$	Description	$\lambda^*(T)$	$\lambda^*(T)$ when $S \equiv 0$
(1)	$x(T) \in Y = X$	Free-end point	$\lambda^*(T) = \Sigma_x[x^*(T)]$	$\lambda^*(T) = 0$
(2)	$x(T) = b \in X$, i.e., $Y = \{b\}$.	Fixed-end point	-	$\lambda^*(T)$ = constant to be determined
(3)	$x(T) \in X \cap [b,\infty)$ i.e., $Y = [b,\infty)$	One-sided constraint	$\lambda^*(T) \geq \Sigma_x[x^*(T)]$ and $(\lambda^*(T)-\Sigma_x[x^*(T)])\ [b - x^*(T)]=0$	$\lambda^*(T) = 0$ and $\lambda^*(T)[b-x^*(T)]=0$
(4)	$x(T) \in Y \subset X$	General constraints	$\{\lambda^*(T)-\Sigma_x[x^*(T)]\}\{y-x^*(T)\} \geq 0$ $\forall\ y \in Y$	$\lambda^*(T)[y-x^*(T)] \geq 0$ $\forall\ y \in Y$

Note 1. $x(T)$ denotes the vector of n state variables, and $\lambda(T)$ denotes vector of n adjoint variables at the terminal time T. $X \subset E^n$ denotes the reachable set of terminal states obtained by using all possible admissible controls: b is an element of X; Y is a subset of X. The function $\Sigma[x(T)]$: $E^n \to E^1$ denotes the salvage value.

In the case of the unspecified terminal time (i.e., when T is free), there is an additional transversality condition

$$H[x^*(T^*),u^*(T^*),\lambda^*(T^*),T^*] - \rho\Sigma[x^*(T^*)] = 0 \tag{3.58}$$

which must be satisfied by an optimal T^* along with the condition in the Table with T^* replacing T everywhere.

The symbol $*$ denotes the optimal values.

Note 2. Table 3.1 will provide transversality conditions for the standard Hamiltonian formulation if we replace Σ by S, and reinterpret λ as being the standard adjoint variable everywhere in the table. Also (3.17) is the standard form of (3.58).

Table 3.1 Summary of Transversality Conditions

$$\text{Maximize } \{J = \int_0^T e^{-\rho t}\log C(t)dt + e^{-\rho T}B\, W(T)\} \tag{3.59}$$

where B is a positive constant, and subject to

$$\dot{W} = rW - C, \quad W(0) = W_0, \quad W(T) \geq 0 \tag{3.60}$$

The Hamiltonian for the problem is given in (3.47) and the adjoint equation is given in (3.48) with the exception that the transversality conditions are from row (3) of Table 3.1:

$$\lambda(T) \geq B, \quad [\lambda(T) - B]\, W(T) = 0 \tag{3.61}$$

In Example 3.3 the value of K, which was the terminal value of the adjoint variable, was

$$K = \frac{1-e^{-rT}}{rW_0} \quad .$$

We now have two cases: (i) $K \geq B$ and (ii) $K < B$.

In case (i) the solution of the problem is the same as that of Example 3. because by setting $\lambda(T) = K$ and recalling that $W(T) = 0$ in that example, it follows that (3.61) holds.

In case (ii) we set $\lambda(T) = B$ and use (3.48) which is $\dot{\lambda} = 0$. Hence $\lambda(t) = B$ for all t. The Hamiltonian maximizing condition remains unchanged. Therefore the optimal consumption is

$$C = \frac{1}{\lambda} = \frac{1}{B} \, .$$

Solving (3.60) with this C gives

$$W(t) = W_0 e^{rt} - \frac{1}{Br}(e^{rt}-1) \, . \tag{3.62}$$

It is easy to show that

$$W(T) = W_0 e^{rt} - \frac{1}{Br}(e^{rT}-1) \geq 0$$

since $K < B$. Note that (3.61) holds for case (ii).

We shall apply the maximum principle next to solve a simple time-optimal problem. It is one of the problems used by Pontryagin et. al. [16] to illustrate the applications of the maximum principle. The problem also elucidates a process of synthesizing the optimal controls.

By the <u>synthesis of optimal controls</u> we mean the procedure of "patching" together various forms of the optimal controls obtained for the Hamilonian maximizing condition. A simple example of synthesis occurs in Example 2.4 where $u^* = 1$ when $\lambda > 0$, $u^* = -1$ when $\lambda < 0$ and when $\lambda = 0$ the control is singular. An optimal trajectory starting at the given initial state variables is synthesized from these. In Example 2.4 this synthesized solution is $u^* = 1$ for $0 \leq t < 1$ and $u^* = 0$ for $1 \leq t \leq 2$. Our next example requires a synthesis procedure which is more complex. In Chapter 4 both the cash management and equity financing models require such synthesis procedures.

<u>Example 3.6</u>. Consider a subway train of mass m (assume $m = 1$), which moves along a smooth horizontal track with negligible friction. The position x of the train along the track at time t is determined by Newton's Law of Motion

$$\ddot{x} = u \tag{3.63}$$

with given initial conditions

$$x(0) = x_0 \quad \text{and} \quad \dot{x}(0) = y_0,$$

and where u is an external controlling force applied to the train. The second order differential equation (3.63) can be changed into a system of two first order differential equations (see Appendix A1.1):

$$\dot{x} = y, \quad x(0) = x_0$$
$$\dot{y} = u, \quad y(0) = y_0. \tag{3.64}$$

Assume further that for the comfort of the passengers the maximum acceleration and deceleration is required to be at most 1 measured in appropriate units. Thus, the control variable constraint is

$$u \in \Omega = [-1,1]. \tag{3.65}$$

The problem is to find a control satisfying (3.65) such that the train stops at the next station, i.e., $x(T) = 0$ and $y(T) = 0$, in a minimum possible time T. We have thus defined the following optimal control problem:

$$\max_{T,u} \{J = \int_0^T -1dt\}$$

subject to

$$\dot{x} = y\ , \quad x(0) = x_0, \quad x(T) = 0$$
$$\dot{y} = u\ , \quad y(0) = y_0, \quad y(T) = 0 \tag{3.66}$$

and the control constraint

$$u \in \Omega = [-1,1].$$

Note that (3.66) is a fixed end point with unspecified terminal time.

The standard Hamiltonian function in this case is

$$H = -1 + \lambda y + \mu u$$

where the adjoint variables λ and μ satisfy

$$\dot{\lambda} = 0 \quad \text{and} \quad \dot{\mu} = -\lambda.$$

Note that there are no terminal conditions on λ and μ because we have a fixed end point problem, see row (2) in Table 3.1. Hence, integrating these equations we have

$$\lambda = c_1 \quad \text{and} \quad \mu = c_2 - c_1(T - t),$$

where c_1 and c_2 are constants to be determined from the maximum principle (2.31), condition (3.17) and the specified initial and terminal values of the state variables. The Hamiltonian maximizing condition yields the form of the optimal control to be

$$u^*(t) = \text{bang}\,\{-1,1;\ c_2 - c_1(T - t)\}. \tag{3.67}$$

The transversality condition (3.17) with $y(T) = 0$ and $S \equiv 0$ yields

$$H + S_T = \mu(T)u(T) = c_2u(T) = 1$$

which together with the bang-bang control policy (3.67), implies either

$$\mu(T) = c_2 = -1 \quad \text{and} \quad u(T) = -1$$

or

$$\mu(T) = c_2 = +1 \quad \text{and} \quad u(T) = +1 .$$

Since the switching function $c_2 - c_1(T - t)$ is a linear function of time remaining, it can change sign at most once.

Therefore we have two cases: (a) $u(\tau) = -1$ in the interval $t \leq \tau \leq T$ for some $t \geq 0$; (b) $u(\tau) = +1$ in the interval $t \leq \tau \leq T$ for some $t \geq 0$. We can integrate (3.64) in each of these cases as shown in Table 3.2. Also in the table we have the curves Γ_- and Γ_+, which are obtained by eliminating t for the expressions for x and y in each case. The parabolic curves Γ^- and Γ^+ are called <u>switching curves</u> and are shown in Figure 3.3.

(a) $u(\tau) = -1$ for $(t \leq \tau \leq T)$	(b) $u(\tau) = +1$ for $(t \leq \tau \leq T)$
$y = T - t$	$y = t - T$
$x = -(T - t)^2/2$	$x = (t - T)^2/2$
Γ^-: $x = -y^2/2$ for $y \geq 0$	Γ^+: $x = y^2/2$ for $y \leq 0$

Table 3.2

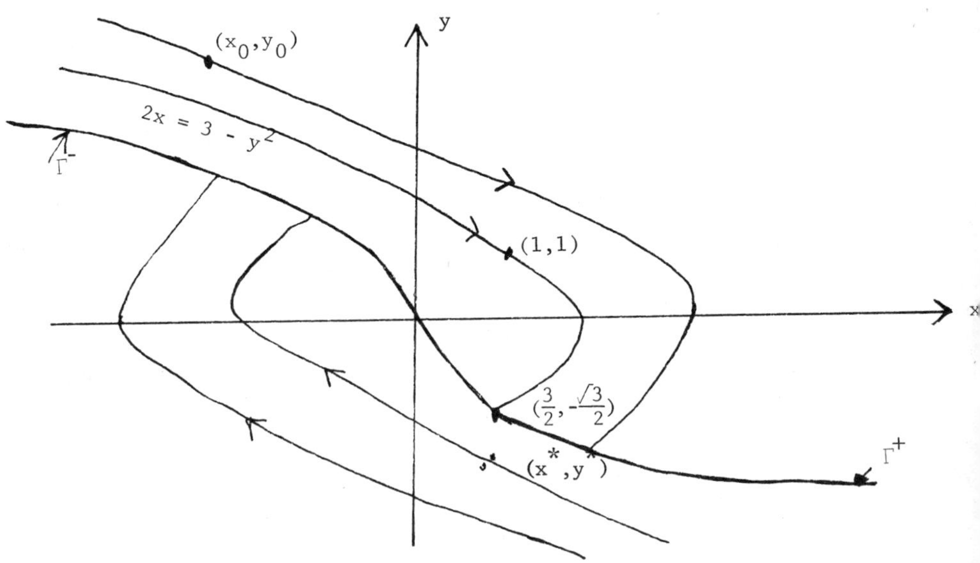

Fig. 3.3 Minimum Time Optimal Response for (3.66)

We can put Γ^- and Γ^+ into a single switching curve Γ as

$$y = \Gamma(x) = \begin{cases} -\sqrt{2x} & , \quad x \geq 0 \\ +\sqrt{-2x} & , \quad x < 0. \end{cases} \tag{3.68}$$

If the initial state (x_0, y_0) lies on the switching curve, then we have $u^* = +1$ ($u^* = -1$) if $x_0 < 0$ ($x_0 > 0$); i.e., if (x_0, y_0) lies on Γ^+ (Γ^-). If, as in most cases, the initial state (x_0, y_0) is not on the switching curve, then $u = \pm 1$ must be chosen to move the system toward the switching curve. By inspection, it is obvious that, above the switching curve, we must choose $u = -1$ and below we must choose $u = +1$.

The other curves in Figure 3.3 are solutions of the differential equations starting from initial points (x_0, y_0). If (x_0, y_0) lies above the switching curve Γ, as shown in Figure 3.3, we use $u = -1$ to compute the curve as follows:

$$\dot{x} = y \quad , \quad x(0) = x_0$$

$$\dot{y} = -1 \quad , \quad y(0) = y_0 \quad .$$

Integrating these equations gives

$$y = -t + y_0$$

$$x = -\frac{t^2}{2} + y_0 t + x_0 \quad .$$

Elimination of t between these two gives

$$x = \frac{y_0^2 - y^2}{2} + x_0 \quad . \tag{3.69}$$

This is the equation of the parabola in Figure 3.3 through (x_0,y_0). The point of intersection of the curve (3.69) with the switching curve Γ^+ is obtained by solving (3.69) and the equation for Γ^+, namely $2x = y^2$, simultaneously, which gives

$$x^* = \frac{y_0^2 + 2x_0}{4}, \quad y^* = -\sqrt{(y_0^2 + 2x_0)/2} \tag{3.70}$$

where the - sign in the expression for y^* in (3.70) was chosen since the intersection occurs when y^* is negative. The time t^* to reach the switching curve, called the <u>switching time</u>, given that we start above it, is then

$$t^* = y_0 - y^* = y_0 + \sqrt{(y_0^2 + 2x_0)/2} \ . \tag{3.71}$$

To find the minimum total time to go from the starting point (x_0,y_0) to the origin (0,0) we substitute t^* into the equation for Γ^+ in Table 3.2(b); this gives

$$T = t^* - y^* = y_0 + \sqrt{2(y_0^2 + 2x_0)} \tag{3.72}$$

As a numerical example let us start at the point (1,1). Then the equation of the parabola (3.69) is $2x = 3 - y^2$; the switching point (3.70) is $(3/4, -\sqrt{3/2})$; finally the switching time is $t^* = 1 + \sqrt{3/2}$ from (3.71). Substituting into (3.72) we find the minimum time to stop is $T = 1 + \sqrt{6}$.

To complete the solution of this numerical example let us evaluate c_1 and c_2, which are needed to obtain λ and μ. Since (1,1) is above the switching curve $u(T) = 1$ and therefore $c_2 = 1$. To compute c_1 we observe that $c_2 - c_1(T - t^*) = 0$ so that $c_1 = c_2/(T - t^*) = 1/\sqrt{3/2} = \sqrt{2/3}$.

In Exercises 3.15 through 3.18 you are asked to work other examples with different starting points above, and below and on the switching curve.

3.4 Infinite Horizon and Stationarity

As a final topic, we briefly discuss the case of $T = \infty$, called the infinite horizon case. This case is especially important in most economic problems and management science problems.

We remark that in the infinite horizon case, the requirement that the transversality conditions of Table 3.1 hold in the limit as $T \to \infty$ are not necessary conditions, in general; see [Arrow and Kurz, 1970]. The limiting transversality conditions are sufficient for optimality, however, provided the other sufficiency conditions stated in Theorem 2.1 are also met.

For the important case of the free-end point, the limiting transversality condition is

$$\lim_{T\to\infty} e^{-\rho T} \lambda^*(T) = 0. \tag{3.73}$$

Another important case is that of one-sided constraints $\lim_{T\to\infty} x(T) \geq 0$. Then the transversality condition is

$$\lim_{T\to\infty} e^{-\rho T} \lambda^*(T) \geq 0 \quad \text{and} \quad \lim_{T\to\infty} e^{-\rho T} \lambda^*(T)x^*(T) = 0. \tag{3.74}$$

In most management science problems, time t does not enter explicitly in the functions Φ, f, and g. The assumption of explicit independence of t is termed the stationarity assumption. More specifically, with respect to problem treated in (3.34), (3.1) and (3.3) where Φ is already independent of time, and without the terminal condition on $x(T)$, the stationarity assumption implies

$$\boxed{\begin{aligned} f(x,u,t) &= f(x,u) \\ g(x,u,t) &= g(x,u) \end{aligned}} \tag{3.75}$$

Note also that for the infinite horizon case we do not need Σ; so that we set $\Sigma \equiv 0$.

This means the state and the current-value adjoint equations, and the current-value Hamiltonian in (3.38) are all explicitly independent of time t. Such a system is termed <u>autonomous</u>.

In case of autonomous systems, considerable attention is focused on equilibrium where all motion ceases, i.e., the values of x and λ for which $\dot{x} = 0$ and $\dot{\lambda} = 0$. The notion is that of <u>optimal long-run stationary equilibrium</u> [Arrow and Kurz, 1970, Chap. 2.]. It is defined by the quadruple $\{\bar{x},\bar{u},\bar{\lambda},\bar{\mu}\}$ satisfying

$$\boxed{\begin{array}{l} f(\bar{x},\bar{u}) = 0 \\ \rho\bar{\lambda} = -L_x[\bar{x},\bar{u},\bar{\lambda},\bar{\mu}], \\ \bar{\mu} \geq 0, \quad \bar{\mu}g(\bar{x},\bar{u}) = 0, \\ \text{and} \\ H(\bar{x},\bar{u},\bar{\lambda}) \geq H(\bar{x},u,\bar{\lambda}) \\ \text{for all} \\ \quad u \in \{u|g(\bar{x},u) \geq 0\}. \end{array}} \tag{3.76}$$

Clearly, if the initial condition $x_0 = \bar{x}$, the optimal control $u^*(t) = \bar{u}$, for all t. If the constraint is not imposed, $\bar{\mu}$ may be dropped from the quadruple.

It may be worthwhile to remark that the <u>optimal long-run stationary equilibrium</u> (which is also called the <u>turnpike</u>) is not the same as the optimal steady-state among the set of possible steady-states. (See Exercises 3.31 - 3.[illegible] The later concept is termed the <u>Golden Rule</u> or <u>Golden Path</u> in economics. Howev[illegible] the two concepts are identical if the discount rate $\rho = 0$.

Example 3.7. Let us return to Example 3.3 and now assume that we have a foundation with initial fund W_0, an infinite horizon, and wants to maximize its total discounted utility subject to the condition that $W(t) \geq 0$ for all t. It suffices for this to impose the condition

$$\lim_{T\to\infty} W(T) \geq 0 . \tag{3.77}$$

For convenience we restate the problem:

$$\text{Maximize} \quad \{J = \int_0^\infty e^{-\rho t} \log C(t)dt\}$$

subject to

$$\dot{W} = rW - C, \quad W(0) = W_0 > 0, \tag{3.78}$$

and (3.77). We again assume $\rho = r$, as in Example 3.3.

By (3.76) we set

$$r\bar{W} - \bar{C} = 0, \quad \bar{\lambda} = k, \text{ where } k \text{ is to be determined.}$$

This gives the optimal control $\bar{C} = r\bar{W}$, and, setting $\bar{\lambda} = 1/\bar{C} = 1/r\bar{W}$, we see all the conditions of (3.76) hold, including the Hamiltonian maximizing condition. Furthermore $\bar{\lambda}$ and $\bar{W} = W_0$ satisfy the transversality conditions (3.74). Therefore, by the sufficiency theorem, the control obtained is optimal. Note that the interpretation of the solution is that the foundation consumes only the interest from its endowment W_0. Note further that the triple $(\bar{W},\bar{C},\bar{\lambda}) = \{W_0, rW_0, 1/rW_0\}$ is an optimal long-run stationary equilibrium for the problem.

3.5 Model Types

We are now able to state the general forms of all the models (with one or two exceptions) that we will use to analyze the applications discussed in the rest of the book. Some other model types will be explained in later chapters.

In Table 3.3 we have listed six different combinations of Φ and f functions. If we assume that the initial value x_0 of x is given we can get a completely specified optimal control model by selecting one of the model types in Table 3.3 together with one of the terminal conditions given in Table 3.1.

The reader will see numerous examples of the uses of Tables 3.1 and 3.3 when we construct optimal control models of various applied situations in later chapters. To help in understanding these we shall give a brief mathematical discussion of the six model types in Table 3.3 together with an indication of where each model types will be used later in the book.

(a) In model type (a) of Table 3.3 we see that both state and objective functions are linear functions of their arguments. Hence it is called the linear-linear case. The Hamiltonian is

$$\begin{aligned} H &= -Cx - Du + \lambda(Ax + Bu + d) \\ &= -Cx + \lambda Ax + \lambda d + (-D + \lambda B)u \end{aligned} \tag{3.79}$$

from the second form of the Hamiltonian it is obvious that the optimal policy is bang-bang with the switching function $(-D + \lambda B)$. Since the adjoint equation is independent of both control variables and state variables it can be solved completely without resorting to two-point boundary value methods. Examples of (a) occur in the cash balance problem of Section 4.1.1 and the maintenance and replacement model of Section 8.1.1.

	Objective Function $\Phi =$	State Equation $\dot{x} = f =$	Current Value Adjoint Equation $\dot{\lambda} =$	Form of Optimal Control Policy
a)	$Cx + Du$	$Ax + Bu + d$	$\lambda(\rho - A) - C$	Bang-bang
b)	$C(x) + Du$	$Ax + Bu + d$	$\lambda(\rho - A) - C_x$	Bang-Bang+Singular
c)	$x'Cx + u'Du$	$Ax + Bu + d$	$\lambda(\rho - A) - 2x'C$	Linear decision rule
d)	$C(x) + Du$	$A(x) + Bu + d$	$\lambda(\rho - A_x) - C_x$	Bang-Bang+Singular
e)	$c(x) + q(u)$	$(ax)b(u) + e(x)$	$\lambda(\rho - ab(u)) - c_x$	Interior or boundary
f)	$c(x)q(u)$	$(ax)b(u) + e(x)$	$\lambda(\rho - ab(u)) - c_x q(u)$	Interior or boundary

Note. The current value Hamiltonian is always used, where $\rho > 0$ is the discount rate; the standard formulation is identical to the current value formulation when $\rho = 0$. Capital letters indicate vector functions and small letters indicate scalar functions or vectors. A function followed by an argument in parentheses indicates a non-linear function; when it is followed by an argument without parenthesis it indicates a linear function, Thus, $A(x)$ and $a(x)$ are non-linear vector and scalar functions, while Ax and ax are linear. The function d is always to be interpreted as an exogeneous function of time only.

Table 3.3 Objective, State, and Adjoint Equations for Various Model Types

(b) Case (b) of Table 3.3 is the same as case (a) except that the function $C(x)$ is non-linear. Thus the term C_x appears in the adjoint equation and two-point boundary value methods are needed to solve the problem. Here there is the possibility of singular control, see section 4.5. A specific example of (b) is the Nerlove-Arrow model in Section 6.1.1.

(c) This case has linear functions in the state equation and quadratic functions in the objective function. Therefore it is sometimes called the linear-quadratic case. A specific example of this case occurs in the production-inventory example of Section 5.1.1; a complete solution of its two-point boundary value problem for an example is asked for in Exercises 11.2 and 11.3. The discrete time version of this problem is solved numerically in Section 11.2.

(d) This case is a more general version of case (b) in which the state equation is non-linear in x. The wheat trading model of Section 5.2.1 illustrates this case.

(e,f) Here the functions are scalar functions, and there is only one state equation so that λ is also a scalar function. In these cases the Hamiltonian function is nonlinear in u. If it is concave in u then the optimal control is usually obtained by setting $H_u = 0$. If it is convex, then the optimal control is as in case (b).

Several examples of case (e) occur in this book: the optimal financing model in Section 4.2.1; the extension of the Nerlove-Arrow model in Section 6.1.; the Vidale-Wolfe advertising model in Section 6.2.1; the nonlinear extension of the maintenance and replacement model in Section 8.1.4; the forestry model in Section 9.2.1; the exhaustible resource model in Section 9.3.1; and all the models of Chapter 10. Model (f) examples are: the Kamien-Schwartz model in Section 8.2.1; and the sole owner fishery resource model in Section 9.1.

Although the general forms of the model are specified in Tables 3.1 and 3.3, there are a number of additional modelling "tricks" that are useful, and that will be employed later. We collect these as a series of remarks below.

Remark 1. We sometimes need to use the absolute value function $|u|$ of a control variable u in forming the functions Φ or f. For example, in the simple cash balance model of Chapter 4, $u < 0$ represents buying, and $u > 0$ represents selling; in either case there is a transaction cost which can be represented as $c|u|$. In order to handle this we write the following equations:

$$u \triangleq u^+ - u^-, \quad u^+ \geq 0, \quad u^- \geq 0 \tag{3.80}$$

$$u^+ u^- = 0 \; . \tag{3.81}$$

Thus we represent u as the difference of two nonnegative variables, u^+ and u^-, together with the quadratic constraint (3.81). We can then write

$$|u| = u^+ + u^- \tag{3.82}$$

and so have written the nonlinear function $|u|$ as a linear function with a quadratic constraint (3.81).

We now observe that we need not impose (3.81) explicitly provided there is a cost associated with each of the variables u^+ and u^-; for because of the transaction cost, no optimal policy would ever choose to make both of them <u>simultaneously</u> positive. (In the cash management example we never simultaneously buy and sell the same security.) Thus by doubling the number of variables and adding inequality constraints we are able to represent $|u|$ as a linear function in the model.

Remark 2. Tables 3.1 and 3.3 are constructed for continuous time models. Exactly the same kinds of models can be developed for the discrete time case, se Chapter 7.

Remark 3. Consider cases (a) and (b) when the control variable constrain are defined by linear inequalities of the form

$$g(u,t) \geq 0 \quad . \tag{3.83}$$

Suppose further that $g(u,t) = g(u)$ is independent of time. Then the problem of maximizing the Hamiltonian function becomes:

$$\begin{aligned} &\text{Maximize} \quad (D + \lambda B)u \\ &\text{Subject to} \\ &\qquad g(u) \geq 0. \end{aligned} \tag{3.84}$$

This is clearly a linear programming problem for each time t since the Hamiltonian function is linear in u.

Further in case (a) the adjoint equation does not contain terms in x and u, so that we can solve for $\lambda(t)$ and hence the objective function of (3.84) varies parametrically with $\lambda(t)$. In this case we can use parametric linear programming techniques to solve the problem over time. Since the optimal solution to the linear program always occurs at an extreme point of the convex set defined by $g(u) \geq 0$, it follows that as $\lambda(t)$ changes, the optimal solution to (3.84) will "bang" from one extreme point of the feasible set to another. This is called a generalized bang-bang optimal policy. Specific examples of this kind of policy is given in the trading models of Chapter 5.

In case (b) the adjoint equation contains terms in x, so that we cannot solve for the trajectory of $\lambda(t)$ without knowing the trajectory of x

It is still true that (3.84) is a linear program for any given t, but the parametric linear programming techniques will not usually work. Instead some type of iterative procedure is needed in general; see Chapter 11.

Remark 4. In Section 3.1 we discussed several ways of handling state space constraints. Another method is to employ penalty functions such as

$$p(x) = x(1 - e^{-x}) . \tag{3.85}$$

The graph of this function is shown in Figure 3.4.

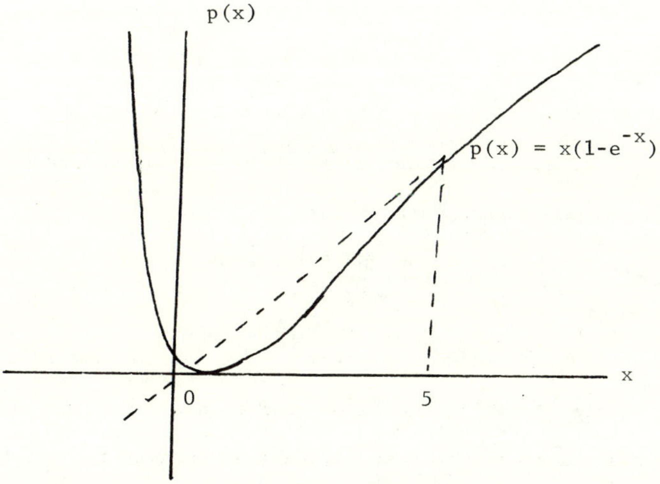

Figure 3.4 Penalty Function

Note that for $x \geq 5$, the functions $p(x)$ and x are so close that they are indistinguishable on the graph; i.e. $5 - p(5) = .034$.
For smaller values of x the function $p(x)$ lies below x. However, for negative x, $p(x)$ is positive and becomes large very rapidly as x goes more negative;

e.g. $p(-2) = 12.8$, $p(-5) = 737$, $p(-10) = 220{,}255$.

We can use p as a penalty function to impose (approximately) the state variable constraint

$$x \geq 0 . \tag{3.86}$$

That is, instead of using a Lagrange multiplier η and adding (3.86) to the Lagrangian we simply add the penalty cost $p(x)$ to the objective function, see [Russell, 1965].

Remark 5. The salvage value part of the objective function, $S(x(T))$, makes sense in two cases:

(a) When T is free, and part of the problem is to determine the optimal stopping time (see Chapter 8).

(b) When T is fixed and we want to maximize the salvage value $S(x(T))$ of the ending state $x(T)$.

The salvage value function for the fixed end point problem and for the infinite horizon case should be omitted.

Remark 6. One important model type that we did not include in the above tables is the impulsive control model of [Bensoussan and Lions, 1975]. In this model a finite control is instantaneously exerted on a state variable. This model is particularly appropriate for the instantaneous reordering of inventory as required in lot-size models, see [Bensoussan, et al., 1974]. Further discussion of impulsive control is given in Section 12.5.

Exercises for Chapter 3

3.1 Consider the constraint set

$$\Omega=\{(u_1,u_2)|0 \leq u_1 \leq x, \quad -1 \leq u_2 \leq u_1\}.$$

Write these in the form shown in (3.3).

3.2 Find the reachable set X defined in Section 3.1 if x and u satisfy

$$\dot{x} = u-1, \quad x_0 = 5, \quad -1 \leq u \leq 1$$

and $T = 3$.

3.3 Check that the solution of Example 3.1 satisfies the sufficiency conditions in Section 2.4.

3.4 Rework Example 3.2 with terminal time $T = 1/2$.

3.5 Derive the transversality condition (3.17) for the case of unspecified terminal time by first solving the problem stated in Mayer form for a fixed time T to obtain the optimal value $J^*(T)$ of the objective function, and then maximize $J^*(T)$ with respect to T by using the condition $dJ^*(T)/dT = 0$.

3.6 Change the objective function of Example 3.2 to be

$$\text{Maximize} \quad \{J = \int_0^2 (-u)dt\}$$

and resolve. Show that the solution is not unique.

3.7 Generalize the maximum principle for the pure state variable inequality constraint of the form

$$h(x,t) \quad \text{for all} \quad t \quad \text{satisfying} \quad 0 \leq t \leq T,$$

in place of $x(t) \geq 0$ in (3.18).

Hint: Define $y = h(x,t)$ as an additional state variable. Assume that $h(x,t) \geq 0$ is a first order constraint.

3.8 Develop the current-value formulation of Section 3.2 for a time varying nonnegative discount rate $\rho(t)$. [Hint. Replace the factor $e^{-\rho\tau}$ in (3.33) by $\alpha(\tau) = e^{-\int_0^\tau \rho(S)dS}$, where $\tau = t$ or T.]

3.9 Starting from (3.17) derive its current value version (3.44).

3.10 Resolve Examples 3.3 and 3.7 when $\rho \neq r$ using the current value formulation.

3.11 Show that the current-value Hamiltonian (3.38) is explicitly independent of time if f is explicitly independent of time. Show that $dH/dt = \rho\lambda f$ and contrast this result with that of Exercise 2.11.

3.12 Rework Example 3.4 with the terminal condition replaced by $x(2) \geq 1$.

3.13 Recall problem 2.12 of the leaking reservoir in Chapter 2. In this probl there was no explicit constraint on the total amount of water available. Suppose we impose the following isoperimetric constraint on that problem:

$$\int_0^{100} u\,dt = K,$$

where $K > 0$ is the total amount of water which must be used. Assume also different the resevoir has infinite capacity. Resolve this problem for various values of K, and the objective functions in parts (a) and (b) of Exercise 2.12.

3.14 Introduce a terminal value in Example 3.4 as follows

$$\text{Maximize } \{J = \int_0^2 (-x)dt + Bx(2)\}$$

$$\text{Subject to} \quad \dot{x} = u, \quad x(0) = 1$$

$$x(2) \geq 0, \text{ i.e., } Y = [0,\infty) \text{ in Table 3.1(3),}$$

$$-1 \leq u \leq 1 \quad .$$

Note that for $B = 0$ the problem is the same as Example 3.4. Solve this problem for $B = 1/2, 1, 3/2, 2, 3$. Conclude that for $B \geq 2$ the solution for the state variable does not change.

3.15 In Example 3.6 determine the optimal control and corresponding trajectory starting at the point (-4,6).

3.16 Carry out the synthesis of the optimal control for Example 3.6 when the starting point (x_0,y_0) lies <u>below</u> the switching curve.

3.17 Use the results of Exercise 3.16 to find the optimal control and corresponding trajectory starting at the point (-1,-1).

3.18 Find the optimal control and the corresponding trajectory for Example 3.6 starting at point (-2,2) which lies on the switching curve.

3.19 What is the shortest time in which a passenger can be transported in a ballistic missile from Los Angeles to New York? Assume that the missile with ultimate mechanical and thermodynamical properties is available but that the passenger imposes the restraint that the maximum acceleration or deceleration is 100 ft/sec^2. The missile starts from rest in Los Angeles and stops at New York. Assume the path is a straight line of length 2400 miles and neglect the rotation and curvature of the earth.

3.20 Solve the following minimum weighted energy and time problem

$$\max_{u,T} \{J = \int_0^T - (1/2)(u^2 + 1)dt\}$$

Subject to

$$\dot{x} = u, \quad x(0) = 5, \quad x(T) = 0.$$

$$|u| \leq 2.$$

[Hint. Use (3.17) to determine the optimal T.]

3.21 Rework Exercise 3.20 with $F = -(1/2)(u^2 + 16)$.

[Hint. Note that use of (3.17) gives an infeasible u. Instead calculate $J^*(T)$ as defined in Exercise 3.5, and choose T to maximize it.]

3.22 Exercise (3.20) becomes a minimum energy problem if we set $F = -1/2\ u^2$. Show that the Hamiltonian maximization condition of the maximum principle implies $u^* = k$, where k is a constant. Note further that the application of (3.17) implies that $k = 0$, which gives $x(t) = 5$ for all $t \geq$ so that the terminal condition $x(T) = 0$ cannot be satisfied.

To see that there exists no optimal control in this situation, let $k < 0$ and compute J^*. It is now possible to see that $\lim_{k \to 0} J^* = 0$. This means that we can make the objective function value as close to zero as we desire, but not equal to zero. Note that in this case there are no feasible solutions satisfying the necessary conditions so we cannot check the sufficiency conditions (see the last paragraph of 2.1.4.).

3.23 Show that every feasible control of the following problem

$$\max_{T,u} \{J = \int_0^T - u\,dt\}$$

$$\dot{x} = u, \quad x = x_0, \quad x(T) = 0$$

$$|u| \leq q, \quad \text{where} \quad q > 0,$$

is an optimal control.

3.24* Let x_0 be the initial velocity of a rocket. Let u be the amount of acceleration (or deceleration) caused by applying a force which consumes fuel at the rate $|u|$. We want to bring the rocket to rest using minimum total amount of fuel. Hence we have the following optimal control problem.

$$\max_{T,u} \{J = \int_0^T - |u| dt\}$$

Subject to

$$\dot{x} = u, \quad x(0) = x_0, \quad x(T) = 0$$

and $$-1 \leq u \leq +1.$$

[Hint. Use (3.80)-(3.82) to deal with $|u|$. Show that for $x_0 > 0$, say $x_0 = 5$, and $T = 10$, every feasible control is optimal.]

3.25 Analyze Exercise 3.24 with state equation

$$\dot{x} = -ax + u,$$

where $a > 0$. Show that no optimal control exists.

3.26 From the transversality conditions for the general terminal constraints in row (4) Table 3.1, derive the transversality conditions in row (1) for the free-end point case, in row (2) for the fixed-end point case and in row (3) for the one-sided constraint case. Assume $\Sigma = 0$, i.e., there is no salvage value and $X = R^1$ for simplicity.

3.27* An example which illustrates that

$$\lim_{t\to\infty} \lambda(t) = 0$$

is not a necessary transversality condition in general, is

$$\max \{J = \int_0^\infty (1-x)u\,dt\}$$

such that

$$\dot{x} = (1-x)u \quad , \quad x(0) = 0$$

and $0 \leq u \leq 1$. Show this by finding an optimum control.

3.28 Consider the regulator problem defined by the scalar equation

$$\dot{x} = u\,, \qquad x(0) = x_0$$

with the cost

$$C(u) = \int_0^\infty \left(\frac{x^4}{4} + \frac{u^2}{2}\right) dt.$$

Show that the optimal feedback control is

$$u^*(x) = \begin{cases} -\,x^2/\sqrt{2} & , \quad x > 0 \\ +\,x^2/\sqrt{2} & , \quad x < 0 \end{cases}$$

where a feedback control determines the control variable as a function of x.

3.29 Assume constraint (3.3) to be of the form $g(u,t) \geq 0$, that is g does not contain x explicitly, and assume $x(T)$ is free. Apply the Lagrangian form of the maximum principle and derive the Hamiltonian form (2.31) with

$$\Omega(t) = \{u \mid g(u,t) \geq 0\}.$$

Assume $g(u,t)$ to be of the form $\alpha \leq u \leq \beta$.

3.30* Consider the problem

$$\text{Maximize} \quad \{J = \int_0^T (-x)dt\}$$

Subject to

$$\dot{x} = -u - 1, \quad x(0) = 1,$$

$$x(t) \geq 0, \qquad 0 \leq u(t) \leq 1.$$

Show that

(a) If $T = 1$ there is exactly one feasible and optimal solution.

(b) If $T > 1$ then there is no feasible solution.

(c) If $0 \leq T < 1$ then there is a unique optimal solution.

(d) If the control constraint is $0 \leq u(t) \leq K$ there is unique optimal for every $K \geq 1$ and $T = 1/2$.

(e) The value of the objective in (d) increases as K increases.

(f) If the control constraint is $u(t) \geq 0$ then the optimal control is an impulse control defined by the limit of the solution in (e).

3.31 Consider the inventory problem

$$\text{Maximize} \quad \{J = \int_0^\infty -e^{-\rho t}[(I-I_1)^2 + (P - P_1)^2]dt\}$$

Subject to $\quad \dot{I} = P - S, \quad I(0) = I_0$

where I denotes inventory level, P denotes production rate, and S denotes a given constant demand rate.

(a) Find the optimal long-run stationary equilibrium, i.e., the turnpike defined in (3.76).

(b) Find the golden rule by setting $\dot{I} = 0$ in the state equation, solve for P, and substitute it into the objective function. Then maximize the integrand with respect to I.

(c) Verify that the golden rule inventory level obtained in (b) is the same as the turnpike inventory level found in (a) when $\rho = 0$.

3.32* Rework Exercise 3.7 by assuming that $h(x,t) \geq 0$ is a kth order constraint when $k \geq 2$.

3.33* Consider a two reservoir system such as shown below

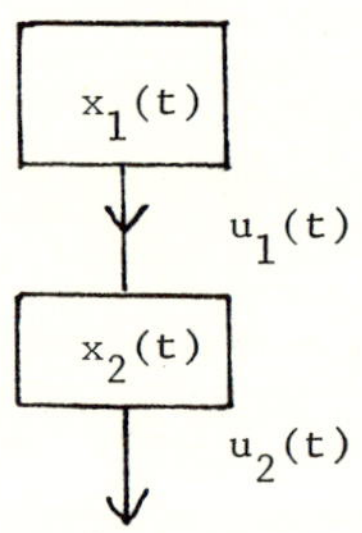

where $x_i(t)$ is the volume of water in reservoir i and $u_i(t)$ is the rate of discharge from reservoir i at time t. Thus

$$\dot{x}_1(t) = -u_1(t) \quad , \quad x_1(0) = 4$$

$$\dot{x}_2(t) = u_1(t) - u_2(t) \quad , \quad x_2(0) = 4 \ .$$

Solve the problem of maximizing

$$J = \int_0^{10} [(10-t)u_1(t) + tu_2(t)]dt$$

subject to the above state equations and the constraints

$$0 \leq u_i(t) \leq 1, \quad x_i(t) > 0 \quad \text{for all} \quad t \in [0,10].$$

[Hint: Guess the optimal solution and verify it by using the Lagrangian form of the maximum principle.]

3.34* Use the Lagrangian form of the maximum principle to obtain the optimal control for the following problem:

$$\text{maximize} \quad \{J = x_1(2)\}$$

subject to

$$\dot{x}_1 = u_1 - u_2 \quad , \quad x_1(0) = 2$$

$$\dot{x}_2 = u_2 \quad , \quad x_2(0) = 1$$

and the constraints

$$x_1(t) \geq 0, \quad x_2(t) \geq 0, \quad u_1(t) \leq x_2(t), \; 0 \leq u_2(t) \leq 2, \text{ for all } t \in [0,5].$$

An interpretation of this problem is that $x_1(t)$ is stock of steel at t, and $x_2(t)$ is the total capacity of steel mills at time t. We can use steel production, u_1 which is bounded by the current steel mill capacity, to either increase steel mill capacity u_2 or add to the stock of steel $u_1 - u_2$. The objective is to build as large a stockpile of steel as possible by time $T = 2$. (It is possible to make the problem more interesting by assuming an exogenous demand d for steel so that $\dot{x}_1 = u_1 - u_2 - d$.)

4 APPLICATIONS TO FINANCE

An important area of finance involves making decisions regarding investment and dividend policies over time and ways to finance them. Among the ways of financing such policies are: issuing equity; retaining earnings; borrowing money, etc. It is possible to model such situations as optimal control theory problems, see [Elton and Gruber, 1975]. Some of these models are simple to analyze and they yield useful insights.

In this chapter we deal with a cash balance problem and a problem of optimal equity financing of a firm. The former, in its simplest form, is the problem of controlling the level of a firm's cash balances to meet its demand for cash at minimum total cost. The latter, a central problem in finance, is that of determining the optimal dividend path along with new equity issued over time, in order to maximize the present value of all future dividends paid to the original stockholders.

Although we only deal with deterministic problems in this chapter, some of the more important problems in finance are stochastic in nature. It is possible to analyze them by using stochastic optimal control theory, which has recently been developed. A brief introduction to this new theory will be provided in Chapter 12, together with applications and references.

In the next section, we introduce a simple cash balance problem as a tutorial. This model is based on [Sethi and Thompson, 1970]. We shall be especially interested in the financial interpretations for the various functions such as the Hamiltonian and the adjoint functions that arise in the course of the analysis.

4.1 The Simple Cash Balance Problem

Consider a firm which has a known demand for cash over time. To satisfy this cash demand, the firm must keep such cash on hand. If the firm keeps too much cash, it loses money in terms of opportunity cost, in that it can earn higher returns by buying securities. On the other hand, if the cash balance is too little, the firm has to sell securities to meet the cash demand and thus incur a broker's commission. The problem, then, is to find the optimal trade-off between the cash and security balances.

4.1.1 The Model

To formulate the optimal control problem, we introduce the following notation:

T = horizon time

$x(t)$ = the cash balance in dollars at time t,

$y(t)$ = the security balance in dollars at time t,

$d(t)$ = the instantaneous rate of demand for cash; $d(t)$ can be positive or negative,

$u(t)$ = the rate of sale of securities in dollars; a negative sales rate means a purchase; $-U_2 \le u(t) \le U_1$, $U_1 \ge 0$ and $U_2 \ge 0$,

$r_1(t)$ = the interest rate earned on the cash balance,

$r_2(t)$ = the interest rate earned on the security balance,

α = the broker's commission in dollars per dollar's worth of securities bought or sold.

Then the state equations are

$$\dot{x} = r_1 x - d + u - \alpha|u|, \quad x(0) = x_0, \tag{4.1}$$

$$\dot{y} = r_2 y - u \quad , \quad y(0) = y_0. \tag{4.2}$$

The objective function is to

$$\text{maximize } \{J = [x(T) + y(T)]\}. \tag{4.3}$$

Note that the problem is in the linear Mayer form.

4.1.2 Solution by the Maximum Principle

Introduce the adjoint variables λ_1 and λ_2 and define the Hamiltonian function

$$H = \lambda_1(r_1x - d + u - \alpha|u|) + \lambda_2(r_2y - u). \tag{4.4}$$

The adjoint variables satisfy the differential equations:

$$\dot{\lambda}_1 = -\partial H/\partial x = -\lambda_1 r_1, \qquad \lambda_1(T) = 1 \tag{4.5}$$

and

$$\dot{\lambda}_2 = -\partial H/\partial y = -\lambda_2 r_2, \qquad \lambda_2(T) = 1. \tag{4.6}$$

It is easy to solve these as

$$\lambda_1(t) = e^{\int_t^T r_1(\tau)d\tau}, \tag{4.7}$$

and

$$\lambda_2(t) = e^{\int_t^T r_2(\tau)d\tau}. \tag{4.8}$$

The interpretations of these solutions are also clear, namely, $\lambda_1(t)$ is the future value (at time T) of one dollar held in the cash account from time t to T; and $\lambda_2(t)$ is the future value of one dollar's worth of securities held as securities from time t to T. Thus, the adjoint variables have natural interpretations as actuarial evaluations of competitive investments at each point of time.

Let us now derive the optimal policy by choosing the control variable u to maximize the Hamiltonian in (4.4). In order to deal with the absolute value function we write the control variable u as the difference of two non-negative variables,

$$u = u_1 - u_2, \qquad u_1 \geq 0, \qquad u_2 \geq 0 \tag{4.9}$$

Recall that this method was suggested in Remark 1, Section 3.5. In order to make $u = u_1$ when u_1 is positive, and $u = u_2$ when u_2 is negative, we also impose the quadratic constraint

$$u_1 u_2 = 0, \tag{4.10}$$

so that at most one of u_1 and u_2 can be nonzero. However, as noted in Remark 1 of Section 3.5, the optimal properties of the solution will automatically cause this constraint to be satisfied. The reason is that the broker's commission must be paid on every transaction, which makes it nonoptimal to simultaneously buy and sell securities. Now, given (4.9) and (4.10), we can write

$$|u| = u_1 + u_2 . \tag{4.11}$$

We now substitute (4.9) and (4.11) into the Hamiltonian (4.4) and reproduce below the part which depends on control variables u_1 and u_2, and denote it by W, thus

$$W = u_1[(1-\alpha)\lambda_1 - \lambda_2] - u_2[(1+\alpha)\lambda_1 - \lambda_2]. \tag{4.12}$$

Obviously, maximizing the Hamiltonian (4.4) with respect to u is same as maximizing W with respect to u_1 and u_2. But W is linear in u_1 and u_2 so that the optimal strategy is bang-bang and is as follows: thus

$$u^* = u_1^* - u_2^* , \tag{4.13}$$

where,

$$u_1^* = \text{bang} \quad [0, U_1; (1-\alpha)\lambda_1 - \lambda_2] , \tag{4.14}$$

and

$$u_2^* = \text{bang} \quad [0, U_2; -(1+\alpha)\lambda_1 + \lambda_2] . \tag{4.15}$$

Thus, $u_1(t)$ represents sale of securities, and (4.14) says that the optimal policy is to sell at the maximum allowable rate if the future value of a dollar less the broker's commission (i.e. the future value of $(1-\alpha)$ dollars) is greater than the future value of a dollar's worth of securities, and not to sell if these future values are in reverse order. In case the future value of a dollar less the commission is exactly equal to the future value of a dollar's worth of securities, then the optimal policy is undetermined. In fact, we are indifferent as to the action taken, and this is called "singular control". Similarly, $u_2(t)$ represents the purchase of securities. Here we buy if the future value of a dollar plus the commission is less than the future value of a dollar's worth of securities, and either do not buy or are indifferent for the reverse inequality and for the equality cases.

Note that if

$$(1-\alpha)\lambda_1(t) \geq \lambda_2(t)$$

then

$$(1+\alpha)\lambda_1(t) > \lambda_2(t)$$

so that if $u_1(t) > 0$, then $u_2(t) = 0$. Similarly, if

$$(1+\alpha)\lambda_1(t) \leq \lambda_2(t)$$

then

$$(1-\alpha)\lambda_1(t) < \lambda_2(t)$$

so that if $u_2(t) > 0$, then $u_1(t) = 0$. Hence, with the optimal policy, (4.10) is always satisfied.

Figure 4.1 illustrates the optimal policy at time t. Notice that the first quadrant is divided into three areas which represent different actions (including no action) to be taken. Also, the dotted lines represent the singular control cases.

A possible path of the vector $(\lambda_1(t),\lambda_2(t))$ of the adjoint variables is also shown in Figure 4.1. Note that on this path there is one period of selling, two periods of buying, and three periods of inactivity. Note also that the final point on the path is (1,1) since the final values of $\lambda_1(T) = \lambda_2(T) = 1$.

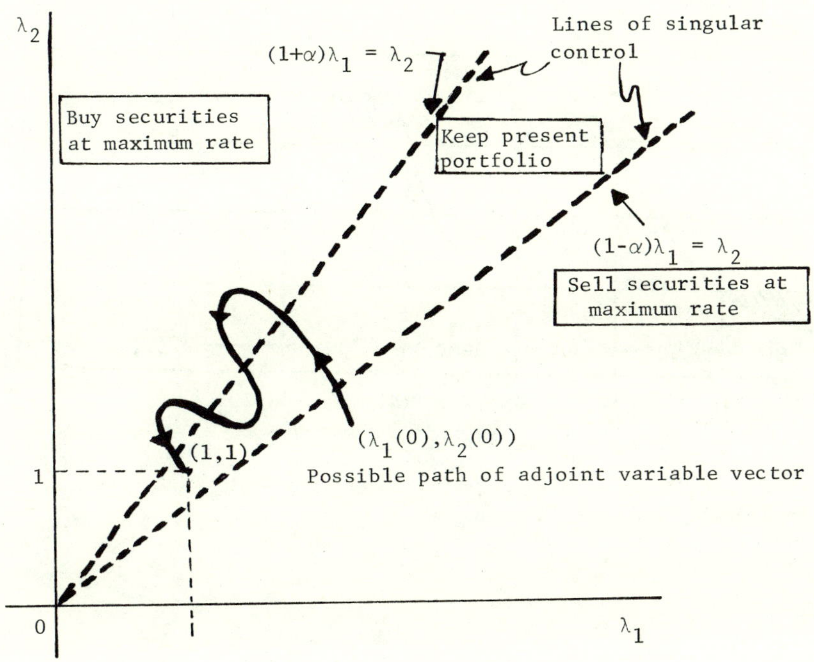

Figure 4.1 Optimal policy Shown in (λ_1,λ_2) Space

Another way to represent the optimal path is in the $(\lambda_2/\lambda_1, t)$ space. The variations in λ_1 and λ_2 shown in Figure 4.1 are also shown in Figure 4.2 below.

4.1.3 An Extension Disallowing Overdraft and Short-Selling

We can also formulate the cash balance problem in which overdrafts and short-sales are disallowed. To do this mathematically we impose the additional constraints

$$x(t) \geq 0 \qquad \text{and} \qquad y(t) \geq 0 \tag{4.16}$$

Furthermore we relax the control constraints, i.e., we set $U_1 = U_2 = \infty$ for simplicity in exposition. In Exercise 4.2 you will be asked to carry out the parallel development when U_1 and U_2 are finite.

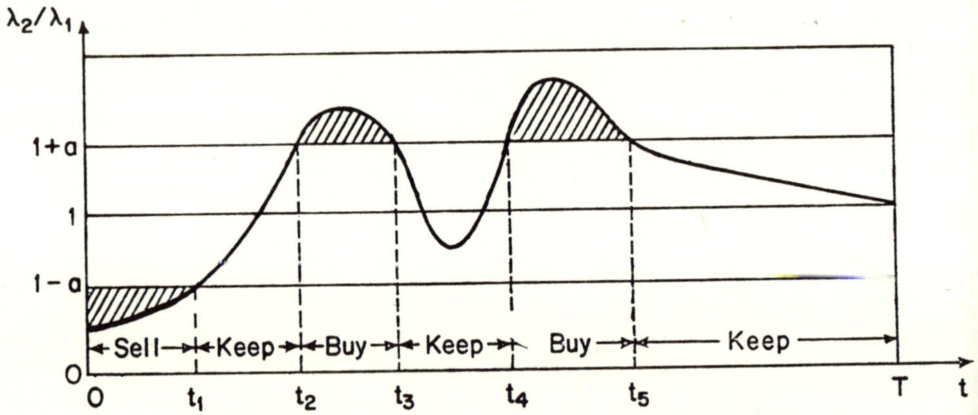

Figure 4.2 Optimal Policy Shown in $(t, \lambda_2/\lambda_1)$ Space

Generally, the presence of (4.16) will require us to form the Lagrangian as in equation (3.20) as

$$L = H + \mu\dot{x} + \nu\dot{y} = \lambda_1(r_1x-d + u - \alpha|u|) + \lambda_2(r_2y-u) +$$

$$\mu(r_1x-d + u - \alpha|u|) + \nu(r_2y-u) \qquad (4.17)$$

where,

$$\dot{\lambda}_1 = -\frac{\partial L}{\partial x} = -(\lambda_1 + \mu)r_1 \quad , \qquad \lambda_1(T) = 1 \qquad (4.18)$$

$$\dot{\lambda}_2 = -\frac{\partial L}{\partial y} = -(\lambda_2 + \nu)r_2 \quad , \qquad \lambda_2(T) = 1 \qquad (4.19)$$

and the Lagrange multipliers μ and ν must satisfy the complementary slackness conditions

$$\mu(t) \geq 0, \quad \mu(t)x(t) = 0, \quad \mu(t)[r_1x(t)-d(t)+u(t)-\alpha|u(t)|] = 0 \quad (4.20)$$

$$\nu(t) \geq 0, \quad \nu(t)y(t) = 0, \quad \nu(t)[r_2y(t) - u(t)] = 0 \qquad (4.21)$$

and

$$\frac{\partial L}{\partial u} = 0 \qquad (4.22)$$

for all $t \in [0,T]$. (See Exercise 4.2.)

In general, the solution of this problem is difficult and requires the use of a computer. However, we illustrate an easy case in which $\alpha = 0$ and r_1 and r_2 are piecewise constant functions of time.

Example 4.1. (We are indebted to C. Norstrom for this example.) Consider the model when $\alpha = 0$, $T = 10$, and r_1 and r_2 vary as follows:

$$r_1(t) = \begin{cases} 0 \text{ for } & 0 \leq t < 5 \\ .3 \text{ for } & 5 \leq t \leq 10 \end{cases} \qquad (4.23$$

$$r_2(t) = .1 \quad \text{for} \quad 0 \leq t \leq 10 \tag{4.24}$$

We also assume $x_o = 0$ and $y_o = 3$. The optimal solution when there is no upper bound on control is easy to guess. It is

$$u^*(t) = 0 \quad \text{for} \quad 0 \leq t < 5 \tag{4.25}$$

$$u^*(t) = 0 \quad \text{for} \quad 5 < t \leq 10 \tag{4.26}$$

At $t = 5$ the optimal action is clearly to sell all the securities instantaneously to take advantage of the higher interest rate on cash; we can do this because there is no upper bound on the rate of sales. Such a control is called an <u>impluse control</u>. It can be conceived to be the result of applying a very large rate of control for a very short time so as to exhaust the entire security balance in that time. The limit of this procedure as the sales rate tends to infinite and the short time interval tends to zero must exhaust the entire security balance instantaneously. That is,

$$\lim_{\substack{\delta t \to 0 \\ u^*(5) \to \infty}} u^*(5) \cdot \delta t = y(5-) = 3e^{.5} \tag{4.27}$$

where $3e^{.5}$ is the security balance at $t = 5-$ (where 5- means time just before 5) obtained by earning interest at rate .1 from 0 to 5. Note that $x(5-) = 0$. Following the application of the impulse sale $x(5+) = 3e^{.5}$ and $y(5+) = 0$, where here 5+ means "just after 5". More discussion of impulse controls will be given in Chapters 6 and 12.

We must now find the adjoint variables λ_1 and λ_2 and Lagrange multipliers μ and ν so that the maximum principle holds. From (4.22) with $\alpha = 0$ we get $\lambda_1 - \lambda_2 + \mu - \nu = 0$ or

$$\lambda_1 + \mu = \lambda_2 + \nu \tag{4.28}$$

We now solve these quantities in three time intervals $0 \leq t < 5$, $t = 5$, and $5 < t \leq 10$. We will start with t in the last interval.

(a) Assume $t \in (5,10]$. Since $x > 0$ we know $\mu = 0$; using (4.28) we can write (4.18) and (4.19) as

$$\dot{\lambda}_1 = -(\lambda_1 + \mu)r_1(t) = -.3\lambda_1 \quad , \quad \lambda_1(10) = 1$$

$$\dot{\lambda}_2 = -(\lambda_2 + \nu)r_2(t) = -.1\lambda_1 \quad , \quad \lambda_2(10) = 1 \; .$$

Solving these we have

$$\lambda_1(t) = e^{.3(10-t)}$$

$$\lambda_2(t) = 1 - \frac{1}{3}[1 - e^{.3(10-t)}] = \frac{2}{3} + \frac{1}{3}\, e^{.3(10-t)}$$

$$\nu(t) = \lambda_1 - \lambda_2 = \frac{2}{3}[e^{.3(10-t)}-1] \quad .$$

Since $\lambda_1(t) \geq \lambda_2(t)$ in this interval and $y(t) = 0$ therefore $u^*(t) = 0$ maximizes the Hamiltonian subject to constraints (4.16).

(b) Assume $t = 5$. As mentioned previously the optimal action is apply impulse $u(5) = \text{Imp}(3e^{.5},0,5)$. In Exercise 4.10 you will be asked to resolve this problem when $-1 \leq u \leq 1$, and compare the solution with (4.25)-(4.28) and Figure 4.3.

(c) Assume $t \in [0,5)$. In this interval $y > 0$ so that $\nu = 0$ and $\mu = \lambda_2 - \lambda_1$. Thus the adjoint equations (4.18) and (4.19) reduce to

$$\dot{\lambda}_1 = -(\lambda_1 + \mu)r_1(t) = 0 \quad , \quad \lambda_1(5) = e^{1.5}$$

$$\dot{\lambda}_2 = -(\lambda_2 + \nu)r_2(t) = -.1\lambda_2 \; , \quad \lambda_2(5) = e^{1.5} \; .$$

Solving these we obtain

$$\lambda_1(t) = e^{1.5}$$

$$\lambda_2(t) = e^{2-.1t}$$

$$\mu(t) = e^{1.5}[e^{.5-.1t}-1] \quad .$$

Since $\lambda_2(t) > \lambda_1(t)$ in this interval, and $x(t) = 0$ it follows that $\lambda_1(5) = \lambda_2(5)$ because 5 is the switching point. Moreover both λ_1 and λ_2 are always nonincreasing. Hence λ_2 must jump at $t = 5$, see Figure 4.3. In the interval $[0,5)$, $\lambda_2 > \lambda_1$ so that it would be optimal to buy if there were cash on hand. Since $x(t) = 0$

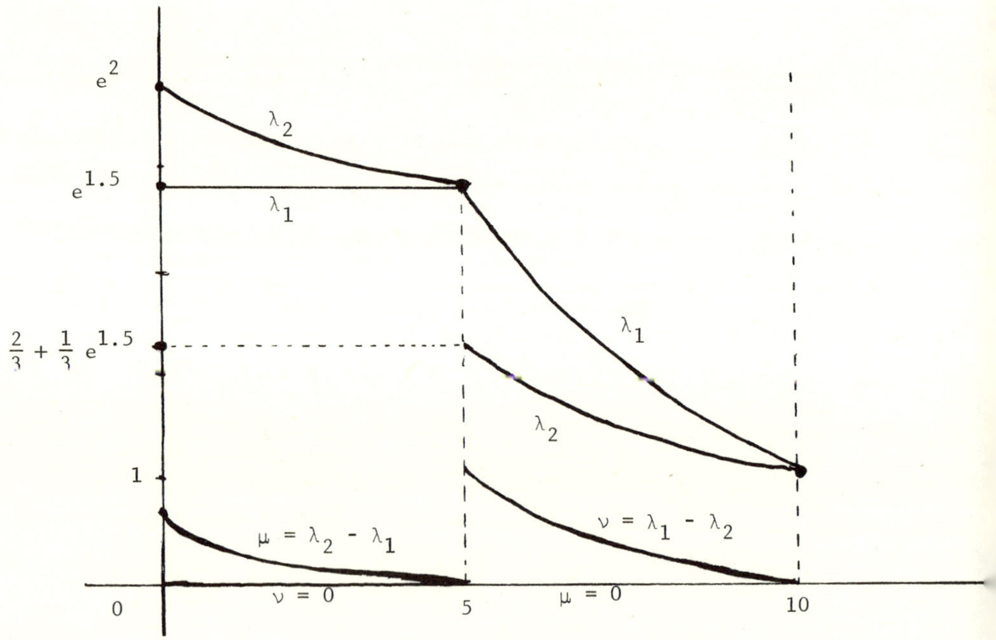

Figure 4.3 Adjoint Variables and Lagrange Multipliers for Example 4.1

it follows that $u^*(t) = 0$ maximizes the Hamiltonian subject to constraints (4.16). Note that if $x(0)$ were positive the optimal policy would (obviously) be to use $x(0)$ to instantaneously buy securities at $t = 0$. A final remark is that $\lambda_2(5-) = \lambda_2(5+) + \nu(5+)$, which is the jump condition specified in Section 3.1.1.

4.2 Optimal Financing Model

In the present section we discuss a model of a firm which must finance its investments by an optimal combination of retained earnings and external equity. The model to be discussed is due to [Krouse and Lee, 1973], with correction and extensions due to [Sethi, 1978].

[Gordon, 1962], [Miller-Modigliani, 1961], [Lintner, 1963] and others have analysed the optimal financing problem of a firm under the convenient assumption of a constant growth rate. If this assumption is removed then the problem of optimal financing of the firm becomes an optimal control problem. The new formulations, such as those of [Davis, 1970], [Krouse, 1972], and [Krouse and Lee, 1973], permit the firm to finance its investments by retained earnings, debt and/or external equity in various proportions which may vary over time.

For reasons of simplicity and ease of its solution, the model analyzed in this section is based on [Krouse and lee, 1973]. It does not permit debt as a source of financing, but does permit retained earnings and external equity to be used in any proportions.

4.2.1 The Model

In order to formulate the model, let

$x(t)$ = the current earnings rate in dollars at time t,

$u(t)$ = external equity financing expressed as a multiple of current earnings; $u \geq 0$,

$v(t)$ = fraction of current earnings retained; $0 \leq v \leq 1$,

c = one minus the (assumed) constant fractional floatation costs for external equity; $0 \leq c < 1$,

ρ = the continuous market-given discount rate (assumed constant) applying to the firm's dividends.

r = the rate of return (assumed constant) on the firm's invested capital, where $r > \rho$,

g = the upper bound on the rate of asset growth the firm can attain.

Given these definitions, the rate of change in current earnings is given by

$$\dot{x} = r(cu + v)x, \qquad x(0) = x_0. \tag{4.29}$$

Furthermore, the upper bound on the rate of growth of the assets implies that there is the same upper bound on the rate of growth in earnings given the assumption of constant rate of return r. Thus we can derive the following constraint on the control variables

$$\dot{x}/x = r(cu + v) \leq g. \tag{4.30}$$

Finally, the objective of the firm is to maximize the present value of the future dividend streams accruing to the firm's initial stockholders. To derive this expression, note that

$$\int_0^T (1-v)x\, e^{-\rho t}\, dt$$

represents the present-value of total dividends issued by the firm. From this we subtract the present-value of the external investments

$$\int_0^T ux\, e^{-\rho t}\, dt,$$

made by the new stockholders. This is because under the assumption of an efficient market, the new stockholders will get a rate of return exactly equal to the discount rate ρ on their investments. Thus, the net present-value of the total future dividends to the initial stockholders is the difference

$$J = \int_0^T e^{-\rho t} (1 - v - u)x \, dt. \tag{4.31}$$

Since the firm at $t = 0$ is completely owned by the initial stockholders J is clearly the objective function which the firm must maximize. Note that in the case of finite horizon, a more realistic objective function would include salvage value or bequest term $S[x(T)]$. This is not very difficult to incorporate. See Exercise 4.9, where the bequest function is linear. We will also solve the infinite horizon problem (i.e., $T = \infty$) after we have solved the finite horizon problem.

The optimal control problem is to choose u and v over time so as to maximize J in (4.31) subject (4.29) and the constraints (4.30), $u \geq 0$, and $0 \leq v \leq 1$. For convenience, we state this problem in the following box.

$$\max_{u,v} \left\{J = \int_0^T e^{-\rho t} (1 - u - v)x \, dt\right\}$$

Subject to

$$\dot{x} = r(cu + v)x, \quad x(o) = x_o, \tag{4.32}$$

and the control constraints

$$cu + v \leq g/r,$$

$$u \geq 0, \quad 0 \leq v \leq 1.$$

4.2.2 Application of the Maximum Principle

This is a bilinear problem with two control variables which is a special case of row (e) of Table 3. The current-value Hamiltonian is:

$$H = (1 - u - v)x + \lambda r(cu + v)x, \tag{4.33}$$

where the current-value adjoint variable λ satisfies

$$\dot{\lambda} = \rho\lambda - (1 - u - v) - \lambda r(cu + v) \tag{4.34}$$

with the transversality condition

$$\lambda(T) = 0 . \tag{4.35}$$

From the Hamiltonian in (4.33) and Section 2.2.2, we know $\lambda(t)$ can be interpreted as the marginal value (in time-t dollars) of a unit change in earnings at time t. Also $(cu + v)x$ is the incremental investment at time t. Thus $\lambda(t)r$ is the marginal value of a unit investment at time t. Therefore the product $\lambda(t)r\,(cu + v)x$ is the value to the stockholders of the incremental investment measured in terms of foregone dividends (in other words in time t dollars). We can also interpret (4.34) as in in Section 2.2 More specifically, if the firm makes an incremental investment of λ (which giv[illegible] an incremental earning of one dollar), $\rho\lambda$ is the expected returns from this investment. In equilibrium this must be equal to the "capital gain" $\dot{\lambda}$, plus the immediate dividend (1-v) less u, the "claim" of the external stockholder and plus the value of the incremental earnings $\lambda r(cu + v)$.

To specify the form of optimal policy, we rewrite the Hamiltonian as

$$H = [W_1 u + W_2 v + 1]x \tag{4.36}$$

where

$$W_1 = cr\lambda - 1 \tag{4.37}$$

and

$$W_2 = r\lambda - 1. \tag{4.38}$$

Note first that the state variable x factors out so that the optimal controls are independent of the state variable. Second, since the Hamiltonian is linear in the two control variables, the optimal policy is a combination of generalized bang-bang and singular controls. Of course, the characterization of these optimal controls in terms of the adjoint variable λ will require solving a linear programming problem at each instant of time t. The Hamiltonian maximization problem can be stated as follows:

$$\boxed{\begin{aligned} &\max_{u,v}[W_1 u + W_2 v] \\ &\text{subject to} \\ &u \geq 0, \quad 0 \leq v \leq 1, \quad cu + v \leq g/r. \end{aligned}} \tag{4.39}$$

Obviously, the constraint $v \leq 1$ becomes redundant if $g/r \leq 1$. Therefore, we have two cases:

Case A: $g/r \leq 1$

Case B: $g/r > 1$

under each of which, we can solve the linear programming problem (4.39) geometrically in a closed form. This is done in Figs. 4.4 and 4.5

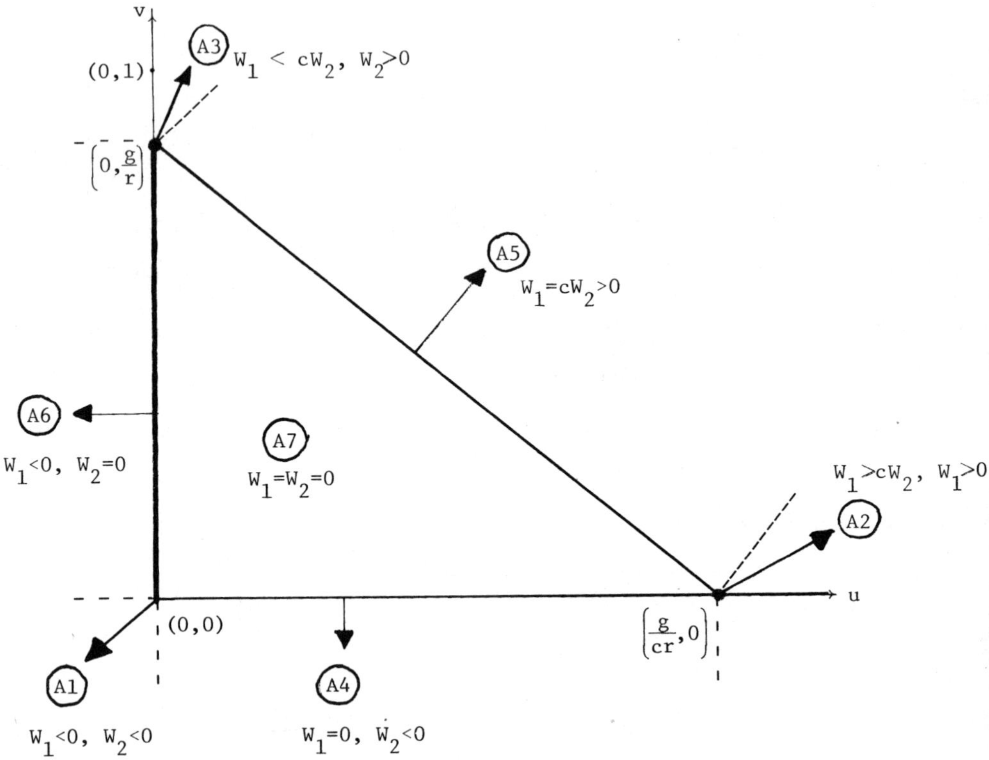

Figure 4.4 Case A: $g/r \leq 1$

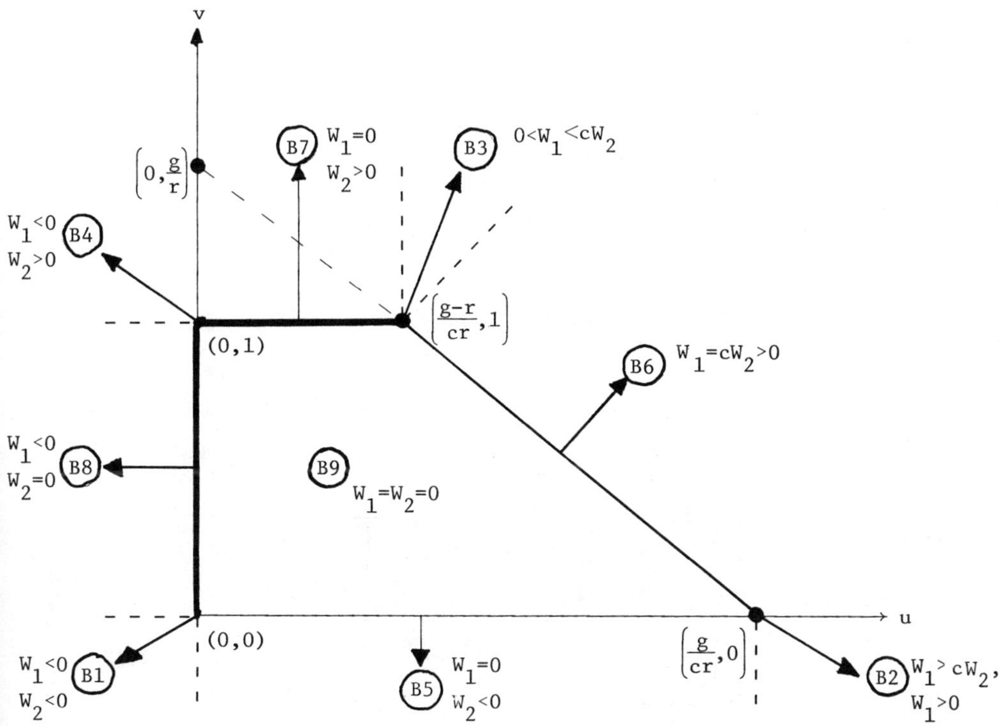

Figure 4.5 Case B: $g/r > 1$

Since $c < 1$ by assumption, the following subcases shown in Figs. 4.4 and 4.5 can be ruled out:

i) $W_1 > cW_2$, $W_1 > 0 \Rightarrow c > 1$ ruling out cases A_2 and B_2

ii) $W_1 = 0$, $W_2 < 0 \Rightarrow c > 1$ ruling out cases A_4 and B_5

iii) $W_1 = cW_2 > 0 \Rightarrow c = 1$ ruling out cases A_5 and B_6

iv) $W_1 = W_2 = 0 \Rightarrow c = 1$ ruling out cases A_7 and B_9.

For the remaining subcases showing adjacent to the darkened lines in the figures, we can characterize the corresponding optimal controls in the following tabular form.

	Condition on W_1 and W_2	SUBCASE		Optimal Controls	Characterization
		Case A: $\frac{g}{r} \leq 1$	Case B: $\frac{g}{r} > 1$		
(1)	$W_1<0$, $W_2<0$	A1	B1	$u^*=0$, $v^*=0$	generalized bang-bang
(2)	$W_1<cW_2$, $W_2>0$	A3	-	$u^*=0$, $v^* = \frac{g}{r}$	"
(3)	$W_1<0$, $W_2=0$	A6	B8	$u^*=0$, $0 \leq v^* \leq \min[1, \frac{g}{r}]$	singular
(4)	$0<W_1<cW_2$	-	B3	$u^* = \frac{g-r}{rc}$, $v^*=1$	generalized bang-bang
(5)	$W_1<0$, $W_2>0$	-	B4	$u^*=0$, $v^*=1$	"
(6)	$W_1=0$, $W_2>0$	-	B7	$0 \leq u^* \leq \frac{g-r}{rc}$ $v^*=1$	singular

Table 4.1 Characterization of Optimal Controls

The catalog of possible optimal control regimes shown in Table 4.1 gives the potential time-paths for the firm. What one needs to do to obtain the optimal path (for a given initial condition) is to synthesize these subcases into an optimal sequence. This is carried out in the following section.

4.2.3 Synthesis of Optimal Control Paths

To obtain an optimal path, we must synthesize an optimal sequence of subcases. The usual procedure employed is that of the reverse-time construction first developed by [Isaacs, 1965]. Reverse time can only be defined for finite horizon problems. However, the infinite horizon solution can usually be inferred from the finite horizon solution if sufficient care is exercised.

Thus, for the time being, assume that T (assumed sufficiently large) denotes the terminal time. Define the reverse time variable τ as

$$\tau = T - t,$$

so that

$$\overset{o}{y} = dy/d\tau = \frac{dy}{dt}\frac{dt}{d\tau} = -\dot{y} .$$

As a consequence, $\overset{o}{y} = -\dot{y}$ and the reverse time version of the state and adjoint equations (4.29) and (4.33), respectively, can be obtained by simply replacing $\dot{y}$ by $\overset{o}{y}$ and changing the signs of the right hand sides. The transversality condition on the adjoint variable

$$\lambda(t = T) = \lambda(\tau = 0) = 0 \tag{4.40}$$

becomes the initial condition in the reverse time sense. Furthermore, let us parametrize the terminal state by assuming that

$$x(t = T) = x(\tau = 0) = \alpha \tag{4.41}$$

where α is a parameter to be determined.

From now on, in this section everything is expressed in reverse time sense, unless otherwise specified. Using the definition of $\overset{o}{x}$ and $\overset{o}{\lambda}$ and the conditions (4.41) and (4.40), we can write reverse time versions of (4.29) and (4.33) as follows.

$$\overset{o}{x} = -r(cu + v)x, \qquad x(0) = \alpha \tag{4.42}$$

and

$$\overset{o}{\lambda} = (1 - u - v) - \lambda\{\rho - r(cu + v)\}, \quad \lambda(0) = 0. \tag{4.43}$$

This is the starting point for our switching point analysis. First we consider Case A.

Case A: $g \leq r$.

Note that the constraint $v \leq 1$ is superfluous in this case and only feasible subcases are A1, A3, and A6. Since $\lambda(0) = 0$, $W_1(0) = W_2(0) = -$ and subcase A1 obtains.

Subcase A1 $(W_1 = cr\lambda - 1 < 0,\ W_2 = r\lambda - 1 < 0)$

From row (1) of Table 4.1 we have $u^* = v^* = 0$, which gives the state equation (4.42) and the adjoint equation (4.43) as

$$\overset{o}{x} = 0 \quad \text{and} \quad \overset{o}{\lambda} = 1 - \rho\lambda. \tag{4.44}$$

With the initial conditions given in (4.40), the solution for λ is:

$$\lambda(\tau) = (1/\rho)[1-e^{-\rho\tau}]. \tag{4.45}$$

It is easy to see that because of the assumption $0 \leq c < 1$ it follows that if $W_2 = r\lambda-1 < 0$ then $W_1 = cr\lambda-1 < 0$. Therefore, to remain in this subcase as τ increases, $W_2(\tau)$ must remain negative for some time as τ increases. From (4.45), however, λ is increasing asymptotically toward a value $1/\rho$ and therefore, $W_2(\tau)$ is increasing asymptotically toward the value $r/\rho-1$. Since, we have assumed $r > \rho$, there exists a τ_1 such that $W_2(\tau_1) = 0$. It is easy to compute

$$\tau_1 = 1/\rho \, \ell n \, [r/(r-\rho)]. \tag{4.46}$$

From this expression, it is clear that the firm leaves subcase A1 provided $\tau_1 < T$. This would usually be the case since we have assumed T to be sufficiently large. (Note that for $r < \rho$, the firm never exits from subcase A1. Obviously, there is no use investing if the rate of return is less than the discount rate.)

At reverse time τ_1, we have $W_2 = 0$ and $W_1 < 0$ and the firm, therefore, is in subcase A6.

Subcase A6 $(W_1 = cr\lambda - 1 < 0, \ W_2 = r\lambda - 1 = 0)$

In this subcase, the optimal control

$$u^* = 0, \quad 0 \le v^* \le g/r \tag{4.47}$$

from row (3) of Table (4.1) is singular with respect to v. This case is termed singular because the Hamiltonian maximizing condition does not yield a unique value for control v. In such cases, the optimal control is obtained by conditions required to sustain $W_2 = 0$ for a finite time interval. This means we must have $\overset{o}{W}_2 = 0$, which in turn implies $\overset{o}{\lambda} = 0$. To compute $\overset{o}{\lambda}$, we substitute (4.53) into (4.39) and obtain

$$\overset{o}{\lambda} = (1 - v^*) - \lambda[\rho - rv^*]. \tag{4.48}$$

Substituting $\lambda = 1/r$ (since $W_2 = 0$) in (4.44) and equating it to zero we obtain

$$r = \rho \tag{4.49}$$

as a necessary condition required to maintain singularity over a finite time interval following τ_1. Condition (4.49) is fortuitous and will not generally hold. In fact we have assumed $r > \rho$. Thus, the firm will not stay in subcase A6 for a finite time interval. Furthermore, since $r > \rho$, $\overset{o}{\lambda}(\tau_1) = (r - \rho)/r > 0$ and therefore W_2 is increasing from zero and becomes positive after τ_1. Thus, at τ_1+, the firm switches to subcase A3.

<u>Subcase A3 ($W_1 = cr\lambda - 1 < cW_2 = cr\lambda - c$, $W_2 = r\lambda - 1 > 0$)</u>

The optimal controls in this subcase from row (2) of Table 4.1 are

$$u^* = 0, \quad v^* = g/r. \tag{4.50}$$

The state and the adjoint equations are

$$\overset{o}{x} = -gx \tag{4.51}$$

and

$$\overset{o}{\lambda} = (1 - g/r) - \lambda(\rho - g). \tag{4.52}$$

Since $\overset{o}{\lambda}(\tau_1) > 0$, λ is increasing at τ_1 from its value of $1/r$. A further examination of the behavior of $\lambda(\tau)$ as τ increases, will be carried out under two different possible conditions.

<u>i) $\rho > g$</u>

Under this assumption, as λ increases, $\overset{o}{\lambda}$ decreases and becomes zero at a value obtained by equating the right-hand-side of (4.52) to zero, i.e., at

$$\bar{\lambda} = \frac{1 - g/r}{\rho - g} . \tag{4.53}$$

This value $\bar{\lambda}$ is, therefore, an asymptote to the solution of (4.52) starting at $\lambda(\tau_1) = 1/r$. Since $r > \rho > g$ in this case.

$$\bar{W}_2 = r\bar{\lambda}-1 = \frac{r(1 - g/r)}{\rho - g} - 1 = \frac{r - \rho}{\rho - g} > 0, \tag{4.54}$$

which implies that the firm continues to stay in subcase A3.

ii) $\rho \leq g$

Under this assumption, as λ increases, $\overset{o}{\lambda}$ increases, Thus $\bar{W}_2 = r\bar{\lambda} - 1$ continues to be greater than zero and the firm continues to remain in subcase A3.

(Note however that $\lambda(\tau)$ in this case increases as τ increases without any bound. This has important implications when we deal later with the solution of the infinite horizon problem.)

Since the optimal path has been found to be independent of α for T sufficiently large we can sketch the solution for Case A in Fig. 4.5. starting with x_o. This also gives the value of α, as shown in Figure 4.6.

In the solution for Case A, there is only one switching point provided T is sufficiently large (i.e., $T > \tau_1$ in this case). The switching time $t_1 = T-\tau_1$ has an interesting economic interpretation. Namely, it requires at least τ_1 units of time for a dollar of retained earnings to be worthwhile. That means, it pays to invest as much earnings as feasible before $T - \tau_1$ and it does not pay to invest any earnings after $T - \tau_1$. Thus, $T - \tau_1$ is the point-of-indifference between retaining earnings or paying dividends out of earnings. To see this directly, let us suppose the firm retains one dollar of earnings at $T - \tau_1$. Since this is the last time any

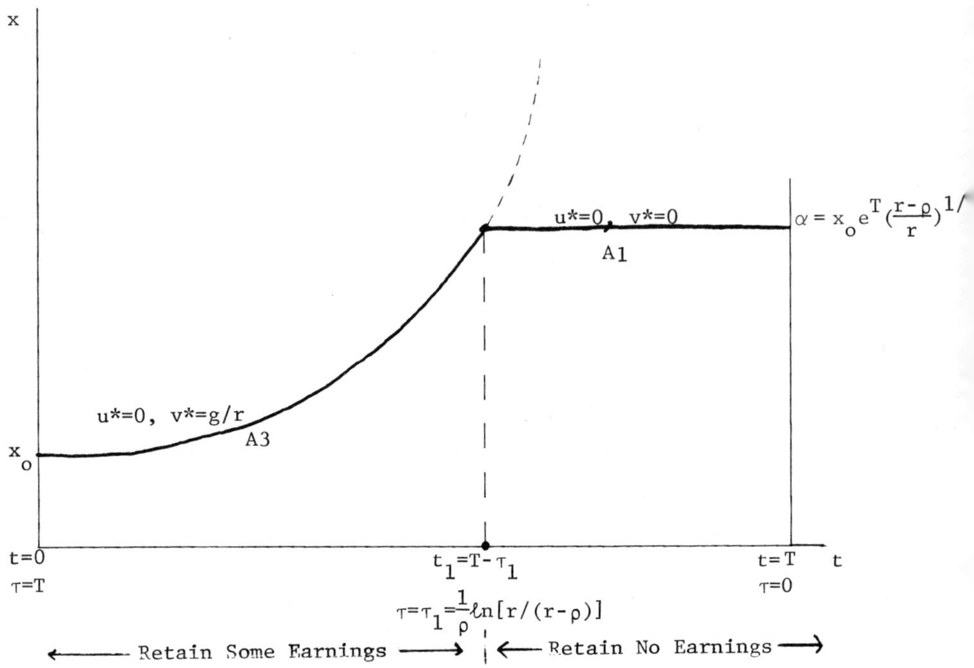

Figure 4.6 Optimal Path for Case A: $g/r \leq 1$.
(Note that equity financing is never used)

earnings invested will be worthwhile, it is obvious (because all earnings are paid out) that the dollar just invested at $T - \tau_1$ yields dividends at the rate r from $T - \tau_1$ to T. The value of this dividend stream in terms of $(T - \tau_1)$-dollars is

$$\int_0^{\tau_1} re^{-\rho t}dt = \frac{r}{\rho}[1-e^{-\rho\tau_1}] \tag{4.55}$$

which must be equated to one dollar to find the point-of-indifference. Equating (4.55) to 1 yields precisely the value of τ_1 given in (4.46).

With this interpretation of τ_1 we conclude that enough earnings must be retained so as to make the firm grow exponentially at the maximum rate of g until $t_1 = T - \tau_1$; then all earnings are paid out and the firm stops growing. Since $g \leq r$ (assumed for Case A), the growth in the first part of the solution can be financed entirely from retained earning. Thus, there is no need to resort to more expensive external equity financing. The latter will not be the case, however, in Case B when $g > r$, which we now discuss.

Case B: $g > r$

Since $g/r > 1$, the constraint $v \leq 1$ in Case B is relevant. The feasible subcases are B1, B8, B4, B7, and B3 shown adjacent to the dark line in Figure 4.5. As in Case A, it is obvious that the firm starts (in the reverse-time sense) in subcase B1.

Subcase B1 $(W_1 = cr\lambda - 1 < 0,\ W_2 = r\lambda - 1 < 0)$

The analysis of this subcase is the same as subcase A1. As in that subcase the firm switches out at time $\tau = \tau_1$ to subcase B8.

Subcase B8 $(W_1 = cr\lambda - 1 < 0,\ W_2 = r\lambda - 1 = 0)$

In this subcase, the optimal control

$$u^* = 0, \quad 0 \leq v^* \leq 1, \tag{4.56}$$

from row (3) of Table (4.1) is singular with respect to v. As before in subcase A6, the singular control cannot be sustained for a finite time because of our assumption $r > \rho$. As in subcase A6 W_2 is increasing at τ_1 from zero and becomes positive after τ_1. Thus, at τ_1^+, the firm finds itself in subcase B4.

Subcase B4 ($W_1 = cr\lambda - 1 < 0,\ W_2 = r\lambda - 1 > 0$)

The optimal controls in this subcase are

$$u^* = 0, \qquad v^* = 1 \tag{4.57}$$

as shown in row 5 of Table 4.1. The state and the adjoint equations are

$$\overset{\circ}{x} = -rx \quad , \quad x(\tau_1) = \alpha \tag{4.58}$$

where α is defined in (4.41), and

$$\overset{\circ}{\lambda} = \lambda(r-\rho), \qquad \lambda(\tau_1) = 1/r. \tag{4.59}$$

Obviously, earnings are growing exponentially at rate r and λ is increasing at rate $(r - \rho)$ as τ increases from τ_1. Since $\lambda(\tau_1) = 1/r$, we have

$$\lambda(\tau) = (1/r)e^{(r-\rho)(\tau-\tau_1)}, \quad \text{for} \quad \tau \geq \tau_1 . \tag{4.60}$$

As λ increases, W_1 increases and becomes zero at a time τ_2 defined by

$$W_1(\tau_2) = cr\lambda(\tau_2) - 1 = ce^{(r-\rho)(\tau_2-\tau_1)} - 1 = 0, \tag{4.61}$$

which, in turn gives

$$\tau_2 = \tau_1 + [1/(r-\rho)]\,\ell n(1/c). \tag{4.62}$$

At τ_2^+ the firm switches to subcase B7.

Subcase B7 ($W_1 = cr\lambda - 1 = 0,\ W_2 = r\lambda - 1 > 0$)

In subcase B7, the optimal controls

$$0 \leq u^* \leq (g-1)/rc, \qquad v^* = 1 \tag{4.63}$$

from row (6) Table (5.1) is singular with respect to u. To maintain this singular control over a finite time interval, we must maintain $W_1 = 0$ in the interval. This means, we must have $\overset{\circ}{W}_1(\tau_2) = 0$, which in turn, implies $\overset{\circ}{\lambda}(\tau_2) = 0$. To compute $\overset{\circ}{\lambda}$, we substitute (4.63) into (4.43) and obtain

$$\overset{\circ}{\lambda} = -u^* - \lambda\{\rho - r(cu+1)\}. \tag{4.64}$$

Substituting $\lambda(\tau_2) = 1/rc$ (since $W_1(\tau_2) = 0$) in (4.64) and equating it to zero, we obtain

$$r = \rho.$$

By our assumption $r > \rho$, a singular path cannot be sustained and the firm will not stay in subcase B7 for a finite amount of time. Furthermore, from (4.64), we have

$$\overset{\circ}{\lambda}(\tau_2) = \frac{r-\rho}{rc} > 0 \tag{4.65}$$

which implies that λ is increasing and therefore, W_1 is increasing. Thus at τ_2^+ the firm switches to subcase B3.

<u>Subcase B3 $(0 < W_1 = cr\lambda - 1 < cW_2 = cr\lambda - c)$</u>

The optimal controls in this subcase from row (4) of Table 4.1 are

$$u^* = \frac{g-r}{rc}, \quad v^* = 1. \tag{4.66}$$

The state and the adjoint equations are

$$\overset{\circ}{x} = -gx \tag{4.67}$$

and

$$\overset{\circ}{\lambda} = -\left(\frac{g-r}{rc}\right) + \lambda(g-\rho). \tag{4.68}$$

Since $\overset{\circ}{\lambda}(\tau_2) > 0$, λ is increasing. Furthermore, in Case B, we assume $g > r$. But $r > \rho$ has been assumed throughout the paper, therefore, $g > \rho$ and the second term in the right-hand side of (4.68) is increasing. That means $\overset{\circ}{\lambda} > 0$ and $\lambda(\tau)$ continues to increase. Therefore, the firm continues to stay in subcase B3.

Note that $\lambda(\tau)$ in Case B increases without bound as τ gets large. This will have important implications when dealing with the infinite horizon problem.

We now can sketch the complete solution for Case B in Figure 4.7.

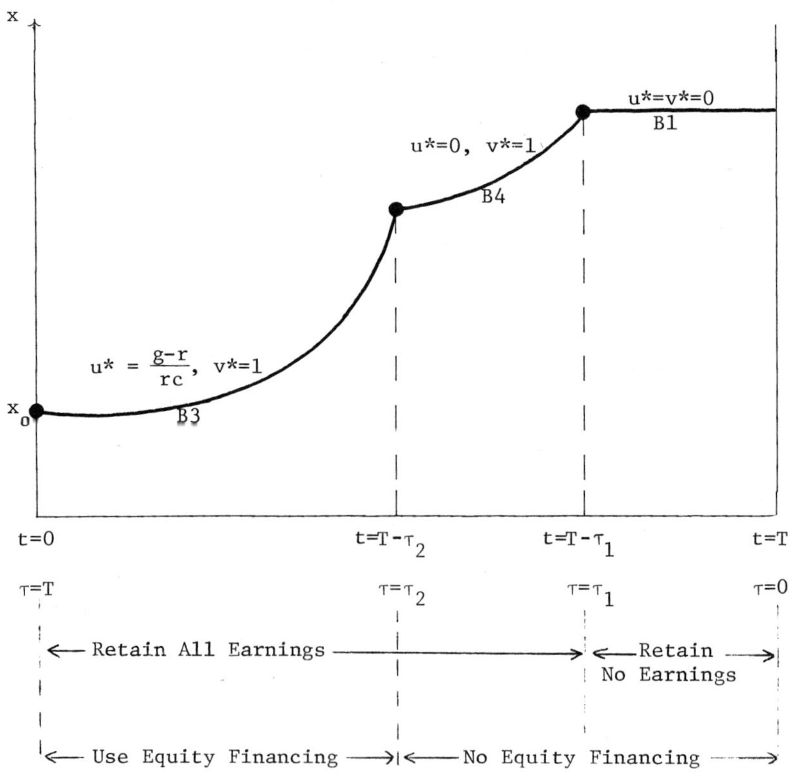

Figure 4.7 Optimal Path for Case FB: g/r > 1

In the solution for Case B, there are two switching points instead of just one as in Case A. The reason for two switching points becomes quite clear when we interpret the significance of τ_1 and τ_2. It is obvious that τ_1 has the same meaning as before. Namely, if τ_1 is the remaining time to horizon, the firm is indifferent between investing a dollar of earnings or paying it out as dividends. Intuitively, it seems that since external equity is more expensive than retained earnings as a source of financing, investment financed by external equity requires more time to be worthwhile. Thus,

$$\tau_2 - \tau_1 = \frac{1}{r-\rho}\,\ell n(1/c) \tag{4.69}$$

should be the time required to compensate for the floatation cost of external equity.

To see this, we suppose that the firm issues a dollars worth of stock at $t = T-\tau_2$. While the cost of this issue is one dollar, the capital acquired is c dollars because of the floatation cost $(1-c)$. Since we are attempting to find the breakeven time for external equity, it is obvious that retaining all of the earnings for investment is still profitable. Thus there is no dividend from $T-\tau_2$ to $T-\tau_1$ and the firm is growing at the rate r. Therefore, the value of this investment at $(T-\tau_2)$ measured in $(T-\tau_2)$-dollars is

$$ce^{(r-\rho)(\tau_2-\tau_1)} = ce^{(r-\rho)\ell n(1/c)/(r-k)} = 1\ . \tag{4.70}$$

Equation (4.70) states that one $(T-\tau_2)$-dollar of external equity at time $(T-\tau_2)$ is equivalent to one $(T-\tau_2)$-dollar investment at $(T-\tau_1)$. But the firm is indifferent between investing or not investing the costless retained earnings at $(T-\tau_1)$. To summarize the firm is indifferent between issuing a dollars worth of stock at $(T-\tau_2)$ or not issuing it before $(T-\tau_2)$, it pays to issue stocks at as large a rate as is feasible; after $(T-\tau_2)$, it does not pay to issue any external equity at all.

Having completely solved the finite horizon case, we now turn to the infinite horizon case.

4.2.4 Solution for the Infinite Horizon Problem

As indicated in Section 3.4, for the infinite dimensional case the transversality condition must be changed to

$$\lim_{t\to\infty} e^{-\rho t}\lambda^*(t) = 0 . \tag{4.71}$$

Furthermore, this condition may no longer be a necessary condition, see Section 3.4. It is a sufficient condition for optimality however, in conjunction with other sufficiency conditions stated in Theorem 2.1.

A usual method of solving an infinite horizon problem is to take the limit as $T\to\infty$ of the finite horizon solution and then proving that the limiting solution so obtained solves the infinite horizon problem. The proof i important because the limit of the solution may or may not solve the infinite horizon problem. The proof is usually based on the sufficiency Theorem 2.1 modified slightly as indicated above for the infinite horizon case.

We now analyze the infinite horizon case following the above procedure. We start with case A.

Case A: $g \leq r$

The limiting solution in this case is given as subcase A3, i.e.,

$$u = 0, \quad v = g/r, \tag{4.72}$$

$$\dot{x} = gx, \quad x(o) = x_o, \tag{4.73}$$

and

$$\dot{\lambda} = -(1-g/r) - \lambda(g-\rho), \quad \lim_{t\to\infty} e^{-\rho t}\lambda(t) = 0. \tag{4.74}$$

To prove that the solution (4.72) is optimal, we must show that there exists a solution of (4.74) which with (4.72) satisfies the maximum principle (i.e., stays in subcase A3).

Going back to subcase A3 in Section 4.2.3 see (4.53), we note that in the reverse-time sense, $\lambda(\tau)$ is increasing asymptotically toward the value

$$\bar{\lambda} = \frac{1-g/r}{\rho-g} \tag{4.75}$$

only in case $\rho > g$. Otherwise, as we noted, $\lambda(\tau)$ increases without bound. Thus, for $\rho > g$, $\lambda = \bar{\lambda}$ clearly satisfies (4.74). Furthermore,

$$\bar{W}_2 = r\bar{\lambda}-1 = \frac{r-\rho}{\rho-g} > 0,$$

and

$$\bar{W}_1 < \bar{W}_2 ,$$

which implies that
the firm stays in subcase A3, that is, the maximum principle holds.

We have now proved the following result: for $g < \rho$ in Case A

$$u^* = 0, \quad v^* = g/r$$

with the corresponding state trajectory

$$x^* = x_o e^{gt}$$

and the adjoint trajectory

$$\lambda^* = \bar{\lambda} = \frac{1-g/r}{\rho-g}$$

represents an optimal path for the infinite horizon problem. Note that the assumption $g < \rho$ together with our overall assumption $\rho \leq r$ gives $g < r$ so that $1 - v^* > 0$ which means a constant fraction of dividends are being paid.

Note that the value of the adjoint variable $\bar{\lambda}$ in this case is constant and its form is reminiscent of the classic Gordon's formula [Gordon, 1962]. In the control theory framework, the value of $\bar{\lambda}$ represents the marginal worth per additional unit of earnings. Obviously, a unit increase in earnings will mean an increase of $1 - v^*$ or $1 - g/r$ units in dividends. This, of course, should be capitalized at a rate equal to the discount rate less the growth rate (i.e., $\rho - g$); which is precisely the Gordon's formula.

For $g \geq \rho$, the reverse time construction in subcase A3 implies that $\lambda(\tau)$ increases without bound as τ increases. Thus, we cannot find any λ which satisfies (4.74). A moment's reflection shows that for $g \geq \rho$ the objective function can be made infinite. For example, any control policy with earnings growing at rate $q \in [\rho,g]$ coupled with a partial dividend payout, i.e. a constant v such that $0 < v < 1$, gives an infinite value for the objective function.

Since there are many policies which give an infinite value to the objective function, the choice among them may be decided on subjective grounds. We shall briefly discuss only the constant optimal policies. If $g < r$ then the rate of growth q may be chosen in the closed interval $[\rho,g]$; if $g = r$ then q may be chosen in the half open interval $[\rho,r)$. In either case the choice of a low rate of growth would mean a higher dividend payout (in dollars) during the early time, but lower payout during later because of the slower growth rate. Similarly, the choice of high growth rate means the opposite in terms of dividend payments.

If $g < r$ in case A, the limiting solution of the finite case is an optimal solution for the infinite horizon problem in the sense that the objective function becomes infinite. However, this will not be the situation for case B.

<u>Case B: $g > r > \rho$</u>

The limit of the finite horizon optimal solution is to grow at the maximum allowable growth rate with

$$u = \frac{g-r}{rc}, \quad v = 1$$

all the way. Since τ_1 disappears in the limit, the stockholders will never collect dividends. The firm has become an infinite sink for investment. In fact, the limiting solution is a <u>pessimal</u> solution because the value of the objective function associated with it is zero. From the point of view of optimal control theory this can be explained as before in case A when $g \geq \rho$. In Case B, we have $g > r$ so that (since $r > \rho$ throughout the chapter) we have $g > \rho$. For this $\lambda(\tau)$ increases without bound as τ increases and therefore (4.70) does not have a solution.

As in case A with $g \geq \rho$, <u>any</u> control policy with earnings growing at rate $q \in [\rho, g]$ coupled with a <u>constant</u> v

$$0 < v < 1$$

has an infinite value for the objective function.

In summary, we note that the only nondegenerate problem in the infinite horizon case is when $g < \rho$. In this case, the policy of maximum allowable growth is optimal. When $g \geq \rho$, whether in Case A or B, the infinite horizon problem is ill-posed; it has nonunique policies with infinite values for the objective function.

Before solving a numerical example we will make a general remark. Let u_T^* be the optimal control for a finite horizon problem where u stands for a vector of all the control variables; let u_∞^* be any optimal control for the infinite horizon problem. Define

$$u_\infty = \lim_{T\to\infty} u_T^*$$

We now note that for our model we have

$$\lim_{T\to\infty} J(u_T^*) = \infty \quad \text{and} \quad J(u_\infty) = 0.$$

Obviously u_∞ is <u>not</u> an optimal control for the infinite horizon problem. Since these two are not equal, we can say that J, regarded as a mapping, is not a <u>closed</u> mapping. We have seen that it is possible in our model to get u_∞^* such that $J(u_\infty^*) = \infty$. If we introduce a salvage value $Bx(T)$, $B > 0$, for the finite horizon problem, then the new objective function

$$J = \int_0^T e^{-\rho t}(1-u-v)x\,dt + Bx(T)e^{-\rho T} \quad \text{for} \quad T < \infty$$

and

$$J = \int_0^\infty e^{-\rho t}(1-u-v)x\,dt + \lim_{T\to\infty} \{Bx(T)e^{-\rho T}\} \quad \text{for } T = \infty,$$

is a closed mapping in the sense that

$$\lim_{T\to\infty} J(u_T^*) = \infty \quad \text{and} \quad J(u_\infty) = \infty$$

for our model.

<u>Example 4.2</u> We will now assign some numbers to various parameters in the optimal financing problem to compute the optimal solution. Let

$$x_o = 1000/\text{month}$$
$$T = 60 \text{ months}$$
$$r = 0.15$$
$$\rho = 0.10$$
$$g = 0.05$$
$$c = 0.98$$

Since $g \leq r$, the problem belongs to Case A. We compute

$$\tau_1 = \frac{1}{\rho} \ell n\ [r/(r-\rho)] = 10\ \ell n\ 3 \approx 11 \text{ months}.$$

The optimal control for the problem is

$$u^* = 0, \quad v^* = g/r = 1/3, \quad t \in [0,49)$$
$$u^* = 0, \quad v^* = 0 \quad , \quad t \in [49,60]$$

and the optimal state trajectory is

$$x(t) = \begin{cases} 1000\ e^{.05t} & t \in [0,49) \\ 1000\ e^{2.45}, & t \in [49,60]. \end{cases}$$

The value of the objective function

$$J^* = \int_0^{49} e^{-.1t}(1-1/3)(1000)e^{0.05t}dt + \int_{49}^{60} 1000\ e^{2.45} \cdot e^{-.1t}\ dt$$
$$= 12,758.4.$$

Note that the infinite horizon problem is well-defined in this case, since $g < \rho$ and $g < r$. The optimal control is

$$u^* = 0, \quad v^* = g/r$$

and

$$J^* = \int_0^{\infty} e^{-.1t}(2/3)(1000)e^{0.05t}\ dt = 2000/0.15 = 13,333\tfrac{1}{3}\ .$$

Exercises for Chapter 4

4.1 Find the optimal policies for the simple cash balance model (Sections 4.1.1 and 4.1.2) with $x_0 = 2$, $y_0 = 2$, $U_1 = U_2 = 5$, $T = 1$, $\alpha = .01$, and the following specifications for interest rates

(a) $r_1(t) = \frac{1}{2}$, $r_2(t) = \frac{1}{3}$

(b) $r_1(t) = \frac{1}{2}t$, $r_2(t) = \frac{1}{3}$.

4.2 Formulate the extension of the model in Section 4.1.3 with finite positive bounds U_1 and U_2 on the control variables for

(a) $\alpha = 0$

(b) $\alpha > 0$.

[Hint. Adjoin the control constraints to the Hamiltonian in forming the Lagrangian. For (b), write $u = u_1 - u_2$ as in (4.9)].

4.3 It is also possible to guess the solution to Example 4.1 when $\alpha > 0$. Show that the optimum policy remains unchanged if $\alpha \le 1-1/e$. [Hint. Use an elementary compound interest argument.]

4.4 Discuss the optimal equity financing model of Section 4.2.1 when $c = 1$. Show that only one control variable is needed. Then solve the problem.

4.5 What happens in the optimal equity financing model when $r < \rho$? Guess the optimal solution (without actually solving it).

4.6 When $g = r$ in case A of the optimal equity financing model, why is the limit of the solution not the solution to the infinite horizon problem?

4.7 Let $g = 0.12$ in Example 4.2. Resolve the finite horizon problem with this new value of g. Also, for the infinite horizon problem, state a policy which yields an infinite value for the objective function.

4.8 Reformulate the simple cash balance problem of Section 4.1.1 and 4.1.2, if the earnings on bonds are paid in cash.

4.9 Introduce a salvage value function

$$B\,x(T)$$

where $B \geq 0$, in the objective function (4.32) and analyze the modified problem. Show how the solution changes when B changes from 0 to $1/rc$.

4.10* Redo Example 4.1 with control constraints $-1 \leq u \leq 1$.

(a) Give reasons why the solution shown in Figure 4.8 is optimal

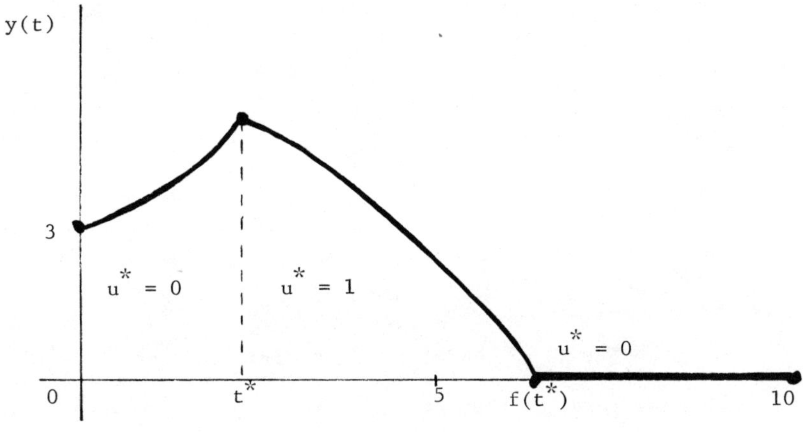

Figure 4.8

(b) Compute $f(t^*)$ in terms of t^*.

(c) Compute J in terms of t^*. Find t^* that maximizes J by setting $\frac{dJ}{dt^*} = 0$. [This is a long and tedious calculus problem; see solutions manual for details.]

4.11* For the solution found in Exercise 4.10 show by using the maximum principle and Exercise 4.10 that the adjoint trajectories are given by: for $0 \le t \le 5$, $\lambda_1(t) = \lambda_1(0) = e^{1.5}$; for $5 \le t \le 10$, $\lambda_1(t) = \lambda_1(0)e^{-.3(t-5)}$; for $0 \le t \le 6.52$, $\lambda_2(t) = \lambda_2(0)e^{-.1t}$; for $6.52 < t \le 10$, $\lambda_2(t) = \lambda_2(6.52) + \frac{1}{3}\lambda_1(0)e^{1.5}[e^{-.3t} - 2^{-1.956t}]$. Sketches of these functions are shown in Figure 4.9.

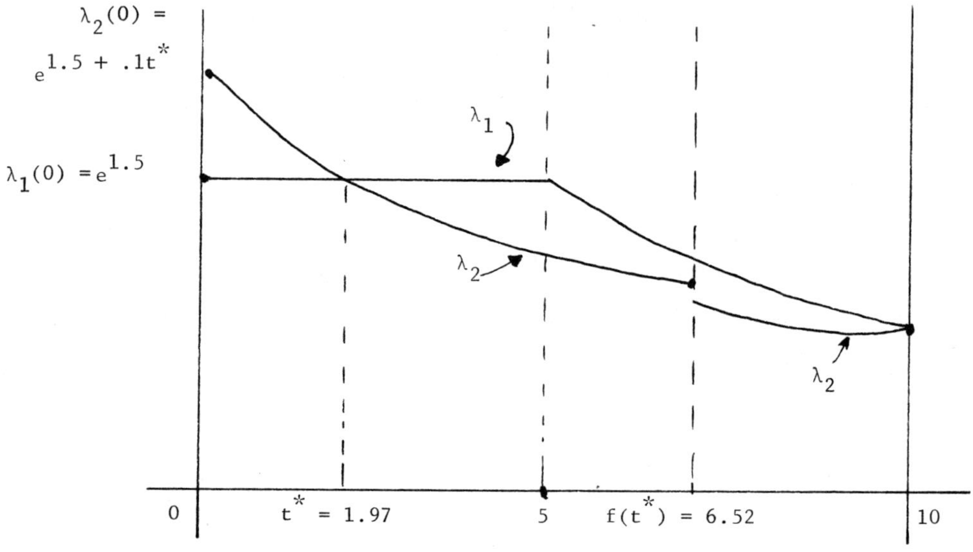

Figure 4.9

4.12* Suppose we extend the model (4.32) to include debt. For this let y denote the total debt at time t and $w \ge 0$ denote the amount of debt issued expressed as a proportion of current earnings. Then

$$\dot{y} = wx, \quad y(0) = y_o.$$

How would you modify the state equation for x and the growth constraint (4.30). Assume i to be the constant interest rate on debt, and $i < r$.

4.13* Find the form of the optimum policy for the following model due to [Davis and Elzinga, 1972]:

$$\max_{u,v} \left\{ J = \int_0^T e^{-\rho t}(1-v)E\, r\, dt + P(T)\, e^{-\rho T} \right\},$$

subject to

$$\dot{P} = k[rE(1-v)-\rho P], \qquad P(0) = P_o,$$

$$\dot{E} = rE[v + u(c-E/P)], \quad E(0) = E_o,$$

and the control constraints

$$u \geq 0, \quad v \geq 0, \quad cu + v \leq g/r.$$

Here P denotes the price of a stock, E denotes equity per stock and $k \geq 0$ is a constant. Also, assume $r > \rho > g$. This example requires the use of the generalized Legendre-Clebsch condition (A4.40).

4.14* Remove the assumption of an arbitrary upperbound g on the growth rate in the financing model of section 4.2.1 by the assumption of an increasing cost function associated with the growth rate. Also assume $c = 1$ as in Exercise 4.4. Formulate the resulting model and apply the maximum principle to find the form of the optimal policy. You may assume the cost function to be quadratic to get an explicit form of the solution.

5 APPLICATIONS TO PRODUCTION AND INVENTORY

Applications of optimization methods to production and inventory problems date back at least to the classical EOQ model and the lot size formula. The EOQ is essentially a static model in the sense that the demand is constant and only a stationary solution is sought. A dynamic version of the lot size model was analyzed by [Wagner and Whitin, 1958]. The solution methodology used there was dynamic programming.

An important dynamic production planning model was developed by [Holt, Modigliani, Muth, and Simon, 1960]. In their model, referred to as the HMMS model, they considered both production costs and inventory holding costs over time. They used calculus of variations techniques to solve the continuous time version of their model.

In Section 5.1 a model [Thompson and Sethi, 1980] similar to the HMMS model is formulated and completely solved using optimal control theory. The turnpike solution is also obtained when the horizon is infinite.

In Section 5.2 the continuous wheat trading model of [Ijiri and Thompson, 1970] is introduced. In this model a wheat speculator must buy and sell wheat in an optimal way in order to take advantage of changes in the price of wheat over time. This model permits short selling of wheat. The model has been further analyzed by [Norstrom, 1978] with short selling disallowed. A simple example which illustrates his results is presented in Section 5.2.4.

In Section 5.3 impose an upper bound warehousing constraint on the wheat trading model. The purpose of doing this is to illustrate the concepts of planning and forecast horizons, by means of examples. This section is expository in nature, but theoretical developments of these ideas are available in the literature.

5.1 A Production Inventory System

Many manufacturing enterprises use a production inventory system to manage fluctuations in consumer demand for the product. Such a system consists of a manufacturing plant and a finished goods warehouse to store those products which are manufactured but not immediately sold. Once a product is made and put into inventory it incurs inventory holding costs of two kinds: first, the costs of physically storing the product, insuring it, etc.; and second the carrying charges of having the firm's money invested in the unsold inventory. The advantages of having products in inventory are: first, they are immediately available to meet demand; second; by using the warehouse to store excess production during low demand periods to be available during high demand periods, it is possible to smooth the production schedules of the manufacturing plant. This usually permits the use of a smaller manufacturing plant than would otherwise be necessary, and also reduces the difficulties of managing the system.

The optimization problem is to balance the benefits of production smoothing versus the costs of holding inventory. Some early references that apply control theory to production and inventory problems are: [Sprzeuzkouski, 1967] and [Huang, Fan, and Erickson, 1967].

5.1.1 The Production Inventory Model

We consider a factory producing a single homogenous good and having a finished goods warehouse. To state the model we define the following quantities:

$I(t)$ = inventory level at time t (state variable)

$P(t)$ = production rate at time t (control variable)

$S(t)$ = sales rate at time t (exogeneous variable); assumed bounded and differentiable for $t \geq 0$.

T = length of planning period

$\hat{I}$ = inventory goal level

I_0 = initial inventory level

$\hat{P}$ = production goal level

h = inventory holding cost coefficient

c = production cost coefficient

ρ = constant nonnegative discount rate.

The interpretation of the inventory goal level, $\hat{I}$, is that it is a "safety stock" that the company wants to keep on hand; e.g., $\hat{I}$ could be two months of sales, one quarter of sales, etc. Similarly, the production goal level, $\hat{P}$, can be interpreted as the desired or most efficient level at which we want to run the factory.

With this notation we can now state the conditions of the model. The first is the stock flow differential equation.

$$\dot{I}(t) = P(t) - S(t); \quad I(0) = I_0 \tag{5.1}$$

which says the inventory at time t is increased by the production rate and decreased by the sales rate. The objective function of the model is:

$$\underset{P\geq 0}{\text{Maximize}}\{J = \int_0^T e^{-\rho t}[\frac{h}{2}(I-\hat{I})^2 + \frac{c}{2}(P-\hat{P})^2]dt\} . \tag{5.2}$$

The interpretation of the objective function is that we want to keep the inventory as close as possible to its goal level, $\hat{I}$, and also keep the production rate, P, as close as possible to its goal level, $\hat{P}$. The quadratic terms $(I-\hat{I})^2$ and $(P-\hat{P})^2$ impose "penalties" for having either I or P not being close to its corresponding goal level.

5.1.2 Solution by the Maximum Principle

We now associate an adjoint function λ with equation (5.1) and can write the current value Hamiltonian function as

$$H = \lambda(P-S) - \frac{h}{2}(I-\hat{I})^2 - \frac{c}{2}(P-\hat{P})^2 . \tag{5.3}$$

Applying the Pontryagin maximum principle, we differentiate (5.3) and set the resulting expression equal to 0, which gives

$$\frac{\partial H}{\partial P} = \lambda - c(P-\hat{P}) = 0.$$

From this we obtain the <u>decision rule</u>

$$P = \hat{P} + \frac{\lambda}{c} \tag{5.4}$$

which is valid unless it drives P negative. In most cases P is constrained to be nonnegative, so that the form of the optimal control is

$$P^* = \text{Maximum } [\hat{P} + \frac{\lambda}{c}, 0] . \tag{5.5}$$

For the rest of this section and the next we shall assume that $\hat{P}$ is large enough so that (5.4) always gives nonnegative production values. (This assumption will be removed in Section 5.1.4.) With this assumption we can substitute (5.4) into (5.1) and obtain

$$\dot{I} = \hat{P} + \lambda/c - S, \; I(0) = I_0 . \tag{5.6}$$

The equation for the adjoint variable is easily found as

$$\dot{\lambda} = \rho\lambda - \frac{\partial H}{\partial I} = \rho\lambda + h(I-\hat{I}), \quad \lambda(T) = 0 . \tag{5.7}$$

We see that (5.6) is an initial value problem and (5.7) a final value problem, so that these two give a two-point boundary value problem. We shall employ a method to solve these two equations simultaneously which works in some special cases including the present case. The method is the well known trick used to solve simultaneous differential equations by differentiation and substutition until one of the variables is eliminated. Specifically, we differentiate (5.6) with respect to t, which creates an equation with $\dot{\lambda}$ in it. We then use (5.7) to eliminate $\dot{\lambda}$ and (5.6) to eliminate λ from the resulting equation, as follows:

$$\ddot{I} = \dot{\lambda}/c - \dot{S} = \rho(\lambda/c) + (h/c)(I-\hat{I}) - \dot{S}$$

$$= \rho(\dot{I} - \hat{P} + S) + (h/c)(I-\hat{I}) - \dot{S} .$$

We rewrite this as

$$\ddot{I} - \rho\dot{I} - \alpha^2 I = -\alpha^2\hat{I} - \dot{S} - \rho(\hat{P}-S) \tag{5.8}$$

where the <u>time constant</u> α is given by

$$\alpha = \sqrt{h/c} . \tag{5.9}$$

We can now solve (5.8) by using the standard method described in Appendix A1. The auxiliary equation for (5.8) is

$$m^2 - \rho m - \alpha^2 = 0$$

which has the two real roots

$$m_1 = (\rho - \sqrt{\rho^2 + 4\alpha^2})/2, \quad m_2 = (\rho + \sqrt{\rho^2 + 4\alpha^2})/2 . \tag{5.10}$$

(Note that $m_1 < 0$ and $m_2 > 0$.)

We can therefore write the general solution to (5.8) as

$$I = a_1 e^{m_1 t} + a_2 e^{m_2 t} + Q(t), \quad I(0) = I_0 \tag{5.11}$$

where $Q(t)$ is a particular integral of (5.8).

We shall say that $Q(t)$ is a <u>special particular integral</u> of (5.8) if it has no additive terms involving the homogeneous solution $e^{m_1 t}$ and $e^{m_2 t}$. From now on we will always assume that $Q(t)$ is a special particular integral.

Although (5.11) has two arbitrary constants, a_1 and a_2, it has only one boundary condition. To get the other boundary condition we differentiate (5.11), substitute the result into (5.6) and solve for λ. We obtain

$$\lambda = c(m_1 a_1 e^{m_1 t} + m_2 a_2 e^{m_2 t} + \dot{Q} + S - \hat{P}), \quad \lambda(T) = 0. \tag{5.12}$$

Note that we have imposed the boundary condition on λ so that we can now determine the constants a_1 and a_2.

To do the latter we define two more constants

$$b_1 = I_0 - Q(0) \tag{5.13}$$

$$b_2 = \hat{P} - \dot{Q}(T) - S(T) \,. \tag{5.14}$$

We now impose the boundary conditions in (5.11) and (5.12) and solve for a_1 and a_2 as follows:

$$a_1 = \frac{b_2 e^{m_1 T} - m_2 b_1 e^{(m_1+m_2)T}}{m_1 e^{2m_1 T} - m_2 e^{(m_1+m_2)T}} \tag{5.15}$$

$$a_2 = \frac{b_1 m_1 e^{2m_1T} - b_2 e^{m_1T}}{m_1 e^{2m_1T} - m_2 e^{(m_1+m_2)T}} \quad . \tag{5.16}$$

If we recall that m_1 is negative, and m_2 is positive we see that for T sufficiently large so that e^{m_1T} and e^{2m_1T} are negligible, we can write approximately

$$a_1 \approx b_1 \tag{5.17}$$

$$a_2 \approx \frac{b_2}{m_2} e^{-m_2T} \quad . \tag{5.18}$$

The reason that we retain the exponential factor in (5.18) is that a_2 is multiplied by e^{m_2t} which, even though it is negligible when t is small, becomes important when t is close to T.

With these values of a_1 and a_2 and (5.4), (5.11) and (5.12), we nc write expressions for I, P, and λ. We will break each expression into three parts: the first part, labelled "Starting Correction, is important only when t is small; the second part, labelled "Turnpike Expression," is significant for all values of t; and the third part, labelled "Ending Correction," is important only when t is close to T.

$$I = \overbrace{(b_1 e^{m_1t})}^{\text{Starting Correction}} + \overbrace{(Q)}^{\text{Turnpike Expression}} + \overbrace{(\frac{b_2}{m_2} e^{m_2(t-T)})}^{\text{Ending Correction}} \tag{5.19}$$

$$P = (m_1 b_1 e^{m_1t}) + (\dot{Q}+S) + (b_2 e^{m_2(t-T)}) \tag{5.20}$$

$$\lambda = c(m_1 b_1 e^{m_1t}) + c(\dot{Q}+S-\hat{P}) + c(b_2 e^{m_2(t-T)}) \tag{5.21}$$

Note that if $b_1 = 0$, which means $I_0 = Q(0)$, then there is no starting correction. In other words, $I_0 = Q(0)$ is a starting inventory that causes the solution to be on the turnpike initially. In the same way, if $b_2 = 0$, then the ending correction vanishes in each of these formulas, and the solution stays on the turnpike until the end.

Note that expressions (5.19) and (5.20) represent closed form solutions for the optimal inventory and production functions I and P that are correct for arbitrarily long time periods $0 \le t \le T$ as long as S is known throughout the same time period. Of course, S must be such that the special particular integral Q can be easily found. For examples see Section 5.1.5.

5.1.3 Infinite Horizon Solution

It is important to show that this solution also makes sense when $T \to \infty$. In other words, we show that for $\rho > 0$, the limit of this finite horizon solution as $T \to \infty$ also solves the infinite horizon problem. Note that as $T \to \infty$, the ending correction disappears, because a_2 defined in (5.16) becomes 0. We now have

$$\lambda(t) = c[m_1 b_1 e^{m_1 t} + \dot{Q} + S - \hat{P}]. \tag{5.22}$$

Clearly $\lim_{t\to\infty} \lambda(t)$ is bounded since $\dot{Q}$ will be bounded if S is bounded. Then for $\rho > 0$

$$\lim_{t\to\infty} e^{-\rho t} \lambda(t) = 0. \tag{5.23}$$

By the sufficiency condition of the maximum principle (Section 2.4), it can be verified that the limiting solution (which is a Pontryagin path, see [Arrow and Kurz, 1970]),

$$I(t) = b_1 e^{m_1 t} + Q, \quad P(t) = m_1 b_1 e^{m_1 t} + \dot{Q} + S \tag{5.24}$$

is optimal. Note that if $I(0) = Q(0)$, the solution is always on the turnpike. Thus, the triple $\{I, P, \lambda\} = \{Q, \dot{Q} + S, c(\dot{Q}+S)\}$ represents a nonstationary turnpike. If $I(0) \neq Q(0)$, then $b_1 \neq 0$ and the expression (5.2.24) imply that the path of inventory and production only approach the turnpike without actually attaining it.

Note that the solution does not satisfy the sufficiency transversality condition for $\rho = 0$. In this case, all solutions give an infinite value (which is the worst value since we are minimizing) for the cost objective function. Our limiting solution also gives an infinite value for the objective function. It makes sense to define the limiting solution for $\rho = 0$ to be also the optimal infinite horizon solution for the undiscounted case.

In the last three sections we ignored the production constraint $P \geq 0$ and used (5.4) as the optimal decision rule. In the next section we shall solve the production inventory problem subject to $P \geq 0$ and use (5.5) as the optimal production rule. For simplicity of analysis and exposition we shall assume also that S is positive constant and $T = \infty$.

5.1.4 A Complete Analysis of the Constant Positive S Case with Infinite Hori

The particular integral $Q(t)$ for the constant positive S case is

$$Q = (\rho/\alpha^2)(\hat{P}-S) + \hat{I} . \tag{5.25}$$

Thus the turnpike is defined by the triple $\{(\rho/\alpha^2)(\hat{P}-S) + \hat{I},\ S,\ c(S-\hat{P})\}$ formed from the turnpike expressions in (5.19), (5.20) and (5.21). Note that we could have obtained the turnpike levels directly by using conditions

$$\dot{\bar{I}} = 0,\ \dot{\bar{\lambda}} = 0, \quad \text{and} \quad \bar{P} = \hat{P} + \bar{\lambda}/c \tag{5.26}$$

as explained in Chapter 3.

If $I_0 = Q$, then the optimal solution stays on the turnpike. If

$I_0 \neq Q$, then the optimal solution is given by

$$P(t) = m_1 b_1 e^{m_1 t} + S = m_1[I_0 - Q]e^{m_1 t} + S. \tag{5.27}$$

Clearly, if $I_0 \leq Q$, this provides a nonnegative optimal production throughout. In case $I_0 > Q$, note first that $P(t)$ increases with t with

$$P(0) = m_1(I_0 - Q) + S . \tag{5.28}$$

Thus if $I_0 - Q > -S/m_1$, we have a negative value for $P(0)$ which is infeasible. By (5.5), $P^*(0) = 0$. We can now depict this situation in Fig. 5.1. The time $\hat{t}$ shown in the figure is the time at which $\hat{P} + \frac{\lambda(\hat{t})}{c} = 0$.

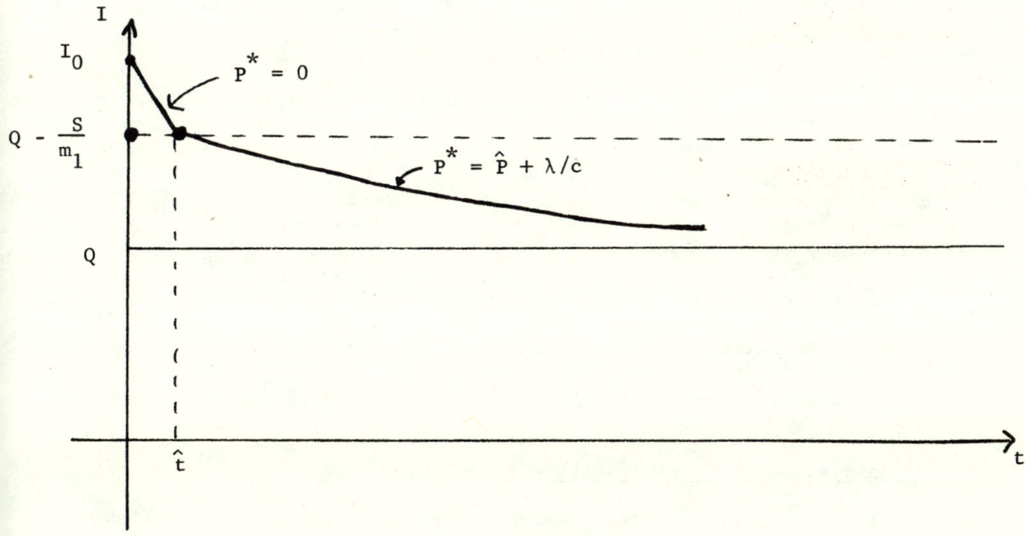

Figure 5.1

It should be obvious that

$$I(\hat{t}) = Q - \frac{S}{m_1} . \tag{5.29}$$

From $\hat{t}$ on, we have the solution of the problem given by (5.19), (5.20) and (5.21). From $t < \hat{t}$, we have

$$\dot{I} = -S \Rightarrow I = I_0 - St \tag{5.30}$$

and

$$\dot{\lambda} = \rho\lambda + h(I-\hat{I}),\ \lambda(\hat{t}) = -c\hat{P} \quad . \tag{5.31}$$

We can substitute $I_0 - St$ for I in equation (5.31) for $\dot{\lambda}$ and solve for λ. Note that we can easily obtain $\hat{t}$ as

$$I_0 - S\hat{t} = Q - \frac{S}{m_1} \Rightarrow \hat{t} = \frac{I_0 - Q}{S} + \frac{1}{m_1} \quad . \tag{5.32}$$

For $t > \hat{t}$, the solution is given by (5.19), (5.20), and (5.21) with t replaced by $(t - \hat{t})$ and $b_1 = I(\hat{t}) - Q = -S/m_1$.

The solution shown in Figure 5.1 represents a situation where there is excess initial inventory, i.e., $I_0 > Q - S/m_1$. For $I_0 \leq Q-S/m_1$ the optimal control given by $\hat{P} + \lambda/c$ is ≥ 0, hence the inventory will be driven asymptotically to the level Q.

In the next section we return to the model of Section 5.1 and examine the behavior of solutions for the case of fluctuating demands, such as polynomial or sinusoidal demand functions.

5.1.5 Special Cases of Time Varying Demands

In this section, we solve some numerical examples of the model of Section 5.1 for $\rho = 0$ and $T < \infty$. As the first class of examples assume $S(t)$ is a polynomial of degree $2p$ or $2p-1$ so that $S^{(2p+1)} \equiv 0$. (We denote the kth time derivative of S with respect to t by $S^{(k)}$). In other words

$$S(t) = C_o t^{2p} + C_1 t^{2p-1} + \ldots + C_{2p} \tag{5.33}$$

where at least one of C_0 and C_1 is not zero. Then it is easy to show that a particluar solution of (5.1.11) is

$$Q(t) = \hat{I} + \frac{1}{\alpha^2} S^{(1)} + \frac{1}{\alpha^4} S^{(3)} + \ldots + \frac{1}{\alpha^{2p}} S^{(2p-1)} . \qquad (5.34)$$

In Exercise 5.3 the reader is asked to verify this by direct substitution.

In the second class of examples we assume that $S(t)$ is sinusoidal of form

$$S(t) = A \sin \pi t + C \qquad (5.35)$$

where A and C are constants. In Exercise 5.4 you are asked to verify that

$$Q(t) = \hat{I} + \frac{\pi A}{\alpha^2 + \pi^2} \cos \pi t \qquad (5.36)$$

is a particular solution of (5.8) for this S.

As is well known in differential equation theory demands which are sums of functions of the form (5.33) and/or (5.35) give rise to solutions which are sums of functions of form (5.34) and/or (5.36).

<u>Example 5.1</u>. Assume $\hat{P} = 30$, $\hat{I} = 15$, $T = 8$, $h = c = 1$ so that $\alpha = 1$ and

$$S(t) = t(t-4)(t-8) + 30$$

$$= t^3 - 12t^2 + 32t + 30 .$$

It is then easy to show from (5.3.2) that

$$Q(t) = 3t^2 - 24t + 53$$

and $b_2 = -24$. In Figure 5.2 the graphs of production, inventory and demand are drawn with $I_0 = 10$ (a low starting inventory) which makes $b_1 = -43$. In

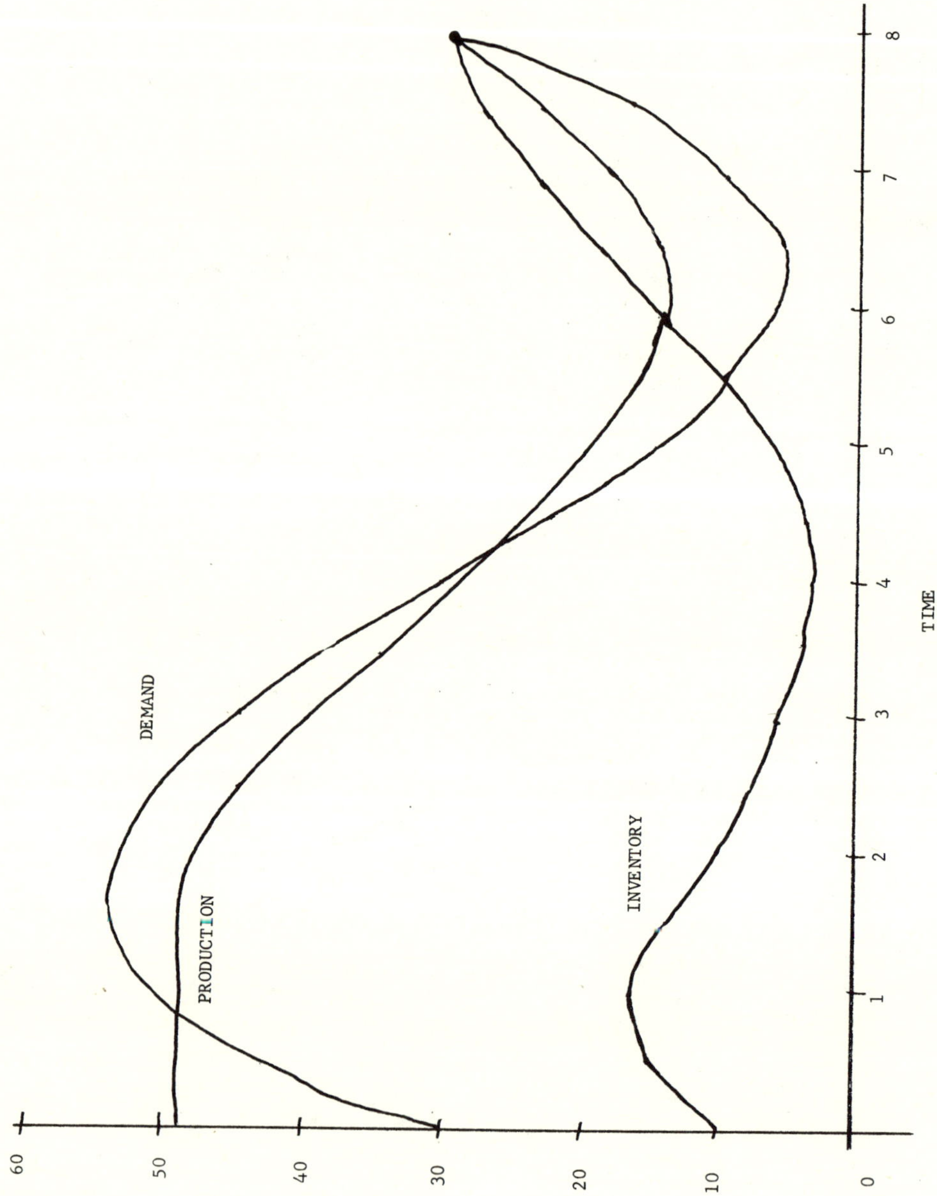

DEMAND
PRODUCTION
INVENTORY
TIME
0
10
20
30
40
50
60
1
2
3
4
5
6
7
8

Figure 5.2

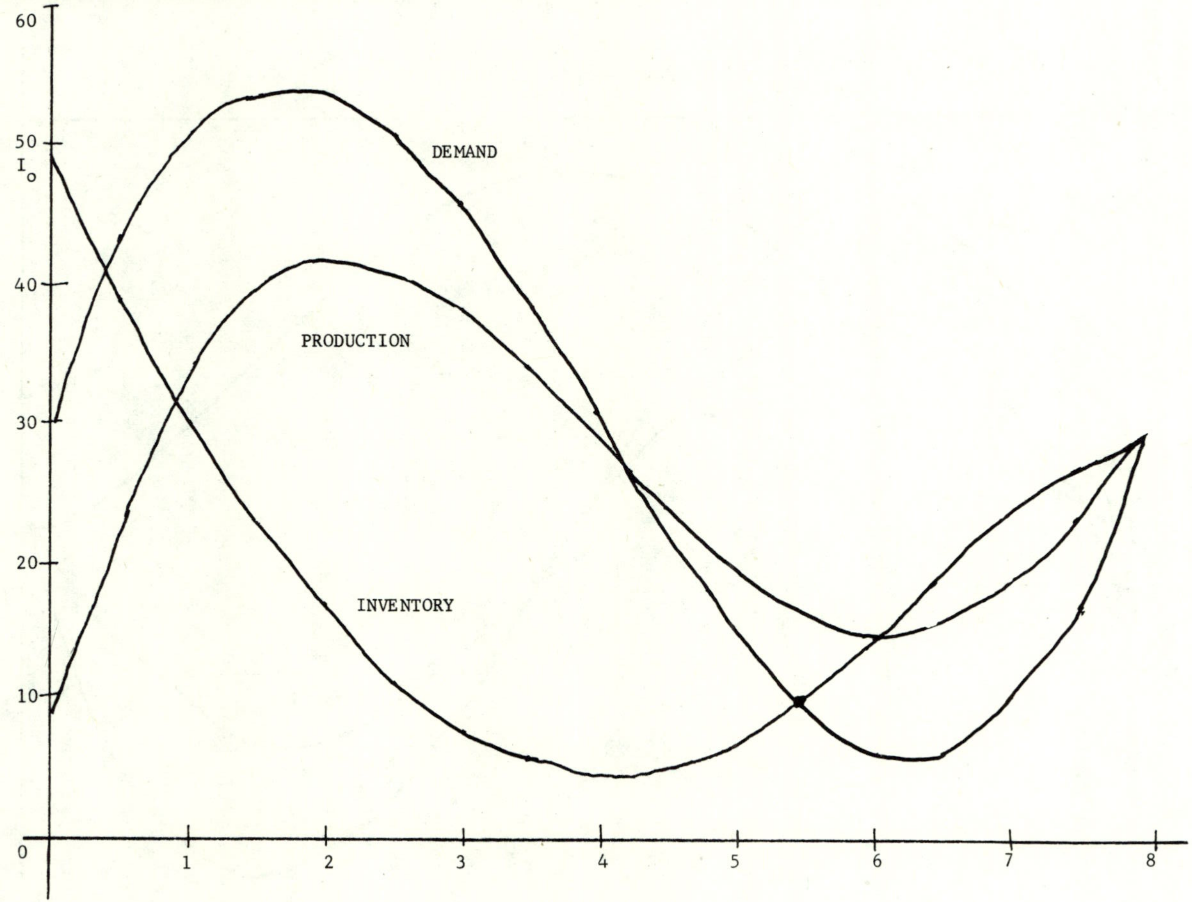

60
50
I_o
40
30
20
10
0
1
2
3
4
5
6
7
8
DEMAND
PRODUCTION
INVENTORY
TIME

Figure 5.3

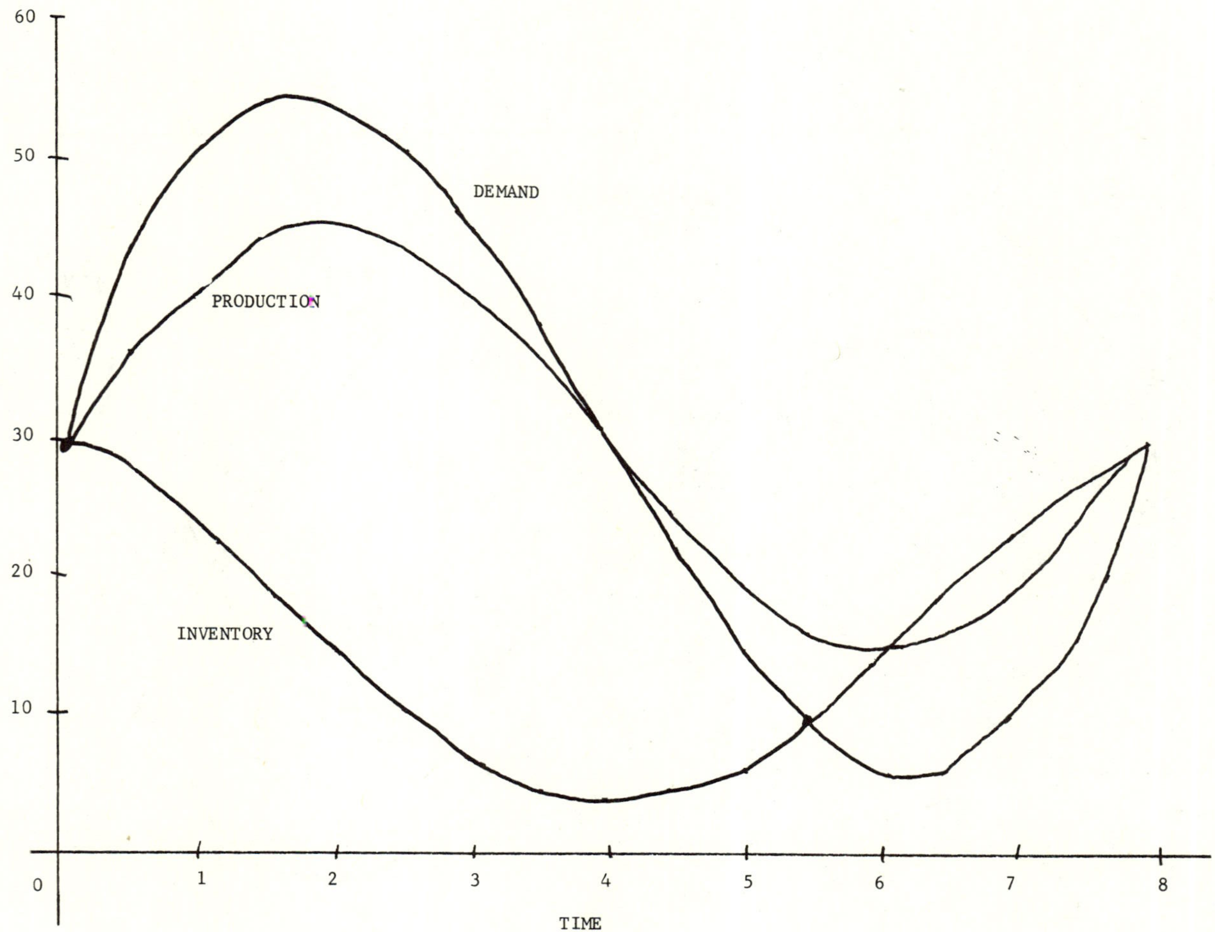
60
50
40
30
20
10
0
1
2
3
4
5
6
7
8
DEMAND
PRODUCTION
INVENTORY
TIME

Figure 5.4

Figure 5.3 the same graphs are drawn with $I_0 = 50$ (a high starting inventory) which makes $b_1 = -3$. In Figure 5.4 the same graphs are drawn with $I_0 = 29$, which is the ending inventory in all three cases.

Example 5.2. As another example assume that

$$S(t) = A + Bt + \sum_{k=1}^{K} C_k \sin(\pi D_k t + E_k) \tag{5.37}$$

where the constants A, B, C_k, D_k, and E_k are estimated from future demand data by means of one of the standard forecasting techniques such as those given in Brown [2]. Making use of formulas (5.34) and (5.36) we have the special particular integral

$$Q(t) = I_1 + \frac{1}{\alpha^2} B + \sum_{k=1}^{K} \frac{\pi C_k D_k}{\alpha^2 + (\pi D_k)^2} \cos(\pi D_k t + E_k) \tag{5.38}$$

5.2 Continuous Wheat Trading Model

Consider a firm that engages in buying and selling of wheat. The firm's only assets are cash and wheat, and the price of wheat over time is given with certainty. The objective of this speculative firm is to buy and sell wheat in order to maximize the total value of its assets at the horizon time T. The problem here is similar to the simple cash balance model of Section 4.1 except that there are nonlinear holding costs associated with storing wheat. An extension to a model having two control variables appears in [Ijiri and Thompson, 1972].

5.2.1 The Model

We introduce the following notation:

T = horizon time.

$x(t)$ = the cash balance in dollars at time t

$y(t)$ = the wheat balance in bushels at time t

$v(t)$ = the rate of purchase of wheat in bushels per unit time; a negative purchase means a sale: $-V_2 \leq v(t) \leq V_1$, where $V_1 \geq 0$ and $V_2 \geq 0$.

$p(t)$ = the price of wheat in dollars per bushel at time t

r = the constant positive interest rate earned on the cash balance

$h(y)$ = the cost of holding y bushels per unit time.

In this section we permit x and y to go negative, meaning borrowing of money and short selling of wheat are allowed. In the next section we disallow the short selling of wheat.

The state equations are:

$$\dot{x} = rx - h(y) - pv, \quad x(0) = x_0 \tag{5.39}$$

$$\dot{y} = v \quad , \quad y(0) = y_0 \ . \tag{5.40}$$

The objective function is to

$$\text{Maximize}\{J = x(T) + p(T)y(T)\} \ . \tag{5.41}$$

Note that the problem is in the linear Mayer form.

5.2.2 Solution by the Maximum Principle

Introducing the adjoint variables λ_1 and λ_2 and defining the Hamiltonian function

$$H = \lambda_1[rx - h(y) - pv] + \lambda_2 v \ , \tag{5.42}$$

The adjoint equations are:

$$\dot{\lambda}_1 = -\lambda_1 r \qquad , \quad \lambda_1(T) = 1 \tag{5.43}$$

$$\dot{\lambda}_2 = \quad h'(y)\lambda_1 \quad , \quad \lambda_2(T) = P(T) \quad . \tag{5.44}$$

It is easy to solve (5.43) as

$$\lambda_1(t) = e^{r(T-t)} \tag{5.45}$$

and (5.44) as

$$\lambda_2(t) = P(T) - \int_t^T h'(y)\, e^{r(T-t)} dt. \tag{5.46}$$

The interpretation of $\lambda_1(t)$ is that it is the future value (at time T) of one dollar held as cash from t to T. Also the interpretation of $\lambda_2(t)$ is the price at time T of a bushel of wheat less the total future value (at time T) of the stream of storage costs incurred to store that bushel of wheat from t to T.

From (5.42) the optimal control is

$$v^*(t) = \text{bang}[-V_2, V_1; \lambda_2(t) - \lambda_1(t)p(t)] \ . \tag{5.47}$$

In Exercise 5.6 you are asked to provide the interpretation of this optimal policy.

Equations (5.39), (5.40), (5.46) and (5.47) determine a two-point boundary value problem which usually requires a numerical solution procedure. In the next section we assume a special form for the storage function $h(y)$ to be able to obtain a closed form solution.

5.2.3 Complete Solution of a Special Case

For this special case we assume $h(y) = \frac{1}{2}|y|$, $r = 0$, $x(0) = 10$, $y(0) = 0$, $V_1 = V_2 = 1$, $T = 6$

$$p(t) = \begin{cases} 3 & 0 \le t \le 3 \\ 4 & 3 < t \le 6 \end{cases} \quad . \tag{5.48}$$

For this problem we have $\lambda_1 = 1$ from (5.45) so that the TPBVP is

$$\dot{x} = -\frac{1}{2}|y| - pv \quad , \quad x(0) = 10 \tag{5.49}$$

$$\dot{y} = v \quad , \quad y(0) = 0 \tag{5.50}$$

$$\dot{\lambda}_2(t) = \frac{1}{2} \operatorname{sgn} y \quad , \quad \lambda_2(6) = 4 \quad . \tag{5.51}$$

Note that $\lambda_1(t) = 1$ for all t. For this simple problem it is easy to guess a solution. From (5.47) the optimal policy is

$$v^*(t) = \text{bang}[-1,1;\ \lambda_2(t)-p(t)] \quad . \tag{5.52}$$

The graph of the price function is shown in Figure 5.5. Since $p(t)$ is increasing short selling is never optimal. Since the storage cost is $\frac{1}{2}$ unit

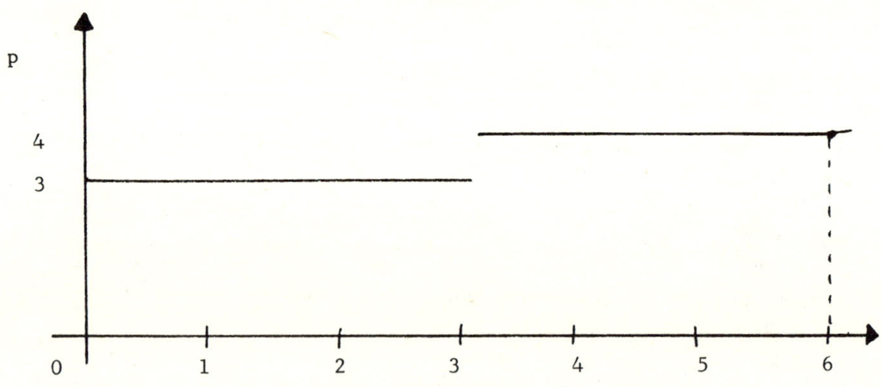

Figure 5.5

per unit time, it never pays to store wheat more than 2 time units, so that

$v^*(t) = 0$ for $0 \leq t \leq 1$. This must be obviously singular control. Suppose we start buying wheat at $t^* > 1$; from (5.52) the rate of buying is 1; clearly buying will continue at this rate until $t = 3$, and no longer. In order not to lose money storing wheat all purchases must be sold within 2 time units from the time of their purchase. Clearly we should start selling at $t = 3+$ at maximum rate 1, and continue until a last sale time t^{**}. In order to exactly sell all the purchases we must have

$$3 - t^* = t^{**} - 3. \tag{5.53}$$

Thus $v^*(t) = 0$ in the interval $[t^{**},6]$, which must also be singular control. With this policy $y(t) > 0$ for all $t \in (t^*,t^{**})$; from (5.51) $\dot{\lambda}_2 = \frac{1}{2}$ in that interval. In order to have singular control in the interval $[t^{**},6]$ we must have $\lambda_2^*(t) = 4$ in that interval. Also, in order to have singular control in $[0,t^*]$ we must have $\lambda_2^*(t) = 3$ in that interval. We can now conclude that

$$t^{**} - t^* = 2 \tag{5.54}$$

and therefore $t^* = 2$ and $t^{**} = 4$ and

$$\lambda_2^*(t) = \begin{cases} 3 & , \quad 0 \leq t \leq 2 \\ 2 + t/2, & 2 \leq t \leq 4 \\ 4 & , \quad 4 \leq t \leq 6 \end{cases} . \tag{5.55}$$

We can now sketch graphs for $\lambda_2^*(t)$, $v^*(t)$, and $y^*(t)$ as shown in Figure 5.6. In Exercise 5.11 you are asked to show that these trajectories are optimal by verifying that the maximum principle necessary conditions hold and that they are also sufficient.

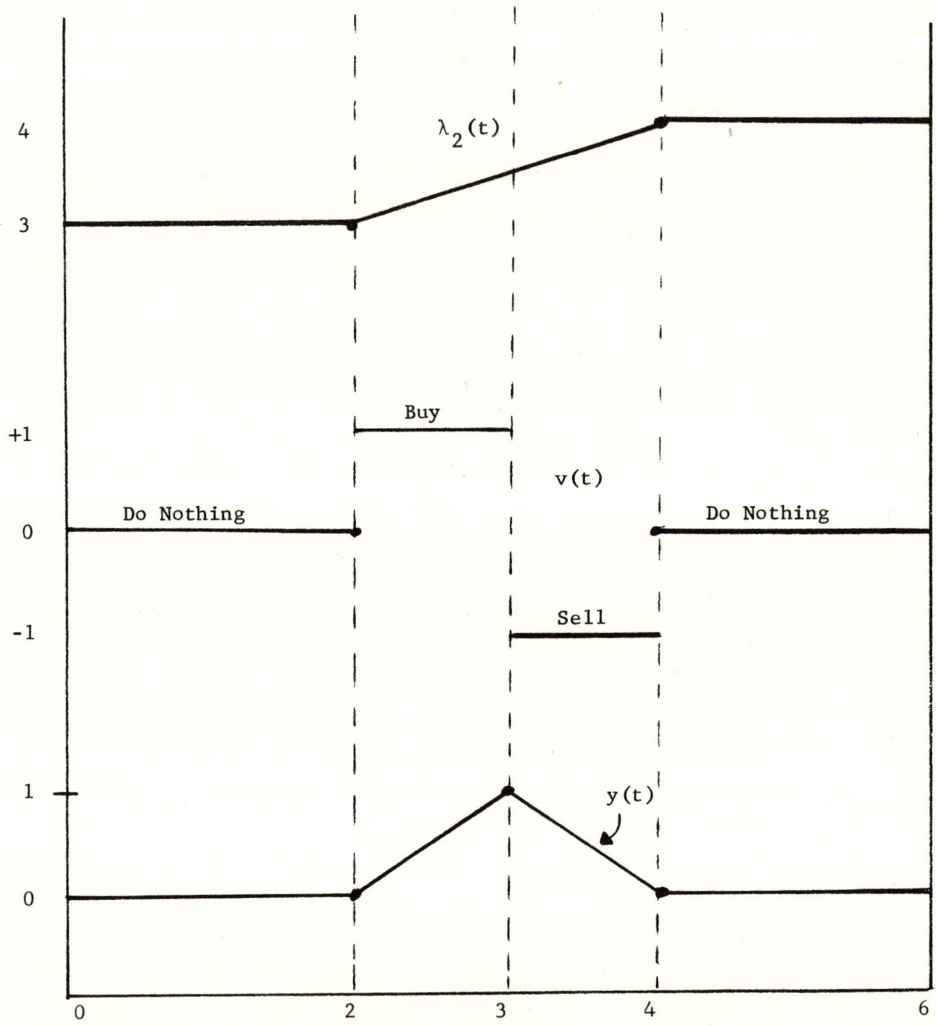

Figure 5.6

5.2.4 The Wheat Trading Model with Short-Selling Not Allowed

We next consider the wheat trading problem for the case in which short selling is not permitted, which requires that we impose the state constraint $y \geq 0$. For simplicity in exposition we consider only a special case due to [Norstrom, 1978] which is a slight modification of the model of 5.2.3.

For the present case we assume $h(y) = \frac{1}{2}y$, $r = 0$, $x(0) = 10$, $y(0) = 1$, $V_1 = V_2 = 1$, $T = 3$, and

$$p(t) = \begin{cases} -2t + 7 & \text{for } 0 \leq t < 2 \\ t + 1 & \text{for } 2 \leq t \leq 3 \end{cases} . \tag{5.56}$$

The statement of the problem is

$$\boxed{\begin{array}{l} \text{Maximize}\{J = x(3) + p(3)y(3) = x(3) + 4y(3)\} \\ \text{Subject to} \\ \qquad \dot{x} = -\frac{1}{2}y - pv, \qquad , \; x(0) = 10 \\ \qquad \dot{y} = v \qquad\qquad\quad , \; y(0) = 1 \\ \qquad v + 1 \geq 0, \; 1 - v \geq 0, \; y \geq 0 \end{array}} \tag{5.57}$$

To solve this problem we use the Lagrangian form of the maximum principle given in Section 3.1. The Hamiltonian is

$$H = \lambda_1(-\frac{1}{2}y - pv) + \lambda_2 v \quad . \tag{5.58}$$

The optimal control, as before, is

$$v^*(t) = \text{bang}[-1,1; \lambda_2(t) - p(t)] \text{ when } \quad y > 0 \quad . \tag{5.59}$$

Whenever $y = 0$ we must impose $\dot{y} = v \geq 0$ in order to insure that no short selling occurs. Therefore

$$v^* = \text{bang}[0,1;\ \lambda_2(t) - p(t)] \quad \text{when} \quad y = 0. \tag{5.60}$$

Next we form the Lagrangian

$$L = H + \mu_1(v+1) + \mu_2(1-v) + \eta v \tag{5.61}$$

where μ_1, μ_2, and η satisfy the complementary slackness conditions

$$\mu_1 \geq 0, \quad \mu_1(v+1) = 0 \tag{5.62}$$

$$\mu_2 \geq 0, \quad \mu_2(1-v) = 0 \tag{5.63}$$

$$\eta \geq 0, \quad \eta y = 0, \quad \eta v = 0 \quad . \tag{5.64}$$

Furthermore the optimal trajectory must satisfy

$$\frac{\partial L}{\partial v} = \lambda_2 - p + \mu_1 - \mu_2 + \eta = 0 \quad . \tag{5.65}$$

From the Lagrangian we get $\lambda_1 \equiv 1$ as before and

$$\dot{\lambda}_2 = \frac{1}{2}, \quad \lambda_2(3) = 4 \quad . \tag{5.66}$$

From this we see that λ_2 is always increasing except possibly at a jump point. Let $\hat{t}$ be a time such that there is no jump in the interval $(\hat{t},3]$, i.e., $\hat{t}$ the time of the last jump; then

$$\lambda_2(t) = \frac{1}{2}t + \frac{5}{2} \quad \text{for} \quad \hat{t} < t \leq 3 \tag{5.67}$$

and the optimal control from (5.59) or (5.60) is $v^* = 1$, i.e., we buy wheat at the maximum rate, 1. The smallest possible value of $\hat{t}$ is obtained by equating $\lambda_2(t) = p(t)$, thus

$$\frac{1}{2} t + \frac{5}{2} = - 2t + 7 \Rightarrow \hat{t} \leq 1.8 \quad . \tag{5.68}$$

Since $p(t)$ is decreasing at the start of the problem, it appears that selling should be optimal at the start. Since the beginning inventory is $y(0) = 1$, selling at the maximum rate can continue only until $t = 1$, when $y(1) = 0$. Suppose that we do nothing, i.e., $v^*(t) = 0$ in the interval $(1,1.8]$. Then $t = 1$ is an entry point (see Section 3.1.1) and $t = 1.8$ is not an entry point. Hence $\lambda_2(t)$ is continuous at $t = 1.8$ and therefore $\lambda_2(t)$ is given by (5.67) in the interval $(1,3]$.

Using (5.65) in the interval $(1,1.8]$ and the fact that $v = 0$ so that $\mu_1 = \mu_2 = 0$, we have

$$\lambda_2 - p + \mu_1 - \mu_2 + \eta = \lambda_2 - p + \eta = 0$$

so that

$$\eta(t) = p(t) - \lambda_2(t) \quad \text{for} \quad t \in (1,1.8]. \tag{5.69}$$

Therefore

$$\eta(1+) = p(1+) - \lambda_2(1+) = 5 - 3 = 2$$

and from the jump condition (3.32)

$$\lambda_2(1-) = \lambda_2(1+) + \eta(1+) = 2 + 3 = 5 = p(1+). \tag{5.70}$$

It is important to note that in the interval $(1,1.8]$ the optimal control condition (5.60) holds, justifying our supposition that $v^* = 0$ in this interval. Furthermore, using (5.70) and (5.66)

$$\lambda_2(t) = \frac{1}{2} t + \frac{9}{2} \quad \text{for} \quad t \in [0,1) \tag{5.71}$$

and the optimal control condition (5.59) holds, justifying our supposition that $v^* = 1$ in this interval. The graphs of $\lambda_2(t)$, $p(t)$, and $v^*(t)$ are displayed in Figure 5.7. To complete the solution of the problem, you are asked to determine the values of μ_1, μ_2 and η in these various intervals.

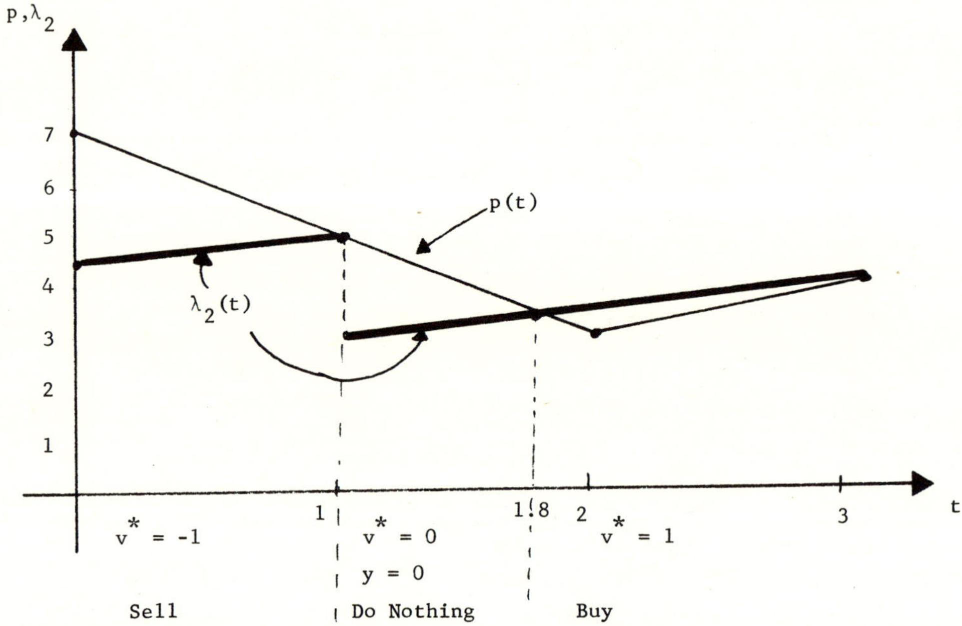

Figure 5.7 Adjoint Trajectory for Wheat Trading Model

5.3 Planning Horizons and Forecast Horizons

In some dynamic problems it is possible to show that the optimal decision during an initial positive time interval are either partially or wholly independent of the data from some future time onwards. In such cases, a forecast of the future data needs to be made only as far as that time to make optimal decisions in the initial time interval. The initial time interval is called the planning horizon and the time up to which data is required to make the optimal decisions during the planning horizon is called the forecast horizon. Whenever they exist, these horizons naturally decompose the problem into a series of smaller problems.

If the optimal decisions during the planning horizon are completely independent of the data beyond the forecast horizon, then the latter is called a strong forecast horizon. If, on the other hand, some mild restrictions on the data after the forecast horizon are required in order that to keep the optimal decisions during the planning horizon unaffected, then it is called a weak forecast horizon.

In this section we shall demonstrate these concepts in the context of the wheat trading model of the previous section. In Section 5.3.1 we shall obtain a planning horizon for the example of Section 5.2.4 which is also a weak forecast horizon. In Section 5.3.2 we modify the wheat trading model by adding a warehousing constraint. For the new problem we obtain a planning horizon and a strong forecast horizon.

In what follows we obtain these horizons and verify them for some examples with different data changes. For more details and proofs in more general situations see [Modigliani and Hohn, 1955], [Leiber, 1973], [Pekelman, 1974,1975, 1979], [Kleindorfer and Leiber, 1979], [Vanthienen, 1975], [Morton, 1978], and [Lundin and Morton, 1975].

5.3.1 Horizons for the Wheat Trading Model

For the example in Section 5.2.4 we will demonstrate that $t = 1$ is a planning horizon as well as a weak forecast horizon. In Figure 5.8 we have redrawn Figure 5.7 with a new price trajectory in the time interval [1,3]. In the figure we have extended the initial λ_2 trajectory and labelled it the "price shield." Its significance is that, as long as the new price trajectory in the interval [1,3] stays below the price shield, the optimal solution in the interval [0,1], which is the planning horizon, remains unchanged, i.e., it is optimal to sell throughout the interval. The restriction that $p(t)$ must stay below the price shield in [1,3] is the reason that $t = 1$ is a weak forecast horizon. The optimality of the control shown in Figure 5.8 can be concluded by obtaining the adjoint trajectory in the interval [1,3] as a straight line with slope 1/2 with its terminal value $\lambda_2(3) = p(3)$. This way of drawing the adjoint trajectory is correct as long as the corresponding policy does not violate the inventory constraint $y(t) \geq 0$. If the inventory constraint is violated, then the $\lambda_2(t)$ trajectory will contain jumps and to obtain it is more complicated. However, the planning horizon and weak forecast horizon still occur at $t = 1$. Moreover, if we let T be any finite horizon > 1 and assume $p(t)$ in the interval $[1,T]$ is always below the price shield line of Figure 5.8 extended to T, then the policy of selling at the maximum rate in the interval [0,1] remains optimal. It is in this sense that the policy of selling in the interval [0,1] can be considered optimal for the infinite horizon problem as long as $p(t)$ stays below the price shield. Suppose for instance that it is known with certainty that $p(t) \leq \bar{P}$ for all t. Then the intersection time, which is $t_1 = 2\bar{P} - 9$ for Figure 5.8, of the price shield and $\bar{P}$ defines a "stronger" forecast horizon than the weak horizon 1.

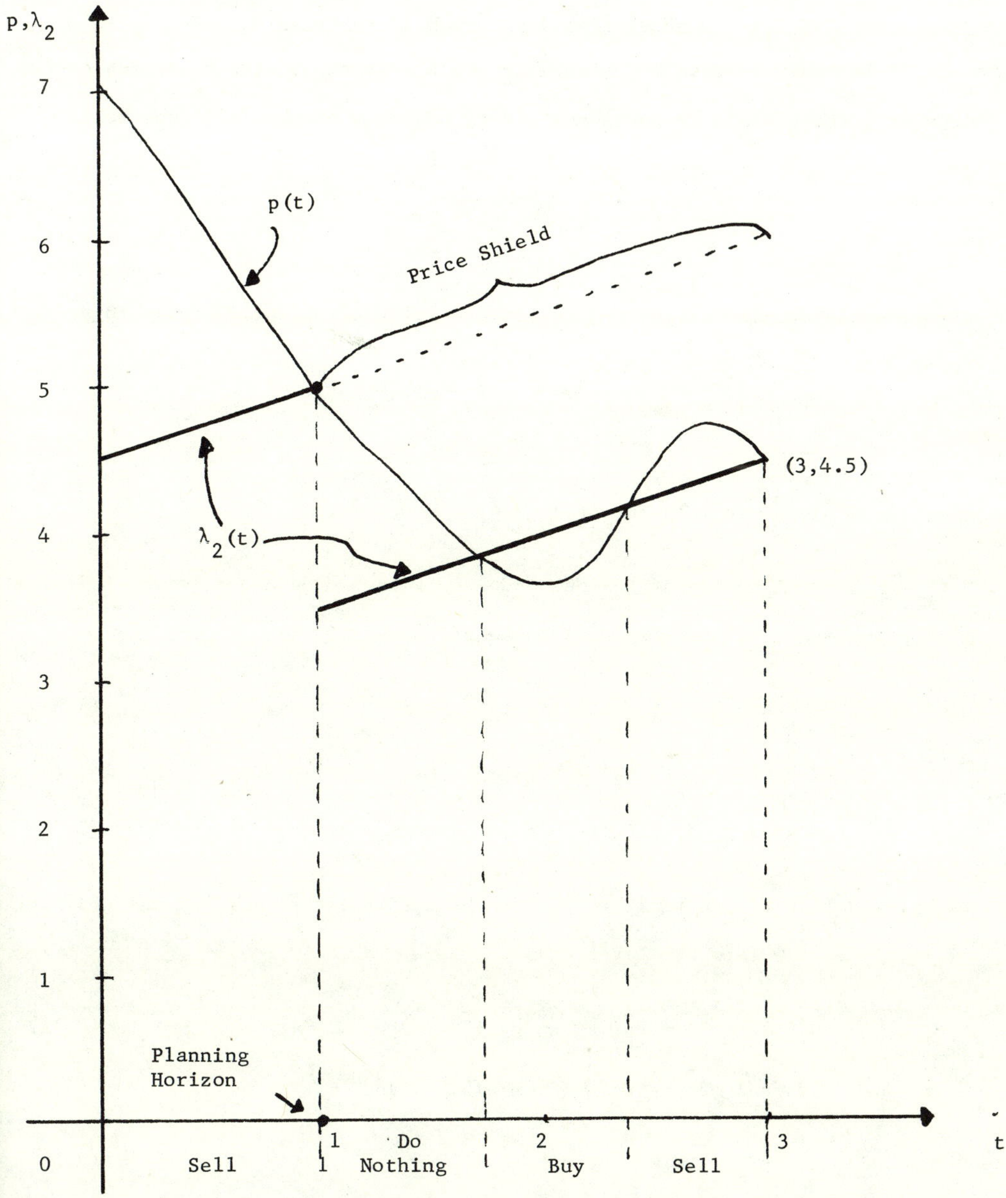

Figure 5.8 Planning Horizon and Optimal Policy for Wheat Trading Model

5.3.2 Horizon for the Wheat Trading Model with Warehousing Constraint

In order to give an example in which a strong forecast horizon occurs we modify the examply of Section 5.2.4 by adding a warehousing constraint $y \leq 1$ or

$$1 - y \geq 0 \tag{5.72}$$

changing the terminal time to $T = 4$, and defining the price trajectory to be

$$p(t) = \begin{cases} -2t + 7 & \text{for} \quad t \in [0,2) \\ t + 1 & \text{for} \quad t \in [2,4] \end{cases} \tag{5.73}$$

The Hamiltonian of the new problem is unchanged and is given in (5.58). The optimal control is defined in three parts as:

$$v^*(t) = \text{bang}[-1,1; \lambda_2(t) - p(t)] \quad \text{when} \quad 0 < y < 1 \tag{5.74}$$

$$v^*(t) = \text{bang}[0,1; \lambda_2(t) - p(t)] \quad \text{when} \quad y = 0 \tag{5.75}$$

$$v^*(t) = \text{bang}[-1,0; \lambda_2(t) - p(t)] \quad \text{when} \quad y = 1 \tag{5.76}$$

Defining a Lagrange multiplier β for the derivative of (5.72), that is for $-\dot{y} = -v \geq 0$, we form the Lagrangian

$$L = H + \mu_1(v+1) + \mu_2(1-v) + \eta v + \beta(-v) \tag{5.77}$$

where μ_1, μ_2, and η satisfy (5.62)-(5.64) and β satisfies

$$\beta \geq 0, \quad \beta(1-y) = 0, \quad \beta v = 0 \quad . \tag{5.78}$$

Furthermore the optimal trajectory must satisfy

$$\frac{\partial L}{\partial v} = \lambda_2 - p + \mu_1 - \mu_2 + \eta - \beta = 0 . \tag{5.79}$$

As before $\lambda_1 \equiv 1$ and λ_2 satisfies

$$\dot{\lambda}_2 = \frac{1}{2}, \quad \lambda_2(4) = p(4) = 5 . \tag{5.80}$$

Let $\hat{t}$ be the time of the last jump discontinuity before $T = 4$ of the adjoint function λ_2. Then

$$\lambda_2(t) = \frac{1}{2} t + 3 \quad \text{for} \quad \hat{t} \leq t \leq 4 . \tag{5.81}$$

The graph of (5.81) intersects the price trajectory at $t = 8/5$ as shown in Figure 5.9. It also stays above the price trajectory in the interval [8/5,4] so that, if there were no warehousing constraints (5.72), the optimal solution in this interval is to buy at maximum rate. However, with constraint (5.72), this is not possible. Thus $\hat{t} > 8/5$ since λ_2 will have a jump in the interval [8/5,4].

To find the actual value of $\hat{t}$ we must insert a line of slope 1/2 above the minimum price at $t = 2$ in such a way that its two intersection points are exactly one time unit apart. Thus using (5.73), $\hat{t}$ must satisfy

$$-2(\hat{t}-1) + 7 + 1/2 = \hat{t} + 1$$

which yields $\hat{t} = 17/6$.

The rest of the analysis for determining λ_2 is similar to that given in Section 5.2.4. Thus

$$\lambda_2(t) = \begin{cases} \frac{1}{2}t + \frac{9}{2} & \text{for} \quad t \in [0,1) \\ \frac{1}{2}t + \frac{19}{12} & \text{for} \quad t \in [1,\frac{17}{6}] \\ \frac{1}{2}t + 3 & \text{for} \quad t \in (\frac{17}{6},4] \end{cases} \tag{5.82}$$

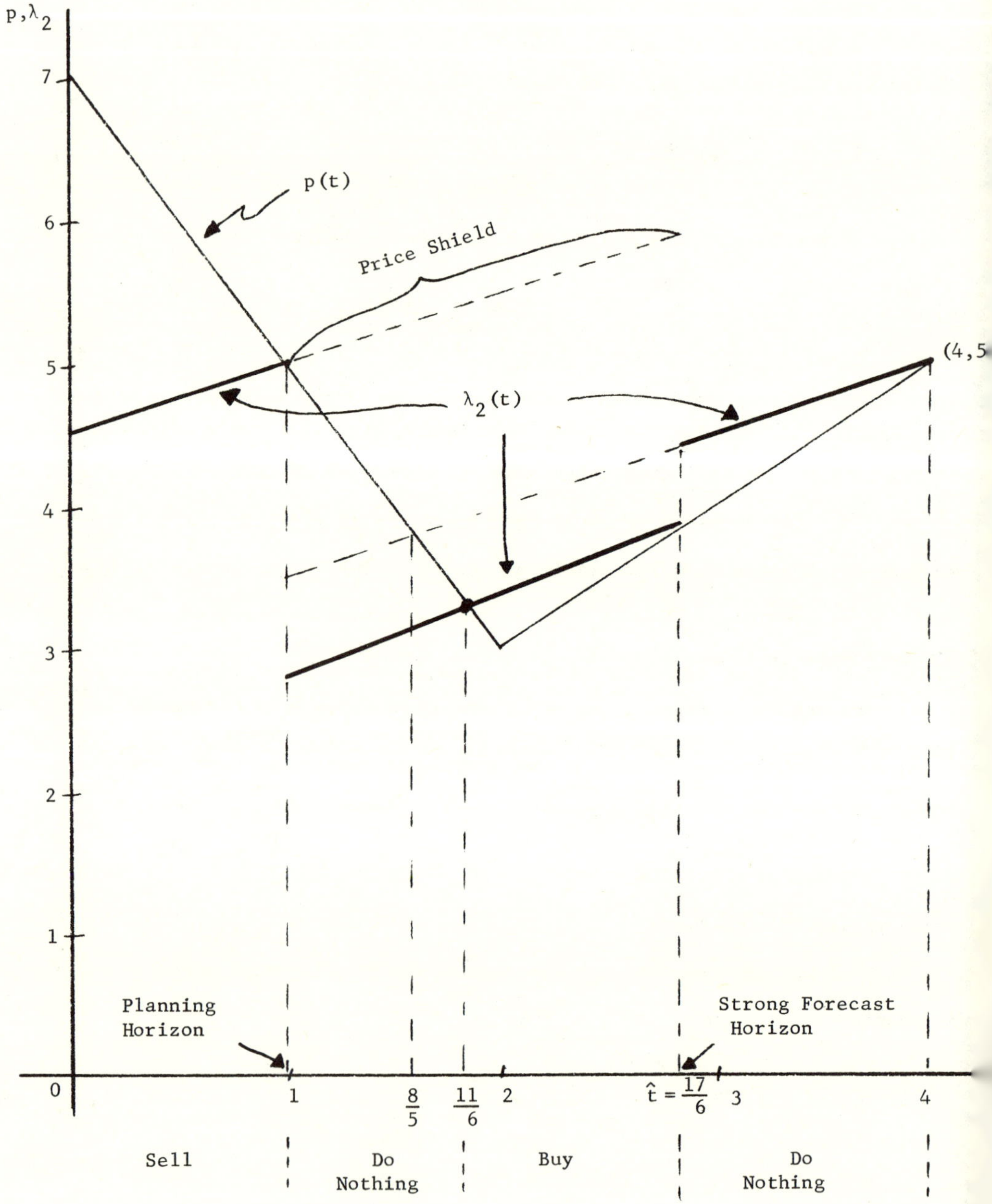

Figure 5.9 Horizons and Optimal Policy for the Wheat Trading Model with Warehouse Constraint

Given (5.82), the optimal policy is given by (5.74)-(5.76) and is marked on Figure 5.9.

To complete the maximum principle we must derive expressions for the Lagrange multipliers in the four intervals shown in Figure 5.9.

Interval [0,1): $\mu_2 = \eta = \beta = 0$, and

$$\mu_1 = p - \lambda_2 > 0, \ v^* = 1, \quad 0 < y^* < 1$$

Interval [1,11/6): $\mu_1 = \mu_2 = \beta = 0$, and

$$\eta = p - \lambda_2 > 0, \ v^* = 0, \quad y^* = 0$$

Interval [11/16,17/6): $\mu_1 = \eta = \beta = 0$, and

$$\mu_2 = \lambda_2 - p > 0, \quad v^* = 1, \quad 0 < y^* < 1$$

Interval [17/6,4]: $\mu_1 = \mu_2 = \eta = 0$, and

$$\beta = \lambda_2 - p > 0, \quad v^* = 0, \quad y^* = 1.$$

In Exercise 5.16 you are asked to work another variant of this problem.

For the example in Figure 5.9 we have labelled $t = 1$ as a planning horizon and $\hat{t} = 17/6$ as a strong forecast horizon. By that we mean that the optimal decision in [0,1] continues to be to sell at maximum rate regardless of the price trajectory $p(t)$ for $t > 17/6$. Because $\hat{t} = 17/6$ is a strong forecast horizon we can terminate the price shield at that time, as shown in the figure.

In order to demonstrate the statements in the previous paragraph, we consider two examples of price changes after $\hat{t} = 17/6$.

Example 5.3. Assume the price trajectory is

$$p(t) = \begin{cases} -2t + 7 & \text{for} \quad t \in [0,1) \\ t + 1 & \text{for} \quad t \in [1,17/6) \\ \frac{25}{7}t - \frac{44}{7} & \text{for} \quad t \in [17/6,4] \end{cases}$$

which is sketched in Figure 5.10. Note that the price trajectory up to 17/6 is the same as before. Note also that the price after 17/6 goes above the extension of the price shield in Figure 5.9. The new λ_2 trajectory is shown in Figure 5.10, which is the same as before up to $\hat{t} = 17/6$, and after that it is $\lambda_2(t) = \frac{1}{2}t + 6$ for $t \in [17/6,4]$. The optimal policy is as shown in Figure 5.10, and as previously asserted, the optimal policy in [0,1) remains unchanged. In Exercise 5.16 you are asked to verify the maximum principle for the solution of Figure 5.10.

Example 5.4. Assume the price trajectory is

$$p(t) = \begin{cases} -2t + 7 & \text{for} \quad t \in [0,1) \\ t + 1 & \text{for} \quad t \in [1,17/6) \\ -\frac{1}{2}t + \frac{21}{4} & \text{for} \quad t \in [17/6,4] \end{cases}$$

which is sketched in Figure 5.11. Again the price trajectory is the same up to 17/6, but the price after 17/6 is declining. This changes the optimal policy in [1,17/6), however the optimal policy will still be to sell in [0,1).

As in the beginning of the section we solve (5.80) to obtain $\lambda_2(t) = \frac{1}{2}t + \frac{5}{4}$ for $\hat{t}_1 \le t \le 4$ where $\hat{t}_1 \ge 1$ is the time of the last jump which is to be determined. It is intuitively clear that some profit can be made by buying and selling to take advantage of the price rise between $t = 2$ and $t = 17/6$. For this the $\lambda_2(t)$ trajectory must cross the price trajectory between 2 and 17/6 as shown in Figure 5.11, and the inventory y must go

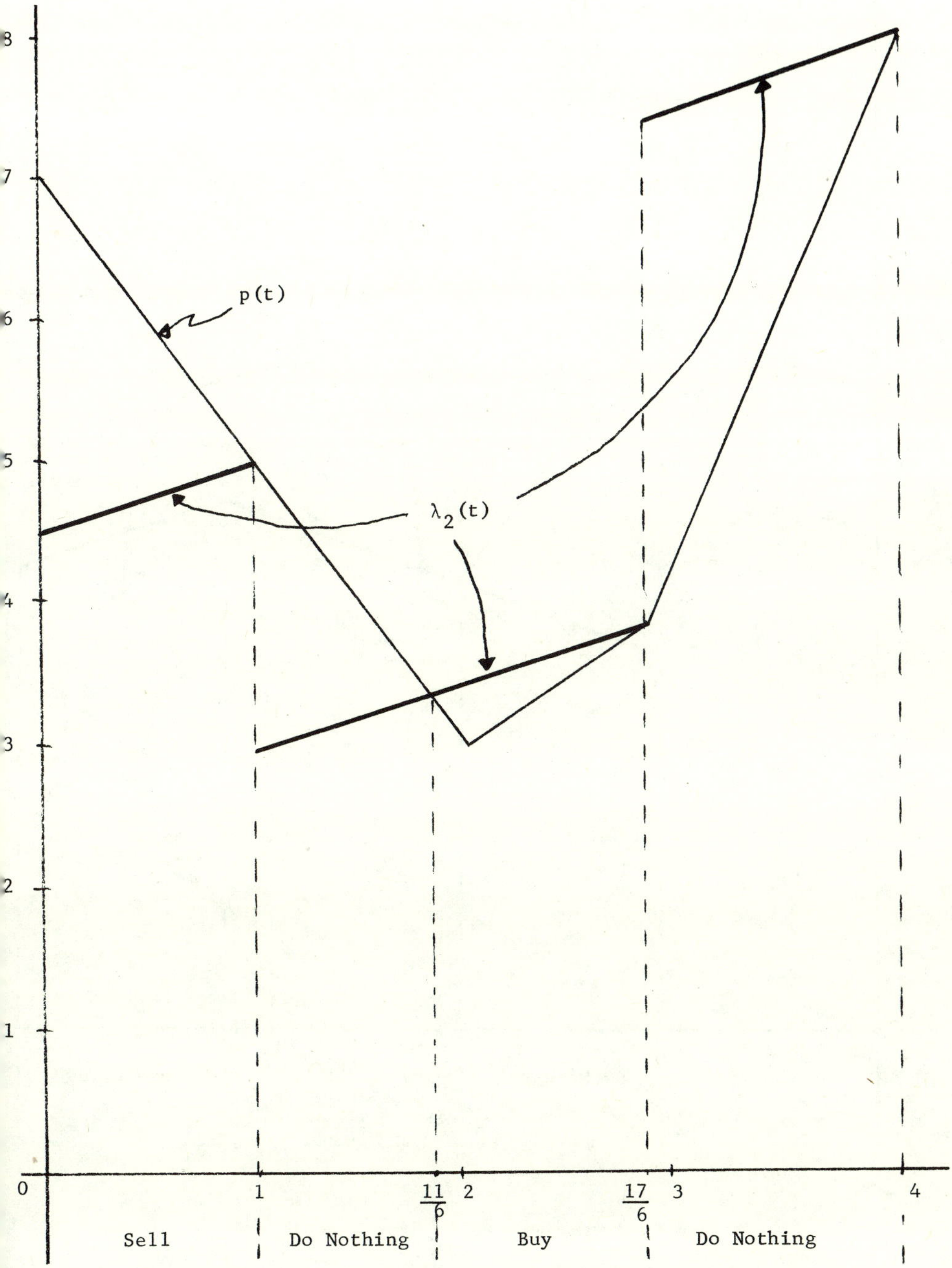

Figure 5.10 Optimal Policy for Example 5.3

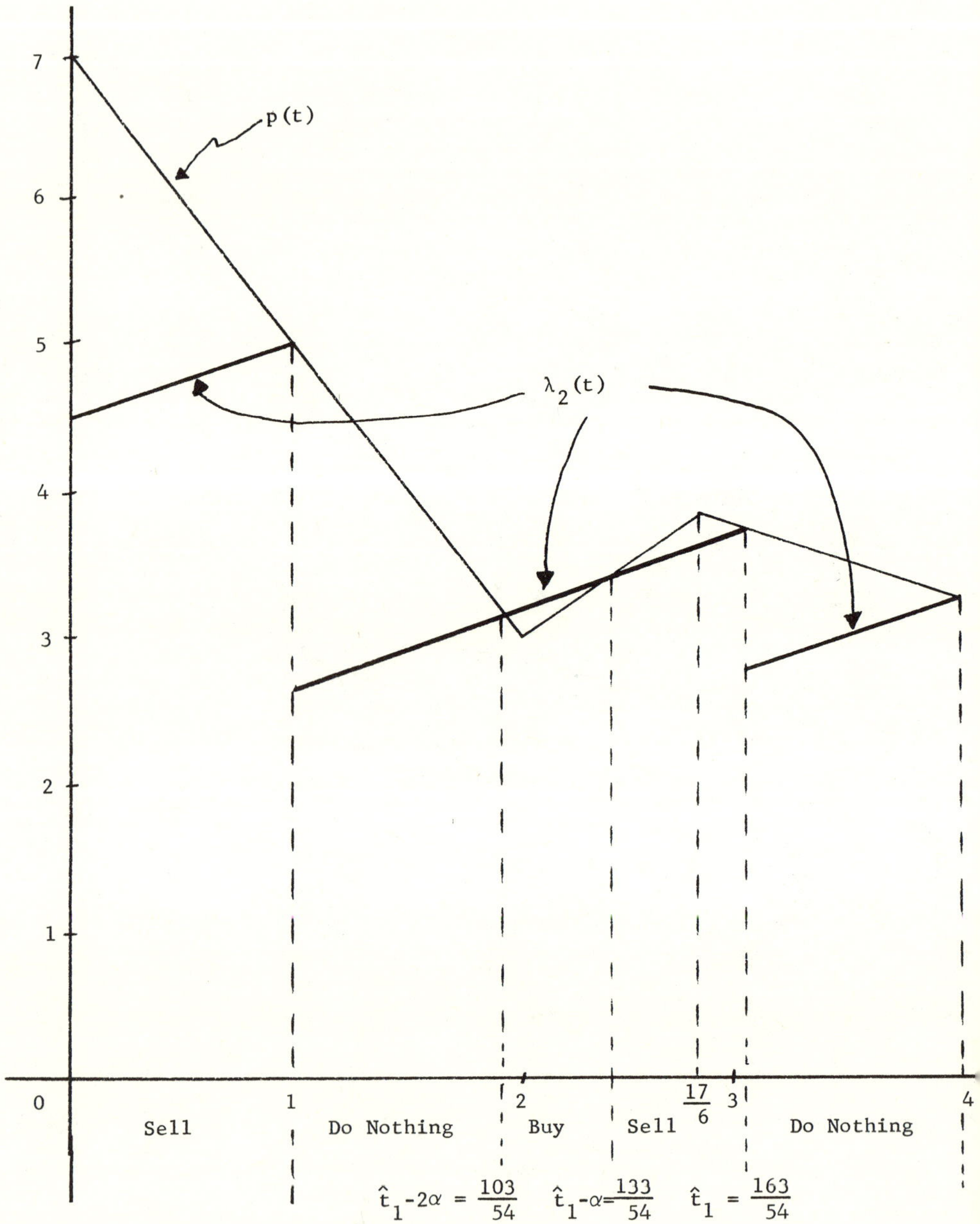

Figure 5.11 Optimal Policy for Example 5.4

to 0 between 17/6 and 4 so that λ_2 can jump downward to satisfy the ending condition $\lambda_2(4) = p(4)$. Since we must buy and sell equal amounts, the point of intersection of the λ_2 trajectory with the rising price segment, i.e., $\hat{t}_1 - \alpha$ must be exactly in the middle of the two other intersection points, $\hat{t}_1$ and $\hat{t}_1 - 2\alpha$, of λ_2 with the two declining price trajectories. Thus $\hat{t}_1$ and α must satisfy

$$-2(\hat{t}_1 - 2\alpha) + 7 + \frac{\alpha}{2} = (\hat{t}_1 - \alpha) + 1$$

$$(\hat{t}_1 - \alpha) + 1 + \frac{\alpha}{2} = -\frac{1}{2}\hat{t}_1 + \frac{21}{4} .$$

The solution to these is $\hat{t}_i = \frac{163}{54}$ and $\alpha = \frac{5}{9}$. Times $\hat{t}_1$, $\hat{t}_1 - \alpha$, and $\hat{t}_1 - 2\alpha$ are shown in Figure 5.11. The $\lambda_2(t)$ trajectory is given by

$$\lambda_2(t) = \begin{cases} \frac{1}{2}t + \frac{9}{2} & \text{for } t \in [0,1) \\ \frac{1}{2}t + \frac{241}{108} & \text{for } t \in [1,\frac{163}{54}) \\ \frac{1}{2}t + \frac{5}{4} & \text{for } t \in [\frac{163}{54},4] \end{cases} . \qquad (5.83)$$

Evaluation of the Lagrange multipliers and verification of the maximum principle is similar to that for the case in Figure 5.9.

In Section 5.3 we have given several examples of planning horizons and weak and strong forecast horizons. In Section 5.3.1 we found a planning horizon which was also a weak forecast horizon. We also introduced the idea of a "price shield" in that section. The planning horizon occurred exactly when $y(t) = 0$. In Section 5.3.2 we obtained the same planning horizon and a strong forecast horizon which occurred when $y(t) = 1$. Note that if we had solved the problem with $T = 1$, then $y^*(1) = 0$; and if we had solved the problem with $T = 17/6$, then $y^*(1) = 0$ and $y^*(17/6) = 1$. The latter problem has the smallest T such that both $y^* = 0$ and $y^* = 1$ occur for $t > 0$, given the price trajectory. This is one of the ways that 17/6 can be found to be a forecast horizon which follows the planning horizon at 1. There are other ways to find strong forecast horizons, see [Pekelman, 1974, 1975, 1979], [Kleindorfer and Lieber, 1979], [Morton, 1978], and [Lundin and Morton, 1975]. More research effort is required to survey and clarify these issues.

Exercises for Chapter 5

5.1 Verify the expressions for a_1 and a_2 given in (5.15) and (5.16).

5.2 For the model of Section 5.1.4 derive the turnpike triple by using the conditions in (5.26).

5.3 Verify (5.34). (Note $\rho = 0$ is assumed in Section 5.1.5).

5.4 Verify (5.36). (Again $\rho = 0$.)

5.5 Given the demand function

$$S = t(t-4)(t-8)(t-12)(t-16) + 30$$

and $\rho = 0$, $\hat{I} = 15$, $T = 16$, $\alpha = 1$,

obtain $Q(t)$ from (5.34).

5.6 Give an intuitive interpretation of (5.47).

5.7 Assume that there is a transaction cost βv^2 when v units of wheat are bought or sold in the model of Section 5.2.1. Derive the form of the optimal policy.

5.8 Set up the two point boundary value problem for Exercise 5.7 with $\beta = .05$, $h(y) = \frac{1}{2} y^2$, and the remaining values of parameters as in the model of Section 5.2.3.

5.9* Use a computer to solve the TPBVP of Exercise 5.8 by using the following flow chart:

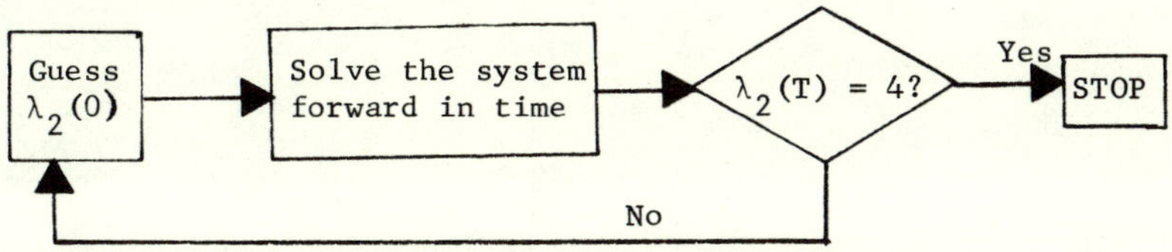

5.9 (continued)

(This method, called the "shooting method," will be described in detail in Chapter 11.)

5.10 For Exercise 5.7 assume $T = 10$, $x(0) = 10$, $y(0) = 0$, $\beta = 1/18$, $h(y) = \frac{1}{2}y^2$, $V_1 = V_2 = \infty$, $r = 0$, $p(t) = 10+t$; solve the TPBVP to obtain the optimal control in closed form.

5.11 Show that the solution obtained for the problem in Section 5.2.3 satisfies the maximum principle necessary conditions. Conclude the optimality of the solution by showing that the maximum principle is sufficient.

5.12 Resolve the problem of Section 5.2.3 with $V_1 = 2$ and $V_2 = 1$.

5.13 Compute the optimal trajectories for $\mu_1 \cdot \mu_2$, and η for the model in Section 5.2.4.

5.14 Solve the model in Section 5.2.4 with each of the following conditions:

(a) $y(0) = 2$

(b) $T = 10$ and $p(t) = 2t-2$ for $3 \leq t \leq 10$

5.15 Resolve the model of Section 5.3.2 with $y(0) = \frac{1}{2}$ and the additional warehousing constraint $y \leq \frac{1}{2}$.

5.16 Verify that the solution shown in Figure 5.10 satisfies the maximum principle.

5.17 Solve the following production planning problem with linear inventory holding costs

$$\text{Maximize}\{J = \int_0^T -[hI + \frac{c}{2} P^2]dt\}$$

$$\text{subject to } \dot{I} = P, \quad I(0) = 0, \quad I(T) = B; \quad 0 < B < \frac{hT^2}{2c}$$

$$P \geq 0 \text{ and } I \geq 0.$$

5.18 Resolve Exercise 5.17 with state equation $\dot{I} = P - S$, $I(0) = I_0 \geq 0$, and $I(T)$ not constrained. Also drop the production constraint $P \geq 0$, for simplicity. (Note that negative production can occur only when initially there is a large inventory I_0 and low S.)

6 APPLICATIONS TO MARKETING

Recently, a considerable number of applications of the optimal control theory have been made to the field of marketing. Most of these applications deal with the problem of finding or characterizing an optimal advertising policy over time; see [Sethi, 1977] for a recent comprehensive survey on dynamic optimal control problems in advertising. In this chapter we discuss optimal advertising policies for two of the well known models called the Nerlove-Arrow model and the Vidale-Wolfe model.

To describe the specific problems under consideration, let us assume that a firm has some way of knowing or estimating the dynamics of sales and advertising. Such knowledge is expressed in terms of a differential equation with good will or rate of sales as the state variable and the rate of advertising expenditures as the control variable. We assume that the firm wishes to maximize an objective function (the criterion function) which reflects its profit motives expressed in terms of sales and advertising rates. The optimal control problem is to find an advertising policy which maximizes the firm's objective function.

The plan of this chapter is as follows. Section 6.1 will cover the Nerlove-Arrow model as well as a nonlinear extension. Section 6.2 deals with the Vidale-Wolfe advertising model and its detailed analysis using the Green's theorem in conjunction with the maximum principle. The switching-point analys for this problem is a good example of the reverse-time construction technique used earlier in Chapters 3 and 4. Extensions of these models to multi-state problems are treated in [Srinivasan, 1976] and [Turner and Neuman, 1975].

6.1 The Nerlove-Arrow Advertising Model

The belief that advertising expenditures by a firm affect its present and future sales and hence, its present and future net revenues, has led a number of economists including [Nerlove and Arrow, 1962], to treat advertising as an investment in building up some sort of advertising capital usually called goodwill. Furthermore, the stock of goodwill depreciates over time. [Vidale and Wolfe, 1957], [Palda, 1964], and others present empirical evidence that the effects of advertising linger on but diminish over time.

Goodwill may be created by adding new customers or by altering the tastes and preferences of consumers and thus changing the shape of the demand function for the firm's product as well as shifting it. The goodwill depreciates over time because consumers "drift" to other brands, as a result of advertising by competing firms, introduction of new products and/or new brands, etc.

6.1.1 The Model

Let $G(t)$ denote the stock of goodwill at time t. The price of (or cost of producing) one unit of goodwill is \$1 so that a dollar spent on current advertising increases goodwill by one unit. It is assumed that the stock of goodwill depreciates over time at a constant proportional rate δ, so that

$$\dot{G} = u - \delta G , \qquad G(0) = G_o, \tag{6.1}$$

where $u = u(t) \geq 0$ is the current advertising effort measured in dollars. In economic terms, equation (6.1) states that the net investment in goodwill is the difference between gross investment, $u(t)$, and depreciation, $\delta G(t)$.

To formulate the optimal control problem for a monopolistic firm, assume that the rate of sales $S(t)$ depends on the stock of goodwill, $G(t)$, the price, $p(t)$, and other exogeneous variables, $Z(t)$, such as consumer incomes, population size, etc. Thus,

$$S = S(p,G,Z). \tag{6.2}$$

Assuming the rate of total production costs is $c(S)$, we can write the total revenue net of production costs as

$$R(p,G,Z) = pS(p,G,Z) - c(S). \tag{6.3}$$

The revenue net of advertising expenditure is, therefore, $R(p,G,Z) - u$. We assume that the firm wants to maximize the present value of net revenue streams discounted at a fixed rate ρ, i.e.,

$$\max_{u\geq 0,\, p\geq 0} \left[J = \int_0^\infty e^{-\rho t}[R(p,G,Z)-u]\,dt \right], \tag{6.4}$$

subject to (6.1).

Since the only place where p occurs is in the integrand, we can maximize J by first maximizing R with respect to price p holding G fixed, and then maximize the result with respect to u. Thus,

$$\partial R(p,G,Z)/\partial p = S + p\partial S/\partial p - c'\partial S/\partial p = 0, \tag{6.5}$$

which implicitly gives the optimal price $p^*(t) = p(G,Z)$. Defining $\eta = -(p/S)(\partial S/\partial p)$ as the elasticity of demand with respect to price, we can rewrite condition (6.5) as

$$p^* = \eta c'/(\eta - 1) \tag{6.6}$$

which is the usual price formula for a monopolist; see any microeconomics text. Defining $\pi(G,Z) = R(p^*,G,Z)$, the objective function in (6.4) can be written as

$$\max_{u\geq 0} \left[J = \int_0^\infty e^{-\rho t}[\pi(G,Z)-u]\,dt \right].$$

For convenience, we restate the optimal control problem which we have just formulated:

$$\max_{u\geq 0} \quad J = \int_0^\infty e^{-\rho t}[\pi(G,Z)-u]dt$$

subject to

$$\dot{G} = u-\delta G, \qquad G(0) = G_0 \tag{6.7}$$

6.1.2 Solution by the Maximum Principle

While Nerlove and Arrow [1962] used calculus of variations we use Pontryagin's maximum principle to derive their results. We form the current-value Hamiltonian

$$H = \pi(G,Z)-u + \lambda[u-\delta G] \tag{6.8}$$

with the current-value adjoint variable λ satisfying the differential equation

$$\dot{\lambda} = \rho\lambda-\partial H/\partial G = (\rho+\delta)\lambda-\partial\pi/\partial G, \tag{6.9}$$

and the condition that $\lim_{t\to\infty} e^{-\rho t}\lambda(t) = 0.$

The adjoint variable $\lambda(t)$ is the shadow price associated with the goodwill at time t. Thus, the Hamiltonian in (6.8) can be interpreted as the dynamic profit rate which consists of two terms: (i) the current net profit rate $(\pi(G,Z)-u)$, and (ii) the value $\lambda\dot{G} = \lambda[u-\delta G]$ of the new goodwill created by advertising at rate u. Also, equation (6.9) corresponds to the usual equilibrium relation for investment in capital goods, see [Arrow and Kurz, 1970] and [Jacquemin, 1973]. It states that the marginal opportunity cost $\lambda(\rho+\delta)dt$ of investment in goodwill should equal the sum of the marginal profit $(\partial\pi/\partial G)dt$ from increased goodwill and the capital gain $d\lambda = \dot{\lambda}dt$.

Defining $\beta = (G/S)(\partial S/\partial G)$ as the elasticity of demand with respect to goodwill and using (6.3), (6.5), (6.6), and (6.9), we can derive (see Exercis 6.3)

$$G^* = \beta pS/\eta[\rho+\delta)\lambda - \dot{\lambda}] \ . \tag{6.10}$$

This is exactly the same result derived by [Jacquemin, 1973] with the interpretation that at the optimum, the ratio of goodwill to sales revenue pS is directly proportional to the goodwill elasticity and inversely proportional to the price elasticity and to the sum of the marginal opportunity cost $\lambda(\rho+\delta)$ of investment and the rate at which the potential contribution of a unit of goodwill to profits becomes its past contribution $(-\dot{\lambda})$.

Using (3.76) to obtain the optimal long-run stationary equilibrium or turnpike $\bar{G}$ we must use $\dot{\lambda} = 0$ and the value for $\bar{\lambda}$ as arrived at by $\partial H/\partial u = 0$, which in the Nerlove-Arrow case gives $\bar{\lambda} = 1$. Thus $\bar{G}$ or also the singular lev G^s is given by

$$\bar{G} = G^s = \beta pS/\eta(\rho+\delta). \tag{6.11}$$

The property of $\bar{G}$ is that the optimal policy is to go to $\bar{G}$ as fast as possible. That is, if $G_o < \bar{G}$, it is optimal to jump instantaneously to $\bar{G}$ by applying an appropriate impulse at $t = 0$ and then set $u^*(t) = \bar{u} = \delta\bar{G}$ for t if $G_o > \bar{G}$, the optimal control $u^*(t) = 0$ until the stock of goodwill depreciates to the level $\bar{G}$ at which time the control switches to $u^*(t) = \delta\bar{G}$ and stays at this level to maintain the level $\bar{G}$ of goodwill. This optimal policy is graphed in Figure 6.1 for these two different cases.

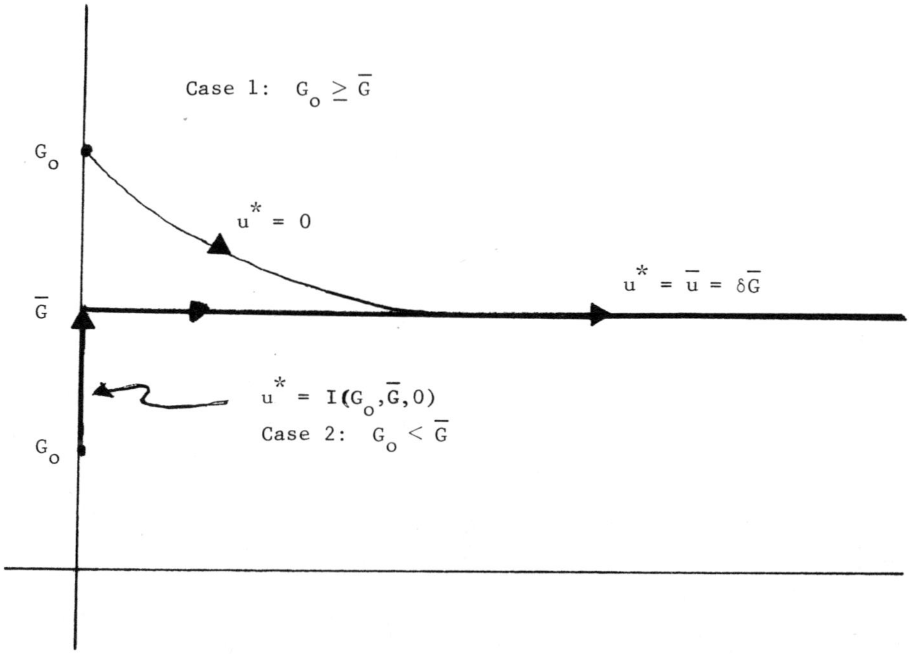

Figure 6.1

However, for a time dependent Z, $\bar{G}(t) = \bar{G}(T) = \bar{G}(Z(t))$ will be a function of time. To maintain this level of $\bar{G}(t)$ the required control is $\bar{u}(t) = \delta\bar{G}(t) + \dot{\bar{G}}(t)$. If $\bar{G}(t)$ is decreasing sufficiently fast, then $\bar{u}(t)$ may become negative, and thus infeasible. If $\bar{u}(t) \geq 0$ for all t then the optimal policy is as before. However, suppose $\bar{u}(t)$ is infeasible in the interval $[t_1,t_2]$, see Figure 6.2. In such a case, it is feasible to set $u = \bar{u}(t)$ for $t \leq t_1$; at $t = t_1$ (which is point A in Figure 6.2) we can no longer stay on the turnpike and must set $u(t) = 0$ until we hit the turnpike again (at point B in Figure 6.2). However, such a policy is not necessarily optimal. For instance, suppose we leave the turnpike at point C anticipating the feasibility at point A. The new path CDE may be better than the old path CAB. Roughly the reason this may happen is that path CDE is "closer" to the turnpike than CAB. The picture in Figure 6.2 illustrates such a case. The optimal policy is the one that is "closest" to the turnpike. This discussion will become clearer in Section 6.2.2, when a similar situation arises in connection with the Vidale-Wolfe model. See also [Breakwell, 1968].

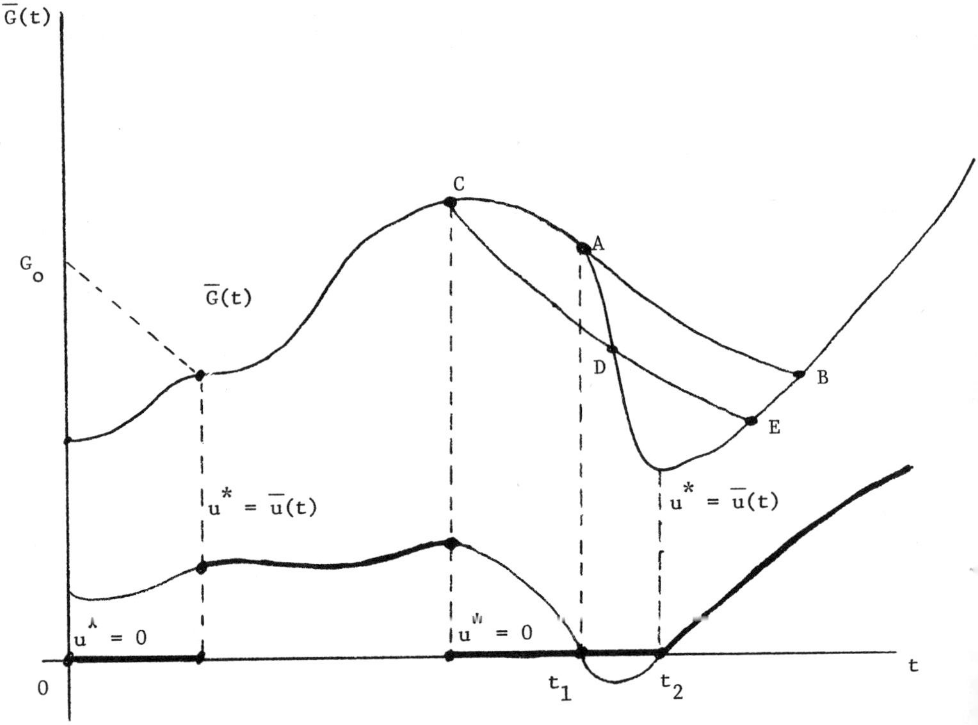

Figure 6.2 Case of Time-Dependent Turnpike with a Possible Optimal Control

From the point of view of control theory, the Nerlove-Arrow model is an example of a bang-bang control followed by a singular control, which arises in a class of optimal control problems which are linear in control. From the point of view of advertising, it is a seminal paper in spite of the fact that its bang-bang policy is not universally accepted as a realistic advertising policy. To avoid such policies we discuss a nonlinear generalization.

6.1.3 A Nonlinear Extension

Nonlinear extensions of Nerlove-Arrow model amount to making the objective function nonlinear in advertising. [Gould, 1970] nonlinearlized the model by assuming a nonlinear cost of 'effective' advertising u. [Jacquemin, 1973] assumed that the demand function S also depends explicitly on advertising effort u; as a result the function π in the objective function of (6.7) becomes

$$\pi = \pi(u,G).$$

Note that we have eliminated the exogenous variable Z for convenience.

We use a simpler version of Jacquemin's model analyzed by [Bensoussan et.al. 1974]. They assumed that the function $\pi(u,G)$ is separable, i.e.

$$\pi(u,G) = \pi_1(u) + \pi_2(G)$$

where π_1 and π_2 are assumed to be increasing strictly concave functions; i.e., $\pi_1' > 0$, $\pi_2' > 0$, and $\pi_1'' < 0$, $\pi_2'' < 0$ for $u \geq 0$ and $G \geq 0$. The new optimal control problem is:

$$\begin{array}{l} \max\limits_{u \geq 0} \; J = \int_0^\infty e^{-\rho t}[\pi_1(u) + \pi_2(G) - u]dt \\ \text{Subject to} \\ \qquad \dot{G} = u - \delta G, \quad G(0) = G_o. \end{array} \tag{6.12}$$

The Hamiltonian for the problem is

$$H = \pi_1(u) + \pi_2(G) - u + \lambda(u-\delta G) \tag{6.13}$$

where

$$\dot{\lambda} = \rho\lambda - \pi_2'(G) + \delta\lambda \; . \tag{6.14}$$

Differentiating H with respect to u gives

$$\frac{\partial H}{\partial u} = \pi_1'(u) - 1 + \lambda = 0 \; . \tag{6.15}$$

Since $\pi_1'' < 0$ we can invert π_1' to solve (6.15) for u as a function of λ, thus

$$u = f_1(\lambda) \tag{6.16}$$

In Exercise 6.9 you will be asked to show f_1 is an increasing function and $f_1(\lambda) \geq 0$ for $\lambda \leq 1$. The function f_1 in (6.16) permits us to rewrite the state equation in (6.12) and the adjoint equation (6.14) as the following initial value problem:

$$\boxed{\begin{array}{ll} \dot{G} + \delta G = f_1(\lambda), & G(0) = G_o \\ -\dot{\lambda} + (\rho+\delta)\lambda = \pi_2'(G), & \lambda(0) = \lambda_o \end{array}} \tag{6.17}$$

where the value $\lambda_o = 1 - \pi_1'(u(0))$ was obtained from (6.15). We also note that $\pi_2'(G) > 0$ for $G \geq 0$ and $\pi_2'(G)$ is a decreasing function.

The system (6.17) can be analyzed geometrically in (λ,G)-space. This method is called the <u>phase diagram method</u>. An alternate phase diagram procedur would be to solve the problem in (u,G)-space by making use of (6.14) and (6.15) to derive a differential equation for u. This derivation is left as Exercise 6.10.

Using the above noted properties of the functions f_1 and π_2'' and setting $\dot{\lambda} = 0$ and $\dot{G} = 0$ we can draw the two equations in (6.17) in the phase diagram of Figure 6.3. The intersection of the two curves $\delta G = f_1(\lambda)$ and $(\rho+\delta)\lambda = \pi_2'(G)$ is the point $(\bar{\lambda},\bar{G})$ with $\bar{\lambda}$ satisfying $0 < \bar{\lambda} < 1$; see Exercise 6.11. These two curves divide the plane into four regions, marked I, II, III, and IV in the figure. Note that $\{\bar{G},\delta\bar{G} = \bar{u},\bar{\lambda}\}$ is the turnpike for this problem (see Section 3.4).

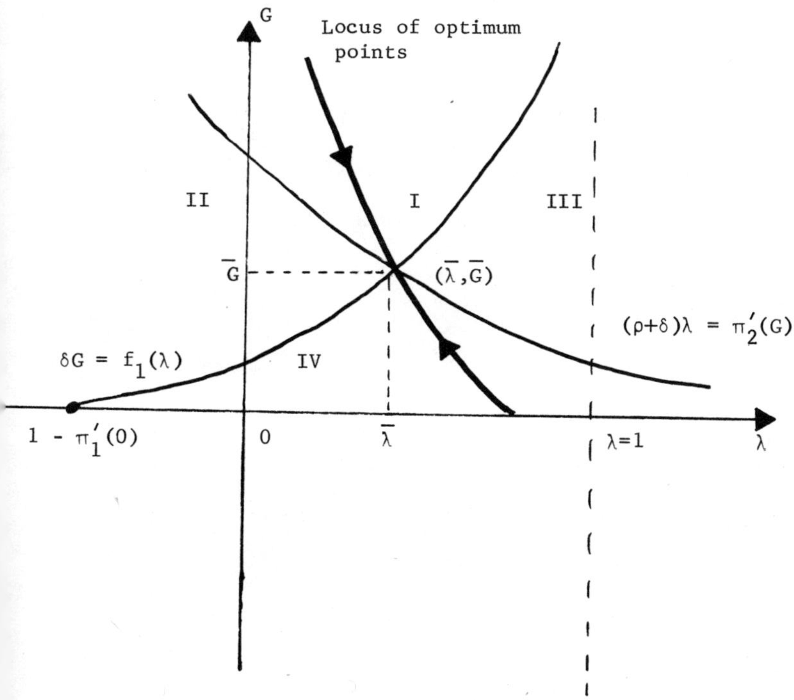

Figure 6.3 Phase Diagram

We note that

$$\left.\begin{array}{ll} \text{in regions I and II} & \delta G - f_1(\lambda) > 0 \\ \text{in regions III and IV} & \delta G - f_1(\lambda) < 0 \\ \text{in regions I and III} & (\rho+\delta)\lambda - \pi_2'(G) > 0 \\ \text{in regions II and IV} & (\rho+\delta)\lambda - \pi_2'(G) < 0 \end{array}\right\},$$

which implies that

$$\left.\begin{array}{lll} \text{in region I} & G \text{ decreasing and } \lambda & \text{increasing} \\ \text{in region II} & G \text{ decreasing and } \lambda & \text{decreasing} \\ \text{in region III} & G \text{ increasing and } \lambda & \text{increasing} \\ \text{in region IV} & G \text{ increasing and } \lambda & \text{decreasing} \end{array}\right\}.$$

Because of these conditions it is clear that for a given G_o, a choice of λ_0 such that (λ_0, G_0) is in regions II and III, will <u>not</u> lead to a path converging to the turnpike point $(\bar{\lambda}, \bar{G})$. On the other hand the choice of (λ_0, G_0) in region I when $G_0 > \bar{G}$ or (λ_0, G_0) in region IV when $G_0 < \bar{G}$, is likely to give a path that converges to $(\bar{\lambda}, \bar{G})$. From a result in [Coddington and Levinson, 1955], it can be shown that there exists a locus of optimum starting points (λ_0, G_0), as shown darkened in Figure 6.3. To find an optimal solution for a given G_0, we find the corresponding λ_0 on the locus and then the optimal path graphed in the (λ, G) space moves along the locus in the direction shown by the arrows.

Given $G_0 > \bar{G}$ we choose λ_0 on the optimum locus in region I of Figure 6.3. Clearly the initial control $u^*(0) = f_1(\lambda_0)$. Furthermore $\lambda(t)$ is increasing and by (6.16), $u(t)$ is increasing, so that in this case the optimal policy is to advertise at a low rate initially and gradually increase advertising to the turnpike level $\bar{u} = \delta\bar{G}$. If $G_0 < \bar{G}$, it can be shown similarly that the optimal policy is to advertise most heavily in the beginning and gradually decrease it to the turnpike level $\bar{u}$ as G approaches $\bar{G}$.

Note that the approach to the equilibrium $\bar{G}$ is no longer via the bang-bang control as in the Nerlove-Arrow model. This, of course, is what one would expect when a model is made nonlinear with respect to u.

6.2 The Vidale-Wolfe Advertising Model

We now present the analysis of the Vidale-Wolfe advertising model (see [Vidale and Wolfe, 1957] and [Sethi, 1973, 1974a]) which, in contrast to the Nerlove-Arrow model, makes no use of the idea of advertising goodwill. Instead the model exploits the closely related notion that the effect of advertising tends to persist, but with diminishing effect, through subsequent time periods. This carryover effect is modelled explicitly by means of a differential equation that gives the relationship between sales and advertising.

Vidale and Wolfe argued that changes in rate of sales of a product depend on two effects, response to advertising which acts (via the response constant a) on the unsold portion of the market, and loss due to forgetting which acts (via the decay constant b) on the sold portion of the market. Assuming a saturation level or market potential M, the Vidale-Wolfe model for a monopolistic firm can be stated as:

$$\dot{S} = au(1 - \frac{S}{M}) - bS. \tag{6.18}$$

The important feature of this equation, which distinguishes it from the Nerlove-Arrow equation (6.1), is the idea of the finite saturation level M. The Vidale-Wolfe model exhibits diminishing returns to the level advertising as a direct consequence of this saturation phenomenon. Note that for large M, the saturation phenomenon disappears reducing (6.18) to an equation (with constant returns to advertising) similar to the Nerlove-Arrow equation (6.1). Nerlove and Arrow, on the other hand, include the idea of diminishing returns to advertising in their model by making the sales S in (6.2) a concave function of goodwill.

Vidale and Wolfe based their model on the results of several experimental studies of advertising effectiveness, which are described in [Vidale and Wolfe, 1957].

6.2.1 Optimal Control Formulation for the Vidale-Wolfe Model

Whereas Vidale and Wolfe offered their model primarily as a description of actual market phenomena represented by cases which they had observed, we obtain the optimal advertising expenditures for the model in order to maximize a certain objective function over the horizon T while also attaining a terminal sales target, see [Sethi, 1973]. For this, it is convenient to transform (6.18) by making the following change of variable:

$$x = \frac{S}{M} . \tag{6.19}$$

Thus, x represent the market share (or more precisely, the rate of sales expressed as a fraction of the saturation level M). Furthermore, we define

$$r = a/M, \qquad \delta = b + \dot{M}/M. \tag{6.20}$$

Now, we can rewrite (6.18) as

$$\dot{x} = ru(1-x)-\delta x, \qquad x(0) = x_0. \tag{6.21}$$

From now on we assume M, and hence δ and r, to be positive constant It would not be difficult to extend the analysis for the case of a changing saturation level, but we do not carry it out here. In Exercise 6.12 you are asked to partially analyze the time dependent case.

To define the optimal control problem arising from the Vidale-Wolfe model, we let π denote the maximum sales revenue corresponding to $x = 1$; assume πx is the revenue function for $x \in [0,1]$. Also let Q be the maximum allowable rate of advertising expenditure and let ρ denote the continuous discount rate. With these definitions the optimal control problem can be stated as:

$$\max \left\{ J = \int_0^T e^{-\rho t}(\pi x-u)dt \right\},$$

subject to

$$\dot{x} = ru(1-x)-\delta x, \qquad x(0) = x_0,$$

the terminal state constraint

$$x(T) = x_T \tag{6.22}$$

and the control constraint

$$0 \leq u \leq Q.$$

Here Q can be finite or infinite and we assume the target x_T is in [0,1]. Recall that the problem is a fixed end point problem. Note that the requirement $0 \leq x_0 \leq 1$ holds, where x_0 is the initial market share.

It is possible to solve this problem by an application of the maximum principle, see Exercise 6.13. However, we will use instead a method based on Green's theorem which does not make use of the maximum principle. This method provides a convenient procedure for solving fixed end point problems having one state variable and one control variable, and where the control variable appears linearly in both the state equation and the objective function see [Miele, 1962] and [Sethi, 1977a]. Problem (6.22) has these properties, and therefore it is also a good example with which to illustrate the method. For the application of Green's theorem we require that Q be large. In particular we can let $Q = \infty$.

6.2.2 Solution by Use of Green's Theorem when Q Is Large

In this section we will solve the fixed end point problem starting with

x_0 and ending with x_T, under the assumption that Q is either unbounded or very large. The places where these assumptions are needed will be indicated.

To make use of Green's theorem it is convenient to consider times a and b where $0 \leq a < b \leq T$ and the problem

$$\text{Max } \{J(a,b) = \int_a^b e^{-\rho t}(\pi x - u)dt\} \tag{6.23}$$

subject to

$$\dot{x} = ru(1-x) - \delta x, \quad x(a) = A, \quad x(b) = B \tag{6.24}$$

$$0 \leq u \leq Q \tag{6.25}$$

To change the objective function in (6.23) into a line integral along any feasible arc Γ_1 from (a,A) to (b,B) in (t,x) space as shown in Figure 6.4, we multiply (6.24) by dt and obtain the formal relation

$$udt = \frac{dx + \delta x dt}{r(1-x)}$$

from (6.24) and substitute into the objective function (6.23); thus

$$J_{\Gamma_1} = \int_{\Gamma_1} \{(\pi x - \frac{\delta x}{r(1-x)})e^{-\rho t}dt - \frac{1}{r(1-x)} e^{-\rho t}dx \}$$

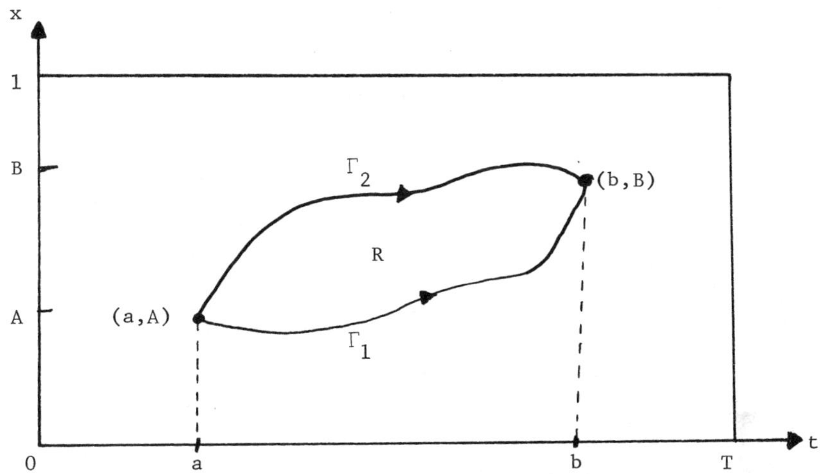

Figure 6.4 Feasible Arcs in (t,x) Space

Consider another feasible arc Γ_2 from (a,A) to (b,B) lying above Γ_1 as shown in Figure 6.4. Let $\Gamma = \Gamma_1 - \Gamma_2$, where Γ is a simple closed curve traversed in the counter-clockwise direction; i.e. Γ goes along Γ_1 in the direction of its arrow, and along Γ_2 in the direction opposite to its arrow. We now have

$$J_\Gamma = J_{\Gamma_1 - \Gamma_2} = J_{\Gamma_1} - J_{\Gamma_2} \tag{6.26}$$

Since Γ is a simple closed curve we can use Green's theorem to express J_Γ as an area integral over the region R enclosed by Γ. Thus treating x and t as independent variables

$$J_\Gamma = \iint_R \left\{\frac{\partial}{\partial t}\left[\frac{-e^{-\rho t}}{r(1-x)}\right] - \frac{\partial}{\partial x}\left[\left(\pi x - \frac{\delta x}{r(1-x)}\right)e^{-\rho t}\right]\right\}dtdx$$

$$= \iint_R \left[\frac{\delta}{(1-x)^2} + \frac{\rho}{(1-x)} - \pi r\right]\frac{e^{-\rho t}}{r}dtdx \tag{6.27}$$

Denote the term in brackets of the integrand of (6.27) by

$$I(x) = \frac{\delta}{(1-x)^2} + \frac{\rho}{(1-x)} - \pi r . \tag{6.28}$$

Note that the sign of the integrand is the same as the sign of $I(x)$.

Lemma 6.1. Let Γ_1 and Γ_2 be the lower and upper feasible arcs as shown in Figure 6.4. If $I(x) \geq 0$ for all $(x,t) \in R$ then the lower arc Γ_1 is more profitable than the upper arc Γ_2. Similarly if $I(x) \leq 0$ for all $(x,t) \in R$ then Γ_2 is more profitable than Γ_1.

Proof. If $I(x) \geq 0$ for all $(x,t) \in R$ then, from (6.27) and (6.28), $J_\Gamma \geq 0$; hence from (6.26) $J_{\Gamma_1} \geq J_{\Gamma_2}$. The proof of the other statement is similar.

To make use of this lemma to find the optimal control for the problem stated in (6.22) we need to find where $I(x)$ is positive and where it is negative. For this note first that $I(x)$ is an increasing function of x in $[0,1]$. Solving $I(x) = 0$ will give that value of x above which $I(x)$ is positive and below which $I(x)$ is negative. Since $I(x)$ is quadratic in $(1-x)^{-1}$ we can use the quadratic formula (see Exercise 6.15) to get

$$x = 1 - \frac{2\delta}{-\rho \pm \sqrt{\rho^2 + 4\pi r\delta}}$$

To keep x in the interval $[0,1]$ we must choose the positive sign before the radical. The optimal x must be nonnegative so we define

$$x^S = \text{Max}[1 - \frac{2\delta}{-\rho + \sqrt{\rho^2 + 4\pi r\delta}} ,0] \tag{6.29}$$

where the superscript S is used because this will turn out to be a singular trajectory. Since x^S is nonnegative the control corresponding to (6.29) is

$$u^S = \frac{\delta x^S}{r(1-x^S)} \tag{6.30}$$

will always be nonnegative. Also since Q is assumed to be large u^S will always be feasible. Moreover, in Exercise 6.16, you will be asked to show that $x^S = 0$ and $u^S = 0$ if and only if $\pi r \leq \delta + \rho$.

We now have enough machinery to obtain the optimal trajectories for (6.22) when Q is assumed to be sufficiently large, i.e., $Q \geq u^S$ where u^S is given in (6.30). We state these in the form of two theorems. Theorem 6.1 refers to the case in which T is "large"; Theorem 6.2 refers to the case in which T is "small." To define these concepts, let t_1 be the shortest time to go from x_0 to x^S; similarly let t_2 be the shortest time to go from x^S to x_T; then we say T is large if $T \geq t_1 + t_2$; otherwise T is small. (Figures 6.5 - 6.8 show cases for which T is large, while Figure 6.10 shows a case for which T is small.) In Exercise 6.18 you are asked to determine whether T is large or small in specific cases. In the statements of the theorems we will assume that x_0 and x_T are such that x_T is reachable from x_0. In Exercise 6.19 you are asked to find the reachable set for any given initial condition x_0.

<u>Theorem 6.1</u>. Let T be large and let x_T be reachable from x_0. For the four cases of inequalities relating x_0 and x_T to x^S, the optimal trajectories are given in the following figures:

(a) Case 1. $x_0 \le x^S$ and $x^S \ge x_T$

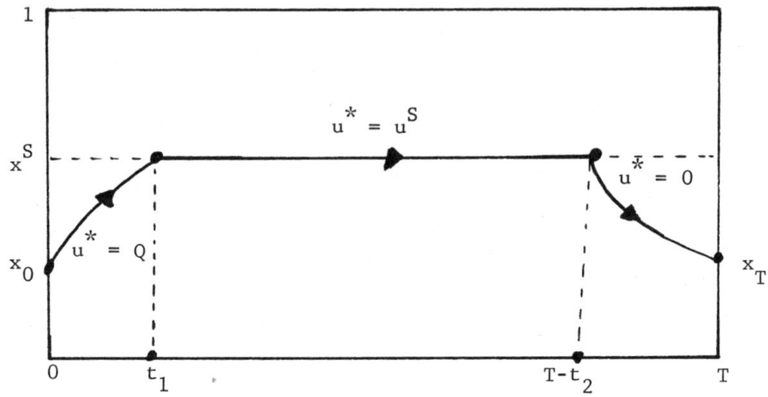

Figure 6.5 Optimal Trajectory for Case 1

(b) Case 2. $x_0 \le x^S$ and $x^S \le x_T$.

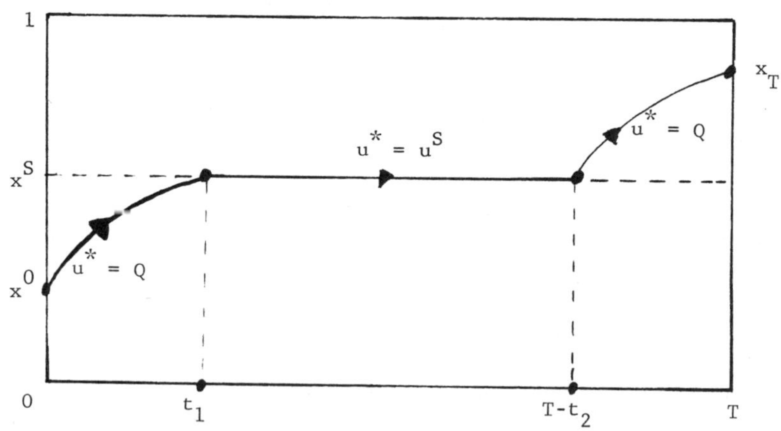

Figure 6.6 Optimal Trajectory for Case 2

(c) Case 3. $x_0 \geq x^S$ and $x^S \geq x_T$

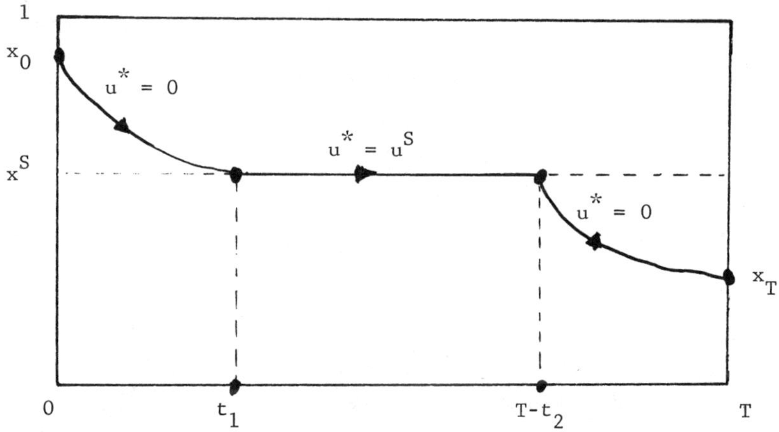

Figure 6.7 Optimal Trajectory for Case 3

(d) Case 4. $x_0 \geq x^S$ and $x^S \leq x_T$

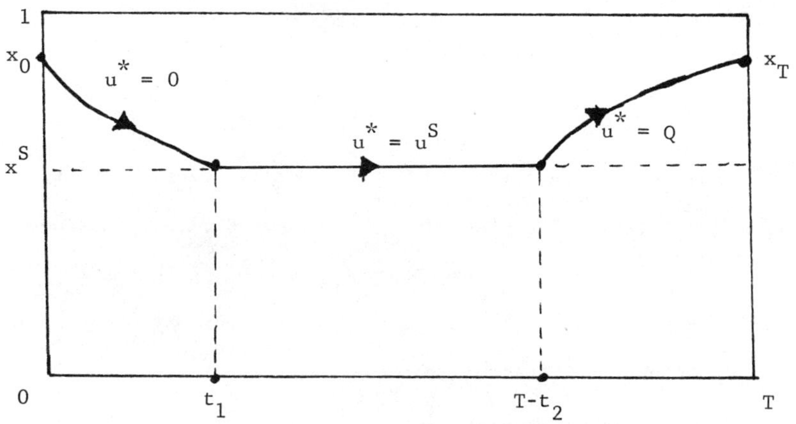

Figure 6.8 Optimal Trajectory for Case 4

(In these figures the quantities t_1 and t_2 are case dependent and not necessarily the same, see Exercise 6.17.)

Proof. We give details for Case 1 only. The proofs for the other cases are similar. Figure 6.9 shows the optimal trajectory of Figure 6.5 together with an arbitrarily chosen feasible trajectory, shown dotted.

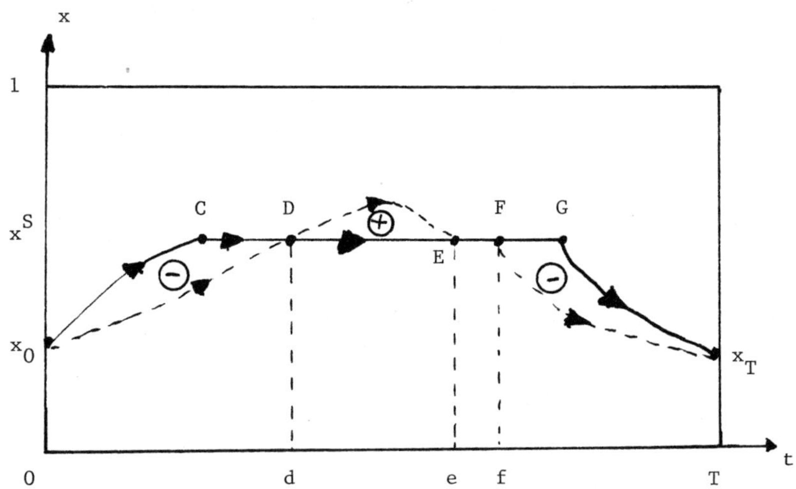

Figure 6.9 Optimal Trajectory (solid) and an Arbitrary Feasible Trajectory (dotted)

It should be clear that the dotted trajectory cannot cross the arc x_0 to A in Figure 6.9, because $u = Q$ on the arc x_0 to A. Similarly the dotted trajectory cannot cross the arc E to x_T because $u = 0$ on that arc.

We subdivide the interval [0,T] into subintervals over which the dotted arc is either above, below, or identical to the solid arc. In Figure 6.9 these subintervals are [0,d], [d,e], [e,f], and [f,T]. Because I(x) is positive for $x > x^S$ and I(x) is negative for $x < x^S$, the regions enclosed by the two trajectories have been marked with a + or - sign. By Lemma 6.1

the solid arc is better than the dotted arc in the subintervals [0,d], [d,e], and [f,T]; in interval [e,f] they have identical values. Hence the dotted trajectory is inferior to the solid trajectory. This proof can be extended to any (countable) number of crossings of the trajectories, see [Sethi, 1977a].

The figures in Theorem 6.1 were drawn for the situation that $T > t_1 + t_2$. You are asked to consider the case $T = t_1 + t_2$ in Exercise 6.23. In the following theorem the case $T < t_1 + t_2$ is dealt with.

<u>Theorem 6.2</u>. Let T be small, i.e. $T < t_1 + t_2$, and let x_T be reachable from x_0. For the two possible cases of inequalities relating x_0 to x_T and x^S, the optimal trajectories are given in Figures 6.10 and 6.11:

(a) Case 1: $x_0 < x^S$ and $x_T < x^S$

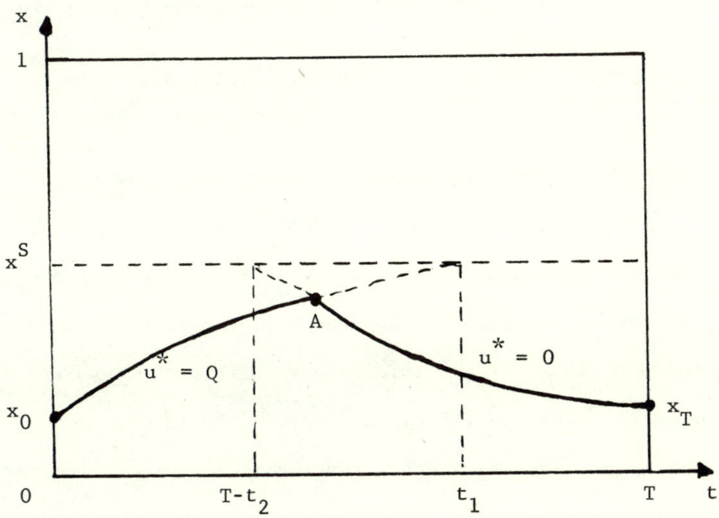

Figure 6.10 Optimal Trajectory for Case 1 of Small T

(b) Case 2: $x_0 > x^S$ and $x_T > x^S$.

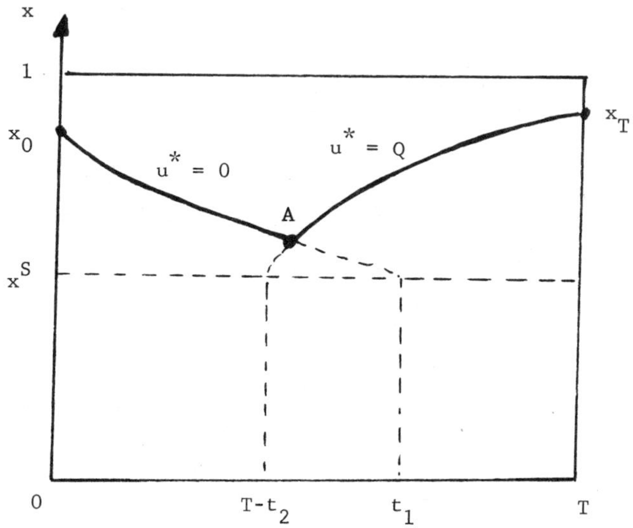

Figure 6.11 Optimal Trajectory for Case 2 of Small T

Proof. The requirement of feasibility when T is small rules out cases where x_0 and x_T are on opposite sides of or equal to x^S. The proofs of optimality of the trajectories shown in Figures 6.11 and 6.11 are similar to the proofs of the parts of Theorem 6.1, and is left as Exercise 6.23. In Figures 6.10 and 6.11, it is possible to have either $t_1 \geq T$ or $t_2 \geq T$. Try sketching some of these special cases.

All the previous discussion has assumed that Q was finite and sufficiently large, but we can easily extend this to the case when $Q = \infty$. This possibility makes the arcs in Figures 6.5 - 6.11 corresponding to $u^* = Q$ become vertical line segments, corresponding to impulse controls. For example Figure 6.6 becomes Figure 6.12 when $Q = \infty$.

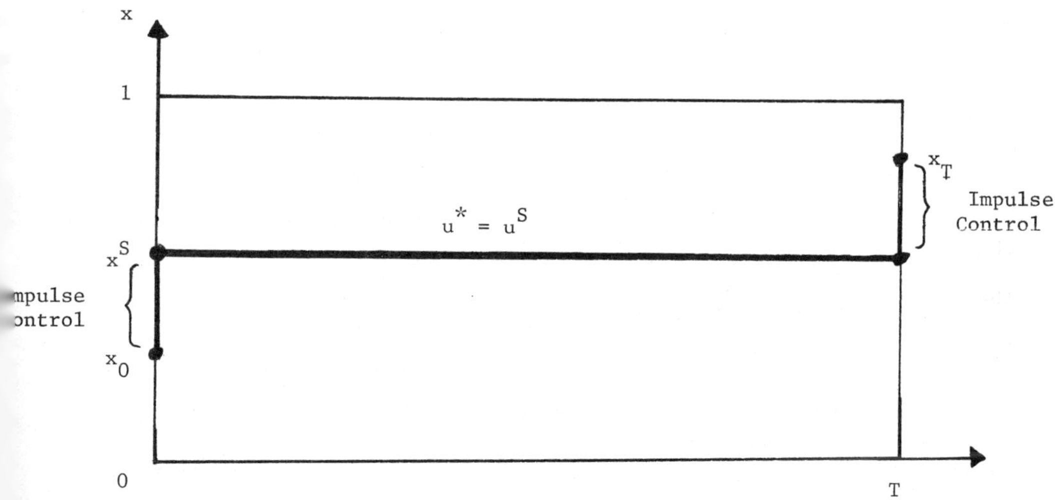

Figure 6.12 Optimal Trajectory for Case 2 of Theorem 6.1 when $Q = \infty$

The effect on the objective function of the impulse control $\mathrm{Imp}(x_0, x^S)$ can be computed as follows: We integrate the state equation in (6.22) from 0 to ε with the initial condition x_0 and u treated as constant. The result is

$$x(\varepsilon) = (x_0 - \frac{ru}{\delta+ru})\, e^{-(\delta+ru)\varepsilon} + \frac{ru}{\delta+ru}$$

According to the procedure at the end of Section 1.4, we must choose $u(\varepsilon)$ so that $x(\varepsilon)$ is x^S. Noting that the right hand side of (1.23) is

$$\lim_{\varepsilon\to 0} -u(\varepsilon)\cdot\varepsilon = \mathrm{Imp}(x_0, x^S) = I$$

so that it is possible to solve for I directly without obtaining $u(\varepsilon)$. This is done by setting $\varepsilon \to 0$, $u(\varepsilon) \to \infty$ and $-u(\varepsilon)\varepsilon \to I$ and $x(\varepsilon) = x^S$ in the expression for $x(\varepsilon)$ obtained above.

$$x(0+) = e^{rI}(x_0 - 1) + 1 \quad .$$

Therefore

$$\mathrm{Imp}(x_0, x^S) = -\frac{1}{r}\ \ln\left(\frac{1-x_0}{1-x^S}\right) \quad . \tag{6.31}$$

We remark that this formula holds for any time t, as well as $t = 0$. Hence it can also be used at $t = T$ to compute the impulse at the end of the period; see Figure 6.12, and Exercise 6.25.

6.2.3 Solution when Q Is Small

When Q is small it is not possible to go along the turnpike x^S, so that the arguments based on Green's theorem become difficult to apply. We therefore return to the maximum principle approach to analyze the problem. By "Q is small" we mean $Q < u^S$ where u^S is defined in (6.30). Another characterization of the phrase "Q is small" in terms of problem parameters is given in Exercise 6.26.

We now apply the maximum principle to the problem under consideration given in (6.22). We form the current value Hamiltonian as

$$\begin{aligned} H &= \pi x - u + \lambda[ru(1-x) - \delta x] \\ &= \pi x - \delta\lambda x + u[-1 + r\lambda(1-x)] \end{aligned} \tag{6.32}$$

and the Lagrangian function

$$L = H + \mu(Q-u). \tag{6.33}$$

From conditions (3.45) the adjoint variable ν satisfies

$$\dot{\lambda} = \rho\lambda - \frac{\partial L}{\partial x} = \rho\lambda + \lambda(ru + \delta) - \pi \tag{6.34}$$

with $\lambda(T)$ being a constant that must be determined. Furthermore, the Lagrang multiplier μ in (6.33) must satisfy

$$\mu \geq 0, \quad \mu(Q-u) = 0. \tag{6.35}$$

From (6.32) we notice that the Hamiltonian is linear in control so the optimal control using Table 3.3 is

$$u^*(t) = \text{bang}[0,Q;W(t)] \tag{6.36}$$

where

$$W(t) = r\lambda(1-x) - 1 \quad . \tag{6.37}$$

We remark that the sufficiency conditions of Chapter 3, which require concavity of the derived Hamiltonian H^0, do not apply here, see Exercise 6.30. However, the sufficiency of the maximum principle for this kind of problem has been established in the literature; see, for example, [Lansdown, 1970].

When $W = r\lambda(1-x) - 1 = 0$ we have the possibility of singular control, provided we can maintain this equality over a finite time interval. For Q large we showed in the previous section that the optimal trajectory contains a segment on which $x = x^S$ and $u^* = u^S$, where $0 \leq u^S \leq Q$. (See Exercise 6.27 for the condition that Q is small.) This can obviously be a singular control. Further discussion of singular control will be given in Appendix A4.3.

A complete solution of problem (6.22) when Q is small requires a lengthy switching point analysis. The details are too voluminous to give here, but an interested reader can find the details in [Sethi, 1973].

6.2.4 Solution when T Is Infinite

In the previous two sections we assumed that T was finite. We now formulate the infinite horizon version of (6.22):

$$\begin{array}{l} \text{Maximize} \quad \left\{ J = \int_0^\infty e^{-\rho t}(\pi x - u)\,dt \right\} \\ \text{Subject to} \\ \qquad \dot{x} = ru(1-x) - \delta x, \quad x(0) = x_0 \\ \text{and} \qquad 0 \leq u \leq Q \quad . \end{array} \tag{6.38}$$

We divide the analysis of this problem into the same two cases defined as before, namely, "Q is large" and "Q is small."

When Q is large, the results of Theorem 6.1 suggest the solution when T is infinite. Because of the discount factor the ending parts of the solutions shown in Figures 6.5 - 6.8 can be shown to be irrelevant (i.e., the discounted profit accumulated during the interval $(T-t_2, T)$ goes to 0 as T goes to ∞). Therefore we only have two cases (a) $x_0 \leq x^S$ and (b) $x_0 \geq x^S$. The optimal control in case (a) is to use $u^* = Q$ in the interval $[0,t_1)$ and $u^* = u^S$ for $t \geq t_1$. Similarly, the optimal control in case (b) is to use $u^* = 0$ in the interval $[0,t_1)$ and $u^* = u^S$ for $t \geq t_1$.

An alternate way to see the above solutions give optimal controls is to check that they satisfy the turnpike conditions (3.76) of Chapter 3. To do this we need to find the values of state, control, and adjoint variables and the Lagrange multiplier along the turnpike. It can be easily shown that $x = x^S$, $u = u^S$, $\lambda^S = \pi/(\rho + \delta + ru^S)$, and $\mu = 0$ satisfy the conditions of the turnpike theorem.

When Q is small, i.e., $Q < u^S$, it is not possible to follow the turnpike $x = x^S$, because we can't impose a control of u^S. Intuitively, it seems clear that the "closest" stationary path which we can follow is the path obtained by setting $\dot{x} = 0$ and $u = Q$, the largest possible control, in the state equation of (6.37). This gives

$$\bar{x} = \frac{rQ}{rQ + \delta} \tag{6.39}$$

correspondingly we obtain

$$\bar{\lambda} = \frac{\pi}{\rho + \delta + rQ} \tag{6.40}$$

by setting $u = Q$ and $\dot{\lambda} = 0$ in (6.34) and solving for λ. More specifically we state the following theorems which give the turnpike and optimal control for Q small. To prove these theorems we need to define two more quantities, namely

$$\hat{x} = 1 - \frac{1}{r\bar{\lambda}} \tag{6.41}$$

and

$$\bar{\mu} = r\bar{\lambda}(1-\bar{x}) - 1 \quad . \tag{6.42}$$

Theorem 6.3. For Q small the following quantities

$$\{\bar{x}, Q, \bar{\lambda}, \bar{\mu}\} \tag{6.43}$$

form a turnpike.

Proof. We show that the conditions in (3.76) hold for (6.43). The first two are obvious. By Exercise 6.28 we know $\bar{x} \le \hat{x}$, which, from definitions (6.41) and (6.42), implies $\bar{\mu} \ge 0$. Furthermore $\bar{u} = Q$ so (6.35) holds and the third condition of (3.76) also holds. Finally because $W = \bar{\mu}$ from (6.37) and (6.42), it follows that $W \ge 0$, so the Hamiltonian maximizing condition of (3.76) holds with $\bar{u} = Q$, and the proof is complete.

Theorem 6.4. For Q small the optimal control at time τ is given by:

(a) If $x(\tau) \le \hat{x}$ then $u^*(\tau) = Q$

(b) If $x(\tau) > \hat{x}$ then $u^*(\tau) = 0$.

Proof. (a) We set $\lambda(t) = \bar{\lambda}$ for all $t \ge \tau$ and note that λ satisfies the adjoint equation (6.34) and the transversality conditions (3.73).

By Exercise 6.28 and the assumption that $x(\tau) \le \hat{x}$ we know that $(t) \le \hat{x}$ for all t. The proof that (6.35) and (6.36) holds for all $t \ge \tau$

relies on the fact that $x(t) \leq \hat{x}$ and an argument similar to the proof of the previous theorem.

Figure 6.13 shows the optimal trajectories for $x(0) < \hat{x}$ and two different starting values $x(0)$, one above and the other below $\bar{x}$. Note that in this figure we are always in case (a) since $x(\tau) \leq \hat{x}$ for all τ, $\tau \geq 0$.

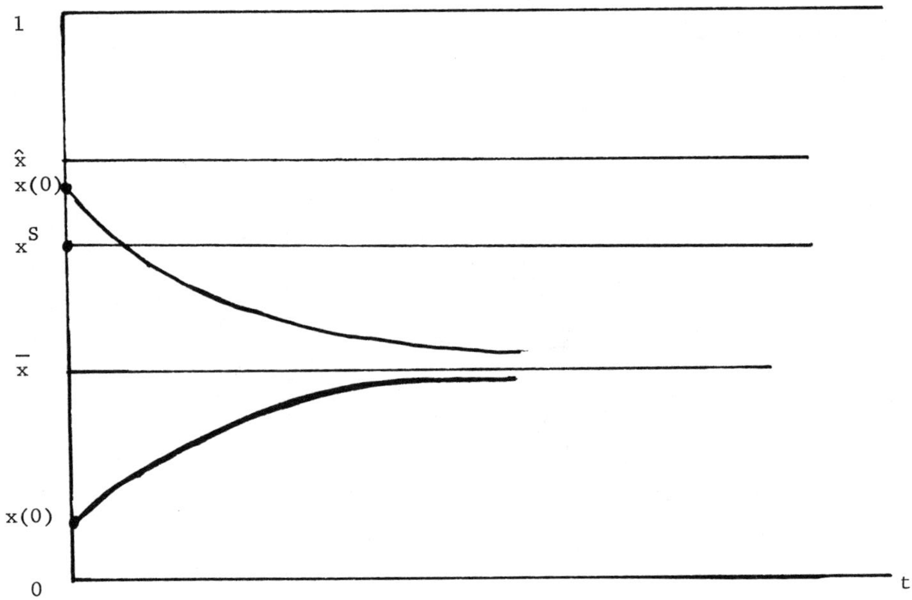

Figure 6.13 Optimal Trajectories when $x(0) < \hat{x}$

(b) Assume $x(0) > \hat{x}$. For this case we will show that the optimal trajectory is as shown in Figure 6.14, which is obtained by applying $u = 0$ u[illegible] $x = \hat{x}$ and $u = Q$ thereafter. Using this policy we can find the time t_1 a[illegible] which $x(t_1) = \hat{x}$ by solving the state equation in (6.38) with $u = 0$. This gives

$$t_1 = \frac{1}{\delta} \log \frac{x(0)}{\hat{x}} \tag{6.44}$$

Clearly for $t \geq t_1$, the policy $u = Q$ is optimal because case (a) applies. We now consider the interval $[0,t_1)$. Let τ be any time in this interval, as shown in Figure 6.14, and let $x(\tau)$ be the corresponding value of the state variable. Consider the following two-point boundary value problem in the interval $[\tau,t_1]$:

$$\left.\begin{aligned} &\dot{x} = -\delta x, \quad x(\tau) \text{ given} \\ &\dot{\lambda} = (\rho+\delta)\lambda - \pi, \quad \lambda(t_1) = \bar{\lambda} \end{aligned}\right\} \quad . \tag{6.45}$$

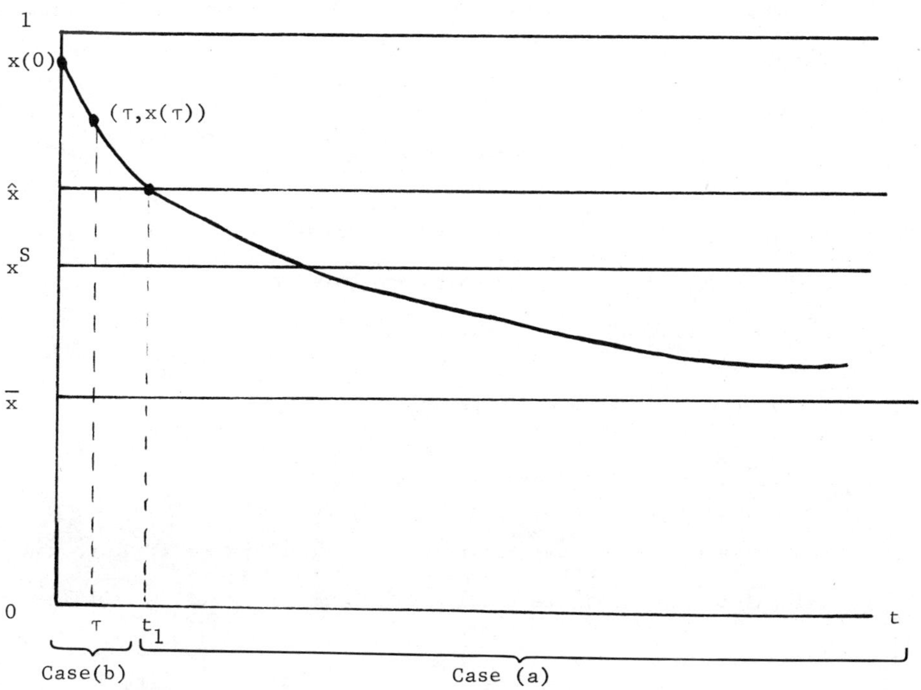

Figure 6.14 Optimal Trajectory for $x(0) > \hat{x}$

In Exercise 6.31 you are asked to show that the switching function $W(t)$ defined in (6.37) is negative in the interval $[\tau, t_1)$ and $W(t_1) = 0$. Therefore by (6.36) the policy $u = 0$ used in deriving (6.45) satisfies the maximun principle. This policy "joins" the optimal policy after t_1 because $\lambda(t_1) = \bar{\lambda}$.

In this book the sufficiency of the transversality condition (3.73) of Section 3.4 was stated under the hypothesis that the derived Hamiltonian was concave; see Theorem 2.1. In the present example this hypothesis does not hold. However, for this simple bilinear problem it can be shown that (3.73) is sufficient for optimality. Because of the technical nature of this issue we omit the details.

Exercises for Chapter 6

6.1 In equations (6.2) and (6.3) assume $S(p,G) = 1000 - 5p + 2G$ and $c(S) = 5S$. Substitute into (6.5) and solve for the optimal price $p^*(t)$ in terms of G.

6.2 Derive the optimal monopoly price formula in (6.6) from (6.5).

6.3 Derive the optimal goodwill level shown in (6.10).

6.4 Show that the only possible singular level of goodwill (which can be maintained over a finite time interval) is the value $\bar{G}$ obtained in (6.11).

6.5 Show that the total cost of advertising required to go from $G_0 < \bar{G}$ to $\bar{G}$ instantaneously (by an impulse) is $\bar{G} - G_0$.
[Hint. Integrate $\dot{G} = u - \delta G$, $G(0) = G_0$ from 0 to ϵ ; equate

$\bar{G} = \lim G(\epsilon)$ where the limit is taken as $\epsilon \to 0$ and $u \to \infty$ in such a way that $u\epsilon \to$ cost; see also the derivation of (6.31).]

6.6 Assume the effect of exogeneous variable $Z(t)$ is seasonal so that the goodwill $\bar{G}(t) = 2 + \sin t$. Assume $\delta = .1$. Sketch the graph of $\bar{u}(t) = \delta\bar{G} + \dot{\bar{G}}$, similar to Figure 6.2, and identify intervals in which maintaining the singular level $\bar{G}(t)$ is infeasible.

6.7 In the Nerlove-Arrow model of Section 6.1.1 and 6.1.2 assume $S(p,A,Z) = \alpha p^{-\eta}G^{\beta}Z^{\gamma}$ and $c(S) = cS$. Show that the optimal stationary policy gives $\bar{u}/pS$ = constant; i.e. that the optimal advertising level is a constant fraction of sales regardless of the value of Z. (Such policies are followed by many industries.)

*6.8 Resolve the advertising model of (6.12) with the objective function replaced by

$$\max_{u\geq 0} \{J = \int_0^\infty e^{-\rho t}[\pi_2(G) - \gamma(u)]dt,$$

where $\gamma(u)$ represents an increasing convex advertising cost function, with $\gamma(u) \geq 0$, $\gamma'(u) \geq 0$, and $\gamma''(u) > 0$ for $u \geq 0$. (This is Gould's model, see [Gould, 1970].)

6.9 Show that $f_1(\lambda)$ in (6.16) is a nonnegative and increasing function for $\lambda \leq 1$.

6.10 Develop a system of equations for the model of Section 6.1.1 similar to those in (6.17) for the phase diagram analysis in (u,G) space for the problem of (6.12)

6.11 Show that the $\bar{\lambda}$ shown in Figure 6.3 satisfies $0 \le \bar{\lambda} \le 1$.

*6.12 In (6.21) assume r and δ are positive, differentiable functions of time. Derive expressions similar to (6.27) - (6.30) in order to get the new turnpike values.

6.13 For problem (6.22) with $\pi r > \delta+\rho$ derive the turnpike $\{\bar{x},\bar{u},\bar{\lambda}\}$ by using the maximum principle. Check your answers with the corresponding values derived by Green's theorem. Show that when $\rho = 0$, $\bar{x}$ reduces to the golden path rule. (See Exercise 6.16 for the case when $\pi r \le \delta+\rho$.)

6.14 Let x^S be such that $A < x^S < B$ in Figure 6.4, and assume that $I(x) \ge 0$ for $x \ge x^S$ and $I(x) \le 0$ for $x \le x^S$. Construct a path Γ_3 going from (a,A) to (b,B) which is better than both Γ_1 and Γ_2. [Hint. Use Lemma 6.1.]

6.15 Solve the quadratic equation $I(x) = 0$ where $I(x)$ is defined in (6.28) to obtain the expression for x shown as the first argument in (6.29).

6.16 Show that both x^S in (6.29) and u^S in (6.30) are 0 if and only if $\pi r \le \delta + \rho$.

6.17 For the problem in (6.22) suppose x_o, x^S, and x_T are given; let t_1 be the shortest time to go from x_0 to x^S, and t_2 be the shortest ti[me] to go from x^S to x_T.

(a) If $x_0 < x^S$ and $x^S > x_T$ show that

$$t_1 = \frac{1}{rQ+\delta}\, \ln\left(\frac{\bar{x} - x_0}{\bar{x} - x^S}\right) \quad \text{and} \quad t_2 = \frac{1}{\delta}\, \ln\left(\frac{x^S}{x_T}\right)$$

where $\bar{x} = rQ/(rQ+\delta)$; assume $\bar{x} > x^S$.

(b) Using the form of the answers in (a) find t_1 and t_2 when $x_0 > x^S$ and $x^S < x_T < \bar{x}$.

6.18 For Exercise 6.17(a) write the condition T large, i.e., $T \geq t_1 + t_2$, in terms of all the parameters of problem (6.22).

6.19 For problem (6.22) find the reachable set for a given initial x_0 and horizon time T.

6.20 (a) For problem (6.22) assume $r = .2$, $\delta = .05$, $\rho = .1$, $Q = 5$, $\pi = 2$, $x_0 = .2$, and $x_T = .3$. Use Exercises 6.17(a) and 6.18 to show that $T > 10.623$ is large, but $T < 10.623$ is small. Sketch the optimal trajectories for $T = 13$ and $T = 18$.

(b) Redo (a) when $x_T = .7$. Show that both $T = 13$ and $T = 8$ are large.

6.21 Prove Theorem 6.1 for Case 3.

6.22 Draw four figures for the case $T = t_1 + t_2$ corresponding to Figures 6.5 - 6.8.

6.23 Prove Theorem 6.2.

6.24 Sketch one or two other possible curves for the case when T is small.

6.25 Obtain the impulse Imp(A,B;t) required to take the state from A up to B instantaneously at time t for the Vidale-Wolfe model in Section 6.2.2

6.26 (a) Resolve Exercise 6.20(a) with $Q = \infty$. Show $T = 10.5$ is no longer small.

(b) Show that $T > 0$ is large for Exercise 6.20(b) when $Q = \infty$. Find the optimal value of the objective function when $T = 8$.

6.27 Show that Q is "small" if and only if

$\pi r\delta/[(\delta + \rho + rQ)(\delta + rQ)] > 1$.

6.28 (a) Show that $\bar{x} < x^S < \hat{x}$ when Q is small, where $\hat{x}$ is defined in (6.41).

(b) Show that $\bar{x} > x^S$ when Q is large.

6.29 Derive (6.44).

*6.30 Show the derived Hamiltonian H^o corresponding to (6.32) and (6.36) is not concave in x for any given λ.

*6.31 Show that the switching function defined in (6.37) is concave in t, hence show that the policy in Figure 6.14 is optimal.

*6.32 Compute the turnpike level for the model

$$\max_{u \geq 0} \{J = \int_0^\infty e^{-\rho t} (\pi x - u^2) dt\}$$

subject to

$$\dot{x} = ru(1-x) - \delta x, \quad x(0) = x_0.$$

Show that the turnpike reduces to the golden path when $\rho = 0$.

Research Exercises for Chapter 6

6.33 Extend the Nerlove-Arrow Model and its results by introducing the additional capital stock variable

$$\dot{K} = v - \gamma K, \; K(0) = K_0$$

where v is the research expenditure. Assume the new cost function to be $C(S,K)$. Note that this model allows the firm to manipulate its cost function. (Hint: see [Dhrymes, 1962].)

6.34 Analyze an extension of a finite-horizon Nerlove-Arrow Model subject to a budget constraint. That is, introduce the following (isoperimetric) constraint

$$\int_0^T ue^{-\rho t}\, dt = B$$

Also assume $\pi(G) = \alpha\sqrt{G}$ where $\alpha > 0$ is a constant, see [Sethi, 1977b].

6.35 Introduce a budget constaint in a different way into the Nerlove-Arrow model as follows: Let $B(t)$ be the budget at time t, and let $\gamma > 0$ be a constant. Assume B satisfies

$$\dot{B} = e^{-\rho t}(-u + \gamma G), \quad B(0) = B_0$$

and $B(t) \geq 0$ for all t. Solve only the infinite horizon model, see [Sethi and Lee, 1979].

6.36 Maximize the present value of total sales in the Nerlove-Arrow model, i.e.

$$\max_{u \geq 0} \left\{ J = \int_0^\infty e^{-\rho t}\, pS(p,G)dt \right\}$$

subject to (6.1) and the isoperimetric profit constraint

$$\int_0^\infty e^{-\rho t}\, [pS(p,G) - C(S) - u]\, dt = \hat{\pi}\, .$$

(see [Tsurumi and Tsurumi, 1973]).

6.37 Consider (6.38) with the state equation replaced by

$$\dot{x} = ru(1-x) + \mu x(1-x) - \delta x, \quad x(0) = x_0$$

where the constant $\mu > 0$ reflects "word-of-mouth" communication between buyers (represented by x) and non-buyers (represented by $(1-x)$) of the product. Assume Q is infinite for convenience. Obtain the turnpike for this problem; see [Sethi, 1974a].

6.38 Solve (6.22) with the state equation replaced by

$$\dot{x} = r \log u - \delta x, \quad x(0) = x_0$$

but without a terminal state constraint. Solve both the finite and infinite horizon versions of the problem , see [Sethi, 1975].

6.39 <u>Ozga Model [Ozga, 1960], Gould, 1970]</u>: Suppose the information spreads by word of mouth rather than by our impersonal advertising medium, i.e., individuals who are already aware of the product inform the indivisuals who are not at a certain rate influenced by advertising expenditure. What we have now is the Ozga model

$$\dot{x} = ux(1-x) - \delta x, \quad x(0) = x_0$$

subject to which we must maximize

$$J = \int_0^\infty e^{-\rho t} [\pi(x) - W(u)] \, dt,$$

where $\pi(x)$ is concave. [See [Sethi, 1979b] for a Green's theorem application to this problem.]

7 THE MAXIMUM PRINCIPLE: DISCRETE TIME

For many purposes it is convenient to assume that time is represented by a discrete variable, $k = 0,1,2,\ldots,T$, rather than a continuous variable, $t\epsilon[0,T]$. This is particularly true when we wish to solve a large control theory problem by means of a computer. It is also desirable, even when solving very small problems which have state or adjoint differential equations whose solutions cannot be expressed in closed form, to formulate them as discrete problems and let the computer solve them in a stepwise manner.

We will see that the maximum principle, which is to be derived in this chapter, is not valid for the discrete problem in as nearly as wide a sense as for the continuous problem. In fact we will reduce it to a constrained nonlinear programming problem and state necessary conditions for its solution by using the well-known Kuhn-Tucker Theorem. In order to follow this procedure we have to make some simplifying assumptions, and hence will obtain only a restricted form of the maximum principle. In Section 7.2.5 we state without proof a more general form of the maximum principle.

7.1 Nonlinear Programming Problems

We begin by stating a general forn of a nonlinear programming problem, see [Mangasarian, 1969]. Let y be an n-component column vector, a be an r-component column vector, and b an s-component column vector. Let h: $E^n \to E^1$, g: $E^n \to E^r$, and w: $E^n \to E^s$ be differentiable functions. Then we consider the nonlinear programming problem

$$\text{Maximize} \quad h(y) \tag{7.1}$$

Subject to

$$g(y) = a \tag{7.2}$$

$$w(y) \geq b \tag{7.3}$$

We will develop necessary conditions, called the Kuhn-Tucker conditions, which a solution y^* to this problem must satisfy. We start with simpler problems and work up to the statement of these conditions in a heuristic fashion. References are given for rigorous developments of these results.

7.1.1 Lagrange Multipliers

Suppose we want to solve (7.1) without imposing constraints (7.2) or (7.3). The problem is now the classical unconstrained maximization problem of calculus, and the first order necessary conditions for its solution are

$$h_y = 0 \tag{7.4}$$

The points satisfying (7.4) are called critical points. Additional conditions, called the second order conditions, are required to conclude whether a critical point is a maximum, minimum, or neither. These are stated in Exercise 7.2. In an important case when the function h is concave, then (7.4) is also a sufficient condition for a global maximum of h.

Next suppose we want to solve (7.1) while imposing just the equality constraints (7.2). The method of Lagrange multipliers permits us to obtain necessary conditions for a solution to the constrained maximization problem (7.1) and (7.2) must satisfy. We define the Lagrangian function

$$L(y,\lambda) = h(y) + \lambda[g(y)-a] \tag{7.5}$$

The necessary condition for y^* to be a maximum solution to (7.1) and (7.2) is that there exists an r-component row vector λ such that

$$L_y = h_y + \lambda g_y = 0 \tag{7.6}$$

$$L'_\lambda = g(y) - a = 0 \tag{7.7}$$

Note that (7.7) is a repetition of the constraints in (7.2); recall that L'_λ is the transpose of L_λ.

The system of $n+r$ equations (7.6) and (7.7) has $(n+r)$ unknowns. Since some or all of the equations are nonlinear, the solution method will, in general, involve nonlinear programming techniques, and may be difficult. In other cases, e.g., when h is linear and g is quadratic, it may only involve the solution of linear equations. Once a solution (y^*,λ^*) is found satisfying the necessary conditions (7.6) and (7.7) the solution must still be checked to see whether it satisfies sufficient conditions for a global maximum. Such sufficient conditions will be stated in Section 7.1.3.

Suppose (y^*,λ^*) is in fact a solution to the constrained problem (7.6) and (7.7), and $h^* = h(y^*)$ is the optimum value of the objective function. The Lagrange multipliers have an important managerial interpretation, namely,

$$\frac{\partial h^*}{\partial a} = -\lambda^* \tag{7.8}$$

In other words λ_i is the negative of the inputed value of "shadow price" of having one unit more or less of the resource a_i. For a proof of (7.8) and interpretations see Exercises 7.4 and 7.5.

<u>Example 7.1</u>. Consider the problem

$$\text{Maximize} \quad h(x,y) = x^2 + y^2$$

Subject to

$$2x + y = 10.$$

<u>Solution</u>. We form the Lagrangian

$$L(x,y,\lambda) = (x^2 + y^2) + \lambda(2x + y - 10) .$$

The necessary conditions found by partial differentiation are:

$$L_x = 2x + 2\lambda = 0$$

$$L_y = 2y + \lambda = 0$$

$$L_\lambda = 2x + y - 10 = 0 \quad .$$

From the first two equations we get

$$-\lambda = x = 2y \quad .$$

Solving this with the last equation yields the quantities

$$x^* = 4, \quad y^* = 2, \quad \lambda^* = -4, \; h^* = 20$$

which can be seen to give a maximum value to h since h is concave, and the constraint set is convex. The interpretation of the Lagrange multiplier $\lambda^* = -4$ can be obtained to verify (7.8) by replacing the constant 10 by $10 + \varepsilon$ and expanding the objective function in a Taylor's series. See Exercise 7.5.

7.1.2 Inequality Constraints

Now suppose we want to solve the inequality constrained problem defined by (7.1) and (7.3), without (7.2). The latter constraints will be appended to the problem in Section 7.1.3.

As before we define the Lagrangian

$$L(y,\mu) = h(y) + \mu[w(y) - b] \quad . \tag{7.9}$$

The Kuhn-Tucker necessary conditions for this problem cannot be as easily derive as for the equality constrained problem in the preceding section. We will write them first, and then give interpretations to make them plausible. The necessary condition for y^* to be a maximum solution of (7.1) and (7.3) is that there exists an s-dimensional row vector μ such that

$$L_y = h_y + \mu w_y = 0 \tag{7.10}$$

$$w \geq b \tag{7.11}$$

$$\mu \geq 0, \quad \mu(w-b) = 0 \quad . \tag{7.12}$$

Note that (7.10) is analogous to (7.6). Also (7.11) repeats the inequality constraint (7.3) in the same way that (7.7) repeated the equality constraint (7.2). However, the inequality constraints (7.12), usually called the Kuhn-Tucker constraints, are new and are particular to the inequality constrained problem. We will see that they include some of the boundary points of the feasible set of points, as well as any unconstrained maximum solution points, as candidates for the solution to the maximum problem. This is best brought out by examples.

Example 7.2. Solve the problem:

$$\text{Maximize} \quad h(x) = 8x - x^2$$

$$\text{Subject to} \qquad x \geq 2 \quad .$$

Solution. We form the Lagrangian

$$L(x,\mu) = 8x - x^2 + \mu(x-2)$$

then the necessary conditions (7.10)-(7.12) become

$$L_x = 8 - 2x + \mu = 0 \tag{7.13}$$

$$x - 2 \geq 0 \tag{7.14}$$

$$\mu \geq 0, \quad \mu(x-2) = 0 \quad . \tag{7.15}$$

Note that the constraint $\mu(x-2) = 0$ in (7.15) can be phrased as: either $\mu = 0$, or $x = 2$. We treat these two cases separately.

Case 1. $\mu = 0$. Then from (7.13) we get $x = 4$, which also satisfies (7.14). Hence this solution which makes $h(4) = 16$ is a possible candidate for the maximum solution.

Case 2. $x = 2$. Here from (7.13) we get $\mu = -4$, which does not satisfy the $\mu \geq 0$ constraint in (7.13).

From these two cases we conclude that the optimum solution is $x^* = 4$ and $h^* = h(x^*) = 16$.

Example 7.3. Solve the problem:

$$\text{Maximize } h(x) = 8x - x^2$$
$$\text{Subject to: } x \geq 6.$$

Solution. The Lagrangian is

$$L(x,\mu) = 8x - x^2 + \mu(x-6).$$

The necessary conditions are:

$$L_x = 8 - 2x + \mu = 0 \qquad (7.16)$$

$$x - 6 \geq 0 \qquad (7.17)$$

$$\mu \geq 0, \quad \mu(x-6) = 0 \quad . \qquad (7.18)$$

Again the constraint $\mu(x-6) = 0$ is an either-or constraint which gives two cases.

Case 1. $\mu = 0$. From (7.16) we obtain $x = 4$ which does not satisfy (7.17), so that this case is infeasible.

Case 2. $x = 6$. Obviously (7.17) holds. From (7.18) we get $\mu = 4$, so that (7.18) holds as well. The optimal solution is then

$$x^* = 6, \quad h^* = h(x^*) = 12,$$

since it is the only solution satisfying the necessary conditions.

The examples above involve only one variable, and are relatively obvious. The next example, which is two-dimensional, will reveal more of the power, and the difficulties of applying the Kuhn-Tucker conditions.

Example 7.4. Find the shortest distance from the point (2,2) and the upper half of the semi-circle of radius one whose center is at the origin. We can state this problem as the following nonlinear programming problem:

$$\text{Maximize} \quad h = -(x-2)^2 - (y-2)^2$$

Subject to

$$x^2 + y^2 \leq 1$$

$$y \geq 0.$$

The Lagrangian function for this problem is:

$$L = -(x-2)^2 - (y-2)^2 + \mu(1-x^2-y^2) + \nu y. \tag{7.19}$$

The necessary conditions are:

$$-2(x-2) - 2\mu x = 0 \tag{7.20}$$

$$-2(y-2) - 2\mu y + \nu = 0 \tag{7.21}$$

$$1 - x^2 - y^2 \geq 0 \tag{7.22}$$

$$y \geq 0 \tag{7.23}$$

$$\mu \geq 0, \quad \mu(1-x^2-y^2) = 0 \tag{7.24}$$

$$\nu \geq 0, \quad \nu y = 0. \tag{7.25}$$

From (7.24) we see that either $\mu = 0$ or $x^2 + y^2 = 1$, i.e., we are on the boundary of the semi circle. If $\mu = 0$ we see from (7.20) that $x = 2$; but $x = 2$ does not satisfy (7.22) for any y; hence we conclude $\mu > 0$ and $x^2 + y^2 = 1$.

From (7.25) we conclude, either $\nu = 0$ or $y = 0$. If $\nu = 0$ then from (7.20), (7.21), and $\mu > 0$ we get $x = y$. Solving the latter with $x^2 + y^2 = 1$ gives the point

(a) $(\sqrt{2}/2, \sqrt{2}/2)$ and $h = (9 - 4\sqrt{2})/4$

If $y = 0$ then solving with $x^2 + y^2 = 1$ gives the two points

(b) $(1,0)$ and $h = 5$

(c) $(-1,0)$ and $h = 13$.

These three points are shown in Figure 7.1. Of the three points found that satisfy the necessary conditions clearly the point $(\sqrt{2}/2,\sqrt{2}/2)$ found in (a) is the nearest point and solves the closest point problem. The point (-1,0) in (c) is in fact the farthest point; and the point (1,0) in (b) is neither the closest nor the farthest point.

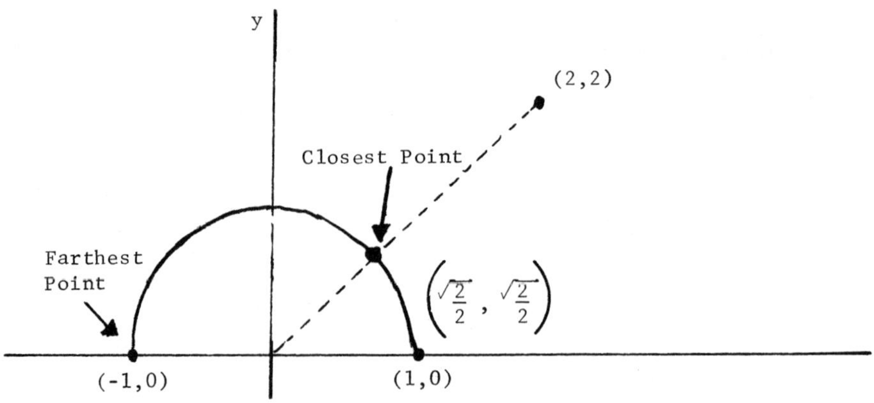

Figure 7.1

The fact that there were three points satisfying the necessary conditions, and only one of them actually solved the problem at hand, emphasizes that they are necessary conditions only. In every case it is necessary to check the solutions to the necessary conditions to see whether they actually solve the desired problem.

Next we work two examples that show some technical difficulties that can arise in the application of the Kuhn-Tucker conditions.

Example 7.5. Consider the problem

$$\text{Maximize} \quad h(x,y) = y \tag{7.26}$$

Subject to

$$-x^{2/3} - y + 1 \geq 0 \tag{7.27}$$

$$y \geq 0 \tag{7.28}$$

The set of points satisfying the constraints is shown shaded in Figure 7.2. From

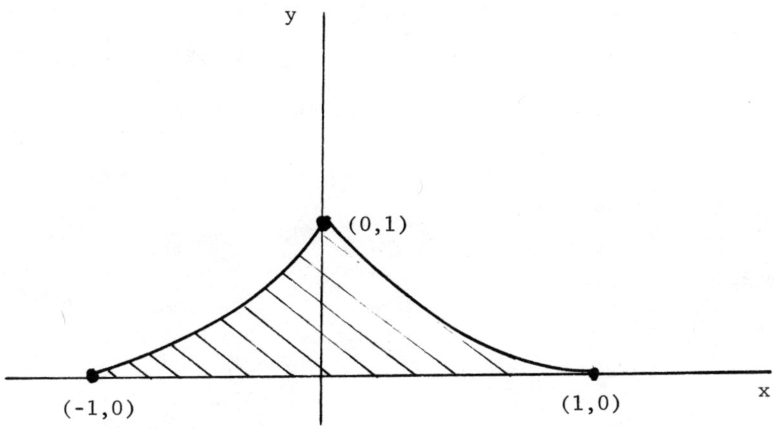

Figure 7.2

the figure it is obvious that the solution point (0,1) maximizes the value of y. Let us see if we can find it using the above procedure.

The Lagrangian is

$$L = y + \lambda(-x^{2/3} - y + 1) + \mu y \quad . \tag{7.29}$$

The necessary conditions can then be stated as:

$$L_x = -\left(\frac{2}{3}\right) \lambda x^{-1/3} = 0 \tag{7.30}$$

$$L_y = 1 - \lambda + \mu = 0 \tag{7.31}$$

$$\lambda \geq 0, \quad \lambda(-x^{2/3} - y + 1) = 0 \tag{7.32}$$

$$\mu \geq 0, \quad \mu y = 0 \tag{7.33}$$

together with (7.27) and (7.28). From (7.30) we get $\lambda = 0$, since $x^{-1/3}$ is never 0 in the range $-1 \leq x \leq 1$. But $\lambda = 0$ in (7.31) gives $\mu = -1 < 0$, which fails to solve (7.33).

You may perhaps think that the reason for the failure of the method is due to the non-differentiability of the constraint (7.27) when $x = 0$. That is part of the reason, but the next example shows that something deeper is involved.

Example 7.6. Consider the problem

$$\text{Maximize} \quad h(x,y) = y \tag{7.34}$$

Subject to

$$(1-y)^3 - x^2 \geq 0 \tag{7.35}$$

$$y \geq 0 \quad . \tag{7.36}$$

The constraints are now differentiable and the set of feasible solutions is exactly the same as for Example 7.5, and is shown in Figure 7.2. Hence the optimum answer is $(x^*,y^*) = (0,1)$ and $h^* = 1$. But once again the Kuhn-Tucker method fails, as we will see. The Lagrangian is

$$L = y + \lambda[(1-y)^3 - x^2] + \mu y \tag{7.37}$$

so that the necessary conditions are:

$$L_x = -2x\lambda = 0 \tag{7.38}$$

$$L_y = 1 - 3\lambda(1-y)^2 + \mu = 0 \tag{7.39}$$

$$\lambda \geq 0, \quad \lambda[(1-y)^3 - x^2] = 0 \tag{7.40}$$

$$\mu \geq 0, \quad \mu y = 0 \tag{7.41}$$

together with (7.35) and (7.36). From (7.41) we get, <u>either</u> $y = 0$ <u>or</u> $\mu = 0$. Since $y = 0$ minimizes the objective function we choose $\mu = 0$. From (7.38) we get, either $\lambda = 0$ or $x = 0$. Since $\lambda = \mu = 0$ gives no solution to (7.39), we choose $x = 0$, $\lambda \neq 0$. But then (7.40) gives $(1-y)^3 = 0$, or $y = 1$. However, $\mu = 0$ and $y = 1$ means there is no solution to (7.39).

The reason for failure of the method in Example 7.6 is that the constraints do not satisfy what is called the <u>constraint qualification</u>, which will be discussed in the next section.

Example 7.6 shows the necessity of imposing some kind of condition to rule out cusp points such as the (0,1) point in Figure 7.2, since the method shown will fail to find the solution when the answer occurs at such points. A complete study of the problem is beyond the scope of this book, but we state here a version of the constraint qualification sufficient for our purposes. For further information see [Mangasarian, 1969].

In order to motivate the definition we illustrate two different situations in Figure 7.3. In Figure 7.3(a) we show two boundary curves $w_1(y) = b_1$ and $w_2(y) = b_2$ intersecting in the boundary point $\bar{y}$. The two tangents to these curves are shown, and v is a vector lying between the two tangents. Starting at $\bar{y}$ there is a differentiable curve $c(\tau)$ drawn so that it lies entirely

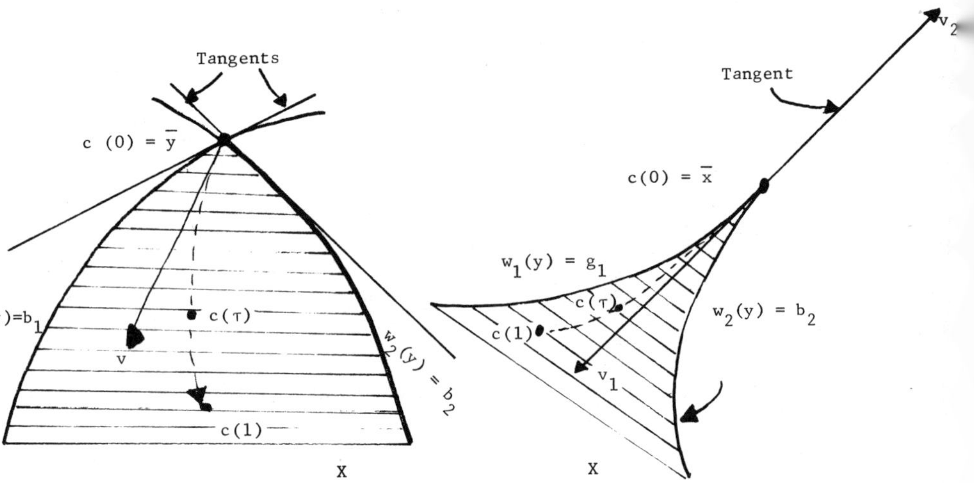

Figure 7.3

within the feasible set Y, and such that its initial slope is equal to v. Whenever such a curve can be drawn from every boundary point $\bar{y}$ in Y and eve v contained between the tangent lines we say that the constraints defining X satisfy the Kuhn-Tucker constraint qualification.

Figure 7.3(b) illustrates a case in which the constraint qualification does not hold. Here the two tangents to the graphs $w_1(y) = b_1$ and $w_2(y) = b$

coincide, so that v_1 and v_2 are vectors lying between these two tangents. Notice that for vector v_1 it is possible to find the differentiable curve $c(\tau)$ satisfying the above conditions, but for vector v_2 no such curve exists. Hence the constraint qualification does not hold for the example in Figure 7.3(b).

In order to state the Kuhn-Tucker conditions analytically we note that the direction of the outward pointing normal to the constraint set $w(y) \geq 0$ is given by the vector $-\frac{\partial w}{\partial y}$. Therefore the vectors v which lie between the tangents at the point $\bar{y}$ must satisfy

$$\left[-\frac{\partial w(\bar{y})}{\partial y}\right] v \geq 0$$

or equivalently

$$\frac{\partial w(\bar{y})}{\partial y} v \leq 0 . \tag{7.42}$$

Note that the set of solutions v to (7.42) forms a convex set V.

The constraint set for Example 7.6, shown in Figure 7.2 also does not satisfy the constraint qualification, but for a slightly different reason. In that example there is a single constraint $w(x,y) = (1-y)^3 - x^2 \geq 0$. The negative of the derivative (gradient) of g is given by the vector

$$\left(-\frac{\partial w}{\partial x}, -\frac{\partial w}{\partial y}\right) = (2x, 3(1-y)^2) .$$

Note that this vector is the zero vector at the point (0,1). Hence every v satisfies the tangent condition (7.42). However, looking at Figure 7.2, we see that there is only one direction (that is the upward vertical direction) along which a differentiable curve lying entirely within X can be constructed. Therefore the constraint qualification does not hold at the point (1,0).

7.1.3 Theorems from Nonlinear Programming

We now state without proof two nonlinear programming theorems which we will use in deriving our version of the discrete maximum principle. For proofs see [Mangasarian, 1969].

We first state the constraint qualification symbolically. For the problem defined by (7.1), (7.2), and (7.3), let Y be the set of all feasible vectors satisfying (7.2) and (7.3). Let $\bar{y}$ be any point of Y and let $z(\bar{y})$ = be the vector of tight constraints at point $\bar{y}$, that is, z includes all the g constraints in (7.2) and those constraints in (7.3) which are satisfied a equalities. Define the set

$$V = \{v \mid \frac{\partial z(\bar{y})}{\partial y} v \leq 0 \} \quad . \tag{7.43}$$

Then we shall say that the constraint set Y satisfies the Kuhn-Tucker constraint qualification at $\bar{y} \in Y$ if z is differentiable at $\bar{y}$ and if, for every $v \in V$, there exists a differentiable curve $c(\tau)$ defined for $0 \leq \tau \leq 1$ such that

(i) $c(0) = \bar{y}$

(ii) $c(\tau) \in Y$ for all τ satisfying $0 \leq \tau \leq 1$

(iii) $\frac{dc(\tau)}{d\tau} = kv$ for some constant $k > 0$.

The interpretation of this condition was given for the two dimensional case in the preceding section.

We now state two theorems giving sufficient and necessary conditions for the problem given by (7.1)-(7.3). The Lagrangian function for this problem is

$$L(y,\lambda,\mu) = h + \lambda(g(y) - a) + \mu(w(y) - b). \tag{7.44}$$

The Kuhn-Tucker conditions at $\bar{y} \in Y$ for this problem are:

$$L_y(\bar{y},\bar{\lambda},\bar{\mu}) = h_j(\bar{y}) + \bar{\lambda}g_y(\bar{y}) + \bar{\mu}w_y(\bar{y}) = 0 \tag{7.45}$$

$$g(\bar{y}) = a \tag{7.46}$$

$$w(\bar{y}) \geq b \tag{7.47}$$

$$\bar{\mu} \geq 0, \quad \bar{\mu}\, w(\bar{y}) = 0 \tag{7.48}$$

where λ and μ are row vectors of multipliers to be determined.

Theorem 7.1. (Sufficient Conditions) If h and g are differentiable, h is concave, g is convex, and $(\bar{y},\bar{\lambda},\bar{\mu})$ solve the conditions (7.45)-(7.48) then $\bar{y}$ is a solution to the maximization problem (7.1)-(7.3).

Theorem 7.2. (Necessary Conditions) If h and g are differentiable, $\bar{y}$ solves the maximization problem, and the constraint qualification holds at $\bar{y}$, then there exist multipliers $\bar{\lambda}$ and $\bar{\mu}$ such that $(\bar{y},\bar{\lambda},\bar{\mu})$ satisfy conditions (7.45)-(7.48).

7.2 A Discrete Maximum Principle

We shall now use the nonlinear programming results of the previous section to derive a special form of the discrete maximum principle. Two references in this connection are: [Luenberger, 1972] and [Mangasarian and Fromovitz, 1967]. A more general discrete maximum principle will be stated in Section 7.5.

7.2.1 The Discrete Optimal Control Problem

In order to state the discrete optimal control problem, we define the following:

T is the terminal time,

k is the time period, $k = 0,1,2,\ldots,T$,

x^k is an n-component column <u>state vector</u>,

u^k is an r-component column <u>control vector</u>,

b^k is an s-component column <u>constant vector</u>,

f: $E^n \times E^m \times E^1 \to E^n$; we assume f is differentiable,

F: $E^n \times E^m \times E^1 \to E^1$; we assume F is differentiable,

g: $E^m \times E^1 \to E^s$; we assume g is differentiable.

Then a discrete time optimal control problem is:

$$\text{Maximize} \sum_{k=0}^{T-1} F(x^k, u^k, k) \tag{7.49}$$

Subject to

$$\Delta x^k = f(x^k, u^k, k), \quad \text{for } k = 0,.,\ldots,T-1, \quad x^0 \text{ given,} \tag{7.50}$$

$$g(u^k, k) \geq b^k, \quad \text{for } k = 0,.,\ldots,T-1 \quad . \tag{7.51}$$

In (7.50) the term on the left hand side is the difference operator $\Delta x^k = x^{k+1} - x^k$.

As in the continuous time case it is convenient to transform the above problem into Mayer form. We do this by adding a variable x^k_{n+1} as an additional component to the vector x^k and a function f^k_{n+1} as an additional component to the vector f defined as follows:

$$f_{n+1}(x^k, u^k, k) = F(x^k, u^k, k) \tag{7.52}$$

$$x^k_{n+1} = \sum_{j=0}^{k-1} F(x^j, u^j, j) \text{ and } x^0_{n+1} = 0 \ . \tag{7.53}$$

Therefore it follows easily that

$$\begin{aligned}\Delta x_{n+1}^{k} &= x_{n+1}^{k+1} - x_{n+1}^{k}\\ &= \sum_{j=0}^{k} F(x^j,u^j,j) - \sum_{j=0}^{k-1} F(x^j,u^j,j)\\ &= F(x^k,u^k,k)\\ &= f_{n+1}(x^k,u^k,k)\end{aligned}$$

as required to keep the form of (7.50).

If we now let c be the $(n+1)$-component constant row vector defined by

$$c = \underbrace{(0,0,\ldots,0,1)}_{n+1} \tag{7.53}$$

we can state the control problem in linear Mayer form as:

Maximize $c \cdot x^T = x_{n+1}^T$ (7.54)

Subject to

$$\Delta x^k = f(x^k,u^k,k), \text{ for } k = 0,.,\ldots,T-1,\ x^0 \text{ given}, \tag{7.55}$$

$$g(u^k,k) \geq b^k, \quad \text{for } k = 0,1,\ldots,T-1 \tag{7.56}$$

We must keep in mind, of course, that the vectors x and f in (7.54) and (7.55) have $n+1$ components.

The objective function (7.54) will become

$$x_{n+1}^T + F(x_1^T,x_2^T,\ldots,x_n^T,u^T,T), \tag{7.57}$$

if the summation in (7.49) ranged from $k = 0$ to $k = T$. Thus the F in (7.57) can be termed the salvage value function. This point is gone into in more detail in Exercise 7.10.

7.2.2 A Discrete Maximum Principle

We now apply the non-linear programming theory of Section 7.1 to find necessary conditions for the solution to the Mayer form of the control problem of Section 7.2.1.

We let λ^{k+1} be an n-component row vector of Lagrange multipliers, which we rename adjoint variables, and which we associate with the constraints (7.55). Similarly, we let μ^k be an s-component row vector of Lagrange multipliers associated with constraints (7.56). These multipliers are defined for each time $k = 0,\ldots,T-1$.

The Lagrangian function of the problem is now

$$L = cx^T + \sum_{k=0}^{T-1} \lambda^{k+1}[\, f(x^k,u^k,k) - x^{k+1} + x^k\,] + \sum_{k=0}^{T-1} \mu^k[g(u^k,k) - b^k]. \tag{7.58}$$

We now define the Hamiltonian function H^k to be

$$H^k = H(x^k,u^k,k) = \lambda^{k+1} f(x^k,u^k,k) \quad . \tag{7.59}$$

Using (7.59) we can rewrite (7.58) as follows:

$$L = cx^T + \sum_{k=0}^{T-1} [\, H^k - \lambda^{k+1}(x^{k+1} - x^k)] + \sum_{k=0}^{T-1} \mu^k[g(u^k,k) - b^k] \; . \tag{7.60}$$

If we now differentiate (7.60) with respect to x^k for $k = 0,1,\ldots,T-1$, we obtain

$$\frac{\partial L}{\partial x^k} = \frac{\partial H^k}{\partial x^k} - \lambda^k + \lambda^{k+1} = 0$$

upon which rearranging terms becomes

$$\Delta\lambda^k = \lambda^{k+1} - \lambda^k = -\frac{\partial H^k}{\partial x^k}, \quad \text{for} \quad k = 0,1,\ldots,T-1 \,. \tag{7.61}$$

If we now differentiate (7.60) with respect to x^T we get

$$\frac{\partial L}{\partial x^T} = c - \lambda^T = 0, \quad \text{or}$$

$$\lambda^T = c. \tag{7.62}$$

The difference equations (7.61) with terminal boundary conditions (7.62) are called the <u>adjoint</u> <u>equations</u>.

Since in the Mayer form c is given by (7.53) we can easily see that boundary conditions (7.62) are simply

$$\lambda_i^T = 0 \quad \text{for} \quad i = 1,\ldots,n \quad \text{and} \quad \lambda_{n+1}^T = 1. \tag{7.63}$$

If we now differentiate L with respect to u^k and state the corresponding Kuhn-Tucker conditions for the multipliers μ^k and constraints (7.56) we have

$$\frac{\partial L}{\partial u^k} = \frac{\partial H^k}{\partial u^k} + \mu^k \frac{\partial g}{\partial u^k} = 0 \qquad \text{or}$$

$$\frac{\partial H^k}{\partial u^k} = -\mu^k \frac{\partial g}{\partial u^k} \tag{7.64}$$

and

$$\mu^k \geq 0, \quad \mu^k[g(u^k,k) - b^k] = 0 \,. \tag{7.65}$$

We note that, provided H^k is concave and $g(u^k,k)$ is convex, and the constraint qualification holds, then the conditions (7.64) and (7.65) are precisely the necessary and sufficient conditions for solving the following Hamiltonian maximization problem:

$$\boxed{\begin{aligned}&\underset{u^k}{\text{Maximize}}\quad H^k\\&\text{subject to}\quad g(u^k,k) \geq b^k\end{aligned}} \quad . \tag{7.66}$$

We have thus derived the following restricted form of the discrete maximum principle.

<u>Theorem 7.3</u>. If for every k, H^k is concave and $g(u^k,k)$ is convex i u^k, and the constraint qualification holds, then necessary conditions for the discrete linear control problem (7.54)-(7.56) are that u^k should solve the Hamiltonian maximization problem (7.66) for every k.

We now state the discrete maximum principle for (7.49)-(7.51) which is in the Lagrange form. First we define the Hamiltonian

$$H^k(x^k,u^k,\lambda^{k+1},k) = F(x^k,u^k,k) + \lambda^{k+1}f(x^k,u^k,k) \ . \tag{7.6}$$

Then with the same assumption as in Theorem 7.3, necessary conditions for u^{k*}, $k = 0,1,\ldots,T-1$, to be an optimal control for the problem in (7.49)-(7.51) are:

$$\boxed{\begin{aligned}&\Delta x^{k*} = f(x^{k*},u^{k*},k), \quad x^0 \ \text{given}\\&\Delta\lambda^{k*} = -\frac{\partial H^k}{\partial x^k}[x^{k*},u^{k*},\lambda^{(k+1)*},k], \ \lambda^{T*} = 0\\&H^k(x^{k*},u^{k*},\lambda^{(k+1)*},k) \geq H^k(x^{k*},u^k,\lambda^{(k+1)*},k)\\&\text{for all } u^k \text{ such that } g(u^k,k) \geq b^k\\&\text{for } k = 0,1,\ldots,T-1.\end{aligned}} \tag{7.6}$$

Section 7.2.3 gives examples of the application of this maximum principle (7.68). In Section 7.3 we state a more general discrete maximization principle.

7.2.3 Examples

Our first example will be similar to Example 2.3, and it will be solved completely. The reader will note that the solutions of the continuous and discrete problems are very similar. The second example is a discrete version of the inventory-production problem of Section 5.1. We shall not try to solve this problem in closed form, but instead leave it for numerical solution in Chapter 11.

Example 7.7. A particle is located at height 5 feet above the x-axis, and we want to bring it down to the x-axis as quickly as possible and keep it there by using a discrete control. We pose the problem as the following discrete optimal control problem:

$$\text{Maximize}\{J = \sum_{k=1}^{T-1} -\frac{1}{2}(x^k)^2\} \tag{7.69}$$

Subject to

$$\Delta x^k = u^k, \quad x^0 = 5 \tag{7.70}$$

$$u^k \in \Omega = [-1,1]. \tag{7.71}$$

We shall solve this problem for various values of T.

Solution. The Hamiltonian is

$$H^k = -\frac{1}{2}(x^k)^2 + \lambda^{k+1}u^k \tag{7.72}$$

from which it is obvious that the optimal policy is bang-bang. Its form is

$$u^k = \text{bang}[-1,1;\lambda^{k+1}] = \begin{cases} 1 & \text{if } \lambda^{k+1} > 0 \\ \text{singular} & \text{if } \lambda^{k+1} = 0 \\ -1 & \text{if } \lambda^{k+1} < 0 \end{cases} \quad . \tag{7.73}$$

Let us assume, as we did in solving the similar problem in Example 2.3, that $\lambda^k < 0$ as long as x^k is positive so that $u^k = -1$. Given this assumption (7.68) becomes $\Delta x^k = -1$ whose solution is

$$x^k = -k + 5 \quad \text{for} \quad k = 1,2,\ldots,T \quad . \tag{7.74}$$

Differentiating (7.70) we obtain the adjoint equation

$$\Delta\lambda^k = -\frac{\partial H^k}{\partial x^k} = x^k, \; \lambda^T = 0 \quad . \tag{7.75}$$

Let us assume $T = 6$. Substitute (7.74) into (7.75) and solve using the table in Appendix (A1). We obtain

$$\Delta\lambda^k = -k + 5, \quad \lambda^6 = 0$$

$$\lambda^k = -\frac{1}{2}(k^2-k) + 5k + c$$

$$\lambda^6 = -\frac{1}{2}(36-6) + 30 + c = 0 \Rightarrow c = -15$$

so that

$$\lambda^k = -\frac{1}{2}k^2 + \frac{11}{2}k - 15 \quad . \tag{7.76}$$

A sketch of the values for λ^k and x^k appears in Figure 7.4. Note that $\lambda^5 = 0$, so that the control for u^4 is singular. However since $x^4 = 1$ we choose $u^4 = -1$ in order to bring x^5 down to 0.

The solution of the problem for $T > 6$ is carried out in the same way we solved Example 2.3. Namely, observe that $x^5 = 0$ and $\lambda^5 = \lambda^6 = 0$, so that control is singular. We simply make $\lambda^k = 0$ for $k > 6$ so that $u^k = 0$

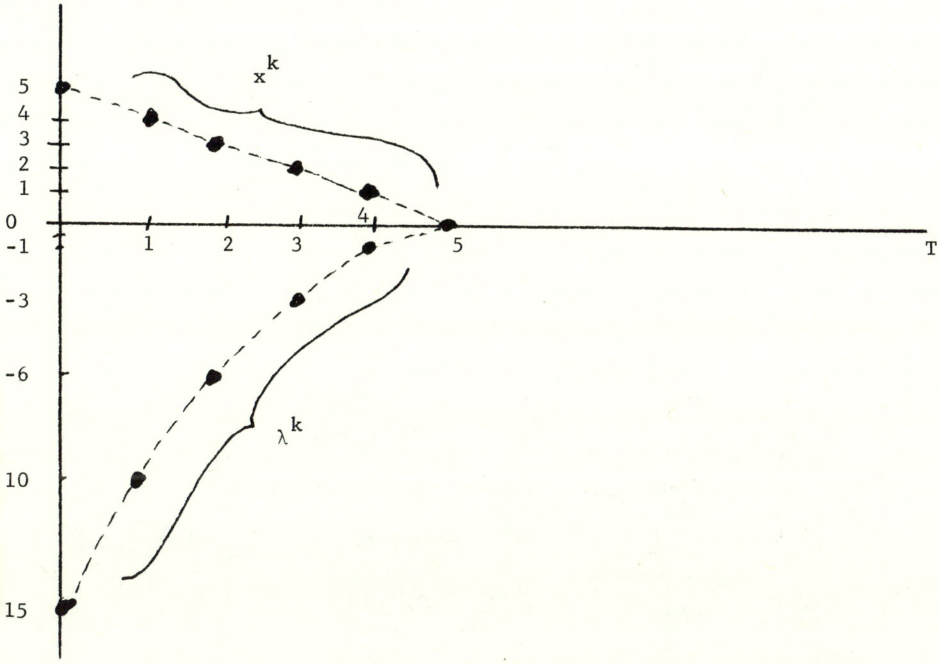

Figure 7.4. Sketch of λ^k and x^k. (Only the blackened dots are relevant. The lightly dotted lines are added to help visualize the two discrete functions.)

or all $k > 6$. It is clear, without a formal proof that this maximizes 7.69).

Example 7.8. Let us consider a discrete version of the production-inventory example of Section 5.1; see also [Kleindorfer et al., 1975]. Let I^k, P^k, and S^k be the inventory, production and demand at time k, let I^0 be the initial inventory, let $\hat{I}$ and $\hat{P}$ be the goal inventory and production amounts, and let h and c be inventory and production cost coefficients. The problem is

$$\underset{P^k \geq 0}{\text{Maximize}}\{J = \sum_{k=0}^{T-1} -\frac{1}{2}[h(I^k-\hat{I})^2 + c(P^k-\hat{P})^2]\} \tag{7.77}$$

subject to

$$\Delta I^k = P^k - S^k, \quad I^0 \text{ given.} \tag{7.78}$$

Forming the Hamiltonian

$$H^k = -\frac{1}{2}[h(I^k-\hat{I})^2 + c(P^k-\hat{P})^2] + \lambda^{k+1}(P^k-S^k) \tag{7.79}$$

where the adjoint variable satisfies

$$\Delta\lambda^k = -\frac{\partial H^k}{\partial I^k} = h(I^k-\hat{I}), \quad \lambda^T = 0. \tag{7.80}$$

The optimal control is obtained by differentiating (7.79)

$$\frac{\partial H^k}{\partial P^k} = -c(P^k - \hat{P}) + \lambda^{k+1} = 0$$

so that

$$P^{k*} = \text{Max}[0, \hat{P} + \frac{1}{c}\lambda^{k+1}] \tag{7.81}$$

in order to keep $P^k \geq 0$.

Expressions (7.78), (7.80), and (7.81) determine a two point boundary value problem. If the constraint $P^k \geq 0$ is dropped it can be solved by the method of Section 5.1, with difference equations replacing the differential equations used there. The numerical solution of this TPBVP will be carried out in Section 11.2.3.

7.3 A General Discrete Maximum Principle

For the maximum principle (7.68) we assumed that H^k was concave and g was convex in u^k so that the set of admissible controls was convex.

These are fairly strong assumptions and can be relaxed. However, this more general maximum principle cannot be proved by using Kuhn-Tucker theory. We state the new maximum principle without proof. The latter can be found in [Canon, Cullum and Polak, 1970[. Other references on discrete maximum principles are: [Halkin, 1966], and [Holtzman, 1966].

The problem to be solved is

$$\text{Maximize}\left\{ J = \sum_{k=0}^{T-1} F(x^k,u^k,k) \right\}$$

Subject to

$$\Delta x^k = f(x^k,u^k,k), \quad x^0 \text{ given}$$

$$u^k \in \Omega_k, \quad k = 0,1,\ldots,(T-1) \quad . \tag{7.82}$$

Assumptions required are:

(i) $F(x^k,u^k,k)$ and $f(x^k,u^k,k)$ are continuously differentiable in x^k for every u^k and k.

(ii) The sets $\{-F(x,\Omega^k,k),\ f(x,\Omega^k,k)\}$ are <u>b -directionally convex</u> for every x and k where $b = (-1,0,\ldots,0)$; i.e., given v and w in Ω^k and $0 \le \lambda \le 1$ implies there exists $u(\lambda) \in \Omega^k$ such that

$$F(x,u(\lambda),k) \ge F(x,v,k) + (1-\lambda)F(x,w,k)$$

and

$$f(x,u(\lambda),k) = \lambda f(x,v,k) + (1-\lambda)f(x,w,k)$$

for every x and k. It should be noted that convexity implies b -directional convexity, but not the converse.

(iii) Ω^k satisfies the Kuhn-Tucker constraint qualification.

With these assumptions replacing the assumptions of Theorem 7.3, the maximum principle (7.68) holds, with control constraint set $g(u^k,k) \ge b^k$ replaced by $u^k \in \Omega^k$.

For a fixed end point problem, i.e. when x^T is also given in (7.82), the more general maximum principle holds with λ^T a constant to be determined. In Exercise 7.14 is an example of a fixed end point problem.

Some final remarks on the uses of discrete optimal control methods are appropriate. Some real life problems are discrete time problems: interest payments on loans; harvesting of crops; production planning for monthly demands; etc. Another reason which makes discrete optimal control important arises from the fact that digital computers are being used increasingly in the control of dynamic systems. Finally, when there are lags in the system dynamics, that is, when the state of the system in a period depends not only on the state and the control in the previous period, but also on the values of these variables in the prior periods, it is easy to adapt the discrete maximum principle to deal with such systems, see [Burdet and Sethi, 1976].

Exercises for Chapter 7

7.1 Determine the critical points of the following functions:

(a) $h = -5y^2 - z^2 + 10y + 6z + 27$

(b) $h = 5y^2 - yz + z^2 - 10y - 18z + 27$.

7.2 Let h be twice differentiable with its Hessian matrix defined to be $H = h_{yy}$. Let y^0 be a critical point, that is, a solution of (7.4). Let H_j be the jth principal minor, that is, the jxj submatrix found in the first j rows and j columns of H; let $|H_j|$ be the determinant of H_j. Then y^0 is a local maximum of h if

$$H_1 < 0, \quad |H_2| > 0, \quad |H_3| < 0, \ldots, |-H_n| = |-H| > 0$$

evaluated at y^0; and y^0 is a local minimum of h if

$$H_1 < 0, \quad |H_2| > 0, \quad |H_3| > 0, \ldots, |H_n| = |H| > 0$$

evaluated at y^0. Apply these conditions to Exercise 7.1.

7.3 During these times of the energy crisis, it is important to economize on fuel consumption. It is learned that when traveling x mile/hour in high gear, a truck burns fuel at the rate of

$$\frac{1}{500}\left[\frac{2500}{x} + x\right] \text{ gal./mile.}$$

If fuel costs 50 cents per gallon, find the speed that will minimize the cost of fuel for a 1000 mile trip. Check the second order condition.

Imposing this optimal speed with respect to the fuel consumption has made the truckers angry to the point of staging <u>the truck blockades</u> on several freeways around the country. To explain these blockades, we find that a crucial figure is the hourly wage rate of the truckers, a realistic estimate being \$3.90 an hour. Recompute the speed that will minimize the total cost of the same trip, this time inclusive of the driver's wages. No check need be made for the second order conditions.

7.4 Derive equation (7.8).

7.5 Verify (7.8) in Example 7.1 by determining $h^*(a)$ and expanding $h^*(10 + \varepsilon)$ in a Taylor's series around 10.

7.6 Maximize $h(x) = \frac{1}{3}x^3 - 6x^2 + 32x + 5$ subject to each of the following constraints

(a) $x \leq 6$

(b) $x \leq 20$

7.7 Rework Example 7.4 with each of the following points

(a) $(0,-1)$

(b) $(\frac{1}{2},\frac{1}{2})$.

7.8 Add the equality constraint $2x = y$ to the problem in Example 7.4 and resolve it.

7.9 Solve the problem

Maximize h

subject to $x^2 \leq (2-y)^3$, $y \geq 0$

for

(a) $h = x + y$

(b) $h = x + 2y$

(c) $h = x + 3y$

Comment on the solution in each case.

7.10 Derive the transversality condition for the objective function in (7.57).

7.11 Show that the necessary conditions for u^k to be an optimal solution for (7.66) are given by (7.64) and (7.65).

7.12 Prove Theorem 7.3.

7.13 Formulate and solve a discrete time version of the cash balance model of Section 4.1.1.

7.14 Minimum Fuel Problem.

$$\text{Minimize}\{J = \sum_{k=0}^{T-1} |u^k|\}$$

Subject to

$$\Delta x^k = Ax^k + bu^k, \quad x^0; \quad x^T \text{ given}$$

$$u^k \in [-1,1], \quad k = 0,1,\ldots,T-1,$$

where A is a given matrix. Obtain the expression for the adjoint variable and the form of the optimal control.

8 MAINTENANCE AND REPLACEMENT

The problem of determining the lifetime of an asset or an activity simultaneously with its management during that life time is an important problem in practice. The most typical example is the problem of optimal maintenance and replacement for a machine, [Rapp, 1974] and [Pierskalla and Voelker, 1976]. Other examples occur in forest management, [Näslund, 1969], and [Clark, 1976]; advertising copy management [Pekelman and Sethi, 1978]; etc.

The first major work dealing with machine replacement problems appeared in 1949 as a MAPI study by [Terborgh, 1954]. For the most part, this study was confined to those problems where the optimization was carried out only with respect to the replacement lives of the machines under consideration. In 1955, Boiteux (see [Massé, 1962]) extended the single machine replacement problem to include the optimal timing of a partial replacement of the machine before its actual retirement. [Näslund, 1966] was the first to solve a generalized version of the Boiteux problem by using the maximum principle. He considered optimal preventive maintenance applied continuously over the entire period instead of a single optimal partial replacement before the machine is retired. [Thompson, 1968] presents a modification of Näslund's model which is described in the following section.

8.1 A Simple Maintenance and Replacement Model

Consider a single machine whose resale value gradually declines over time. Its output is assumed to be proportional to its resale value. By applying preventive maintenance it is possible to slow down the rate of decline of resale value. The control problem consists of simultaneously determining the optimal rate of preventive maintenance, at the sale date of the machine. Clearly this is an optimal control problem with unspecified terminal time.

8.1.1 The Model

In order to define Thompson's model, we use the following notation:

T = the sale date of machine to be determined,

ρ = the continuous discount rate,

$x(t)$ = the resale value of machine at time t; assume $x(0) = x_o$,

$u(t)$ = the preventive maintenance rate at time t (maintenance here means money spent over and above the minimum required for necessary repairs),

$g(t)$ = the maintenance effectiveness function at time t (measured in dollars added to the resale value per dollar spent on preventive maintenance),

$d(t)$ = obsolescence function at time t (measured in terms of dollars subtracted from x at time t),

π = constant production rate in dollars per unit resale value; (assume $\pi > \rho$ or else it doesn't pay to produce).

It is assumed that $g(t)$ is a nonincreasing function of time and $d(t)$ is a nondecreasing function of time, and that for all t

$$u(t) \in \Omega = [0,U], \tag{8.1}$$

where U is a positive constant.

The present value of the machine is the sum of two terms, the discounted production stream during its life, plus the discounted resale value at T.

$$J = \int_0^T [\pi x(t) - u(t)]e^{-\rho t}dt + x(T)e^{-\rho T}. \tag{8.2}$$

The state variable x is affected by the obsolescence factor, the amount of preventive maintenance, and the maintenance effectiveness function. Thus

$$\dot{x}(t) = -d(t) + g(t)u(t). \tag{8.3}$$

In the interests of realism, we assume that

$$-d(t) + g(t)U \leq 0. \tag{8.4}$$

The assumption implies that preventive maintenance is not so effective as to enchance the resale value of the machine over previous values; rather it can at most slow down the decline of resale value, even when it is performed at the maximum rate U. A modification of (8.3) is given in [Arora and Lele, 1970].

The optimal control problem is to maximize (8.2) subject to (8.1) and (8.3).

8.1.2 Solution by the Maximum Principle

This problem is similar to model type (a) of Table 3.3 with constraint conditions (1) of Table 3.1. Therefore, we follow the steps for solution by the maximum principle stated for it in Chapter 3.

The Hamiltonian is

$$H = (\pi x - u)e^{-\rho t} + \lambda(-d + gu), \tag{8.5}$$

where the adjoint variable λ satisfies

$$\dot{\lambda} = -\pi e^{-\rho t}, \quad \lambda(T) = e^{-\rho T}. \tag{8.6}$$

Since T is unspecified, the additional terminal condition of (3.17) becomes

$$-\rho e^{-\rho T}x(T) = -H, \tag{8.7}$$

which must hold on the optimal path at time T^*.

The adjoint variable λ can be easily obtained by integrating (8.6) i.e.,

$$\lambda(t) = e^{-\rho T} + \frac{\pi}{\rho} [e^{-\rho t} - e^{-\rho T}]. \tag{8.8}$$

The interpretation of $\lambda(t)$ is as follows. It gives in present value terms, the marginal value per dollar additional resale value at time t. The first term represents the present value of one dollar additional resale value at T brought about by one dollar additional resale value at time t. The second term represents the present value of incremental production from t to T brought about by the extra productivity of the machine due to the additional one dollar resale value at time t.

Since the Hamiltonian is linear in the control variable u, the optimal control for a problem with any fixed T is, as in model type (a) in Table 3.3, bang-bang. Thus

$$u^*(t) = \text{bang}[0, U; \{e^{-\rho T} + \frac{\pi}{\rho} (e^{-\rho t} - e^{-\rho T})\}g - e^{-\rho t}]. \tag{8.9}$$

To interpret this optimal policy, we see that the term $\{e^{-\rho T} + \frac{\pi}{\rho} (e^{-\rho t} - e^{-\rho T})\}g$ is the present value of the marginal return of increasing the preventive maintenance by one dollar at time t. The term $e^{-\rho t}$ is the present value of that one dollar spent for preventive maintenance at time t. Thus, in words, the optimal policy means the following: If the marginal return of one dollar additional preventive maintenance is more than one dollar, then perform the maximum possible preventive maintenance, otherwise do not carry out any at all.

To find how the optimal control switches, we need to examine the last expression inside the bang function. Rewriting this expression as

$$e^{-\rho t}[\{\frac{\pi}{\rho} - (\frac{\pi}{\rho} - 1)e^{\rho(t-T)}\}g - 1], \tag{8.10}$$

we can conclude that the expression inside the square bracket is monotonically decreasing with time on account of the assumptions that $\pi/\rho > 1$ and that g is nonincreasing with time. We can therefore conclude that there will not be any singular control for any finite interval of time. Furthermore, since $e^{-\rho t} > 0$ for all t, we can conclude that the first expression in the bang function can only go from positive to negative and not vice versa. Thus, the optimal control will either be U or zero, or U followed by zero. The switching time t^s is obtained by equating (8.10) to zero and solving for t. If the solution is negative, let $t^s = 0$, if the solution is greater than T, let $t^s = T$, otherwise set t^s equals to the solution. It is clear that the optimal control in (8.9) can now be rewritten as

$$u^*(t) = \begin{cases} U & t \le t^s \\ 0 & t > t^s. \end{cases} \tag{8.11}$$

Note that all the above calculations were made on the assumption that T was fixed, or without imposing the condition (8.7). For an optimal path, this condition, which uses (8.5) and (8.7), can be restated as

$$-\rho e^{-\rho T^*} x^*(T^*) = -[\{\pi x^*(T^*) - u^*(T^*)\}e^{-\rho T^*} + e^{-\rho T^*}\{-d(T^*) + g(T^*)u(T^*)\}]. \tag{8.12}$$

That means, if $u^*(T^*) = 0$ (i.e., $t^s < T^*$) we have

$$x^*(T^*) = \frac{d(T^*)}{\pi - \rho}, \tag{8.13}$$

and when $u^*(T^*) = U$ (i.e., $t^s = T^*$), we have

$$x^*(T^*) = \frac{d(T^*) - [g(T^*) - 1]\,U}{(\pi - \rho)} . \tag{8.14}$$

Since $d(t)$ is nondecreasing, $g(t)$ is nonincreasing and $x(t)$ is nonincreasing, equations (8.13) and (8.14) will have solutions for T^*.

8.1.3 A Numerical Example

It is instructive to work an example of this model in which specific values are assumed for the various functions. Other examples that illustrate other kinds of qualitatively different behavior that can occur in the model are left as Exercises 8.2 - 8.4.

Suppose $U = 1$, $x(0) = 100$, $d(t) = 2$, $\pi = .1$, $\rho = .05$ and $g(t) = 2/(1 + t)^{1/2}$. First we write the condition on t^s by equating (8.10) to 0, which gives

$$\pi - (\pi - \rho)e^{-\rho(T-t^s)} = \frac{\rho}{g} . \tag{8.15}$$

Substituting, we have

$$.1 - .05e^{-.05(T-t^s)} = .025\ (1+t^s)^{1/2}$$

which simplifies to

$$(1 + t^s)^{1/2} = 4 - 2e^{-.05(T-t^s)}. \tag{8.16}$$

Then, integrating (8.3), we find

$$x(t) = -\ 2t + 4(1 + t)^{1/2} + 96, \quad \text{if} \quad t < t^s,$$

and hence

$$x(t) = -\ 2t^s + 4(1 + t^s)^{1/2} + 96 - 2(t - t^s)$$

$$= 4(1 + t^s)^{1/2} + 96 - 2t, \quad \text{if} \quad t > t^s.$$

Hence, substituting $x(T)$ into (8.13) we have

$$4(1 + t^s)^{1/2} + 96 - 2T = 2/.05 = 40$$

which yields

$$T = 2(1 + t^s)^{1/2} + 28. \tag{8.17}$$

We must solve (8.16) and (8.17) simultaneously. Substituting (8.17) into (8.16) we find that t^s must be zero of the function

$$h(t^s) = (1 + t^s)^{1/2} - 4 + 2e^{-[2(1+t^s)^{1/2} - t^s + 28]/20}. \tag{8.18}$$

A simple binary search program was written to solve this equation which obtained the value $t^s = 10.6$. Substitution of this into (8.17) yielded $T = 34.8$. Thus, the optimal solution is to maintain at the maximum rate during the first 10.6 time periods, and thereafter not at all. The sale date is at 34.8 after purchase. Figure 8.1 gives the functions $x(t)$ and $u(t)$ for this optimal maintenance and sale policy.

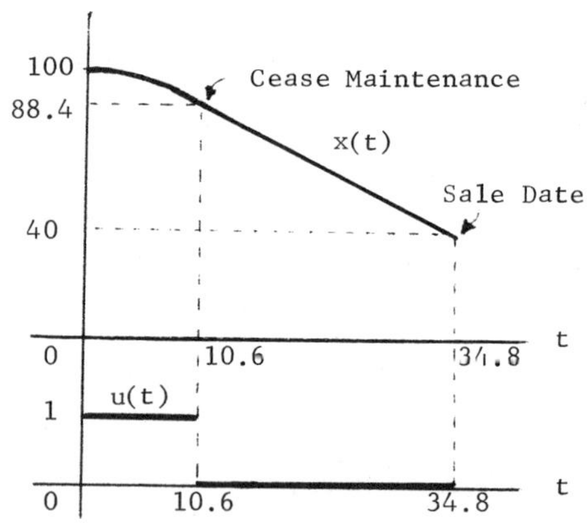

Figure 8.1

8.1.4 An Extension

The pure bang-bang result in the model developed above is as a result of the linearity in the problem. The result can be enriched see [Sethi, 1973a] if we generalize the salvage value equation (8.3) as follows,

$$\dot{x}(t) = -d(t) + g(u(t),t), \tag{8.19}$$

where g is nondecreasing and concave in u. For this section, we will assume g to be strictly concave in u, i.e., $g_u \geq 0$ and $g_{uu} < 0$ for all t. Also, $g_t \leq 0$, and $g(0,t) = 0$.

The Hamiltonian is

$$H = (\pi x - u)e^{-\rho t} + \lambda[-d + g(u,t)] \tag{8.20}$$

where λ is given in (8.8). To maximize the Hamiltonian, we differentiate it with respect to u and equate the result to zero. Thus,

$$H_u = -e^{-\rho t} + \lambda g_u = 0. \tag{8.21}$$

If we let $u^o(t)$ denote the solution of (8.21) then $u^o(t)$ maximizes the Hamiltonian (8.20) on account of the concavity of g in u. Thus, for a fixed T, the optimal control is

$$u^*(t) = \text{sat}[0, U; u^o(t)]. \tag{8.22}$$

To determine the direction of change in $u^*(t)$, we use (8.21) and the value of λ from (8.8) to obtain

$$g_u = \frac{e^{-\rho t}}{\lambda(t)} = \frac{1}{\frac{\pi}{\rho} - e^{\rho(t-T)}(\frac{\pi}{\rho} - 1)}. \tag{8.23}$$

The denominator on the right hand side of (8.23) is the same as the term in the square bracket of (8.10). Since we had already concluded that this expression is decreasing with time, it is clear that the right-hand side of (8.23) is increasing with time.

Taking the time derivative of (8.23), we have

$$g_{uu}\dot{u} > 0.$$

But $g_{uu} < 0$, it is obvious that $\dot{u}^o(t) < 0$. To sketch the optimal control u^*, let us define $0 \leq t_1 \leq t_2 \leq T$ such that $u^o(t) \geq U$ for $t \leq t_1$ and $u^o(t) \leq 0$ for $t \geq t_2$, then we can rewrite the sat function in (8.22) as

$$u^*(t) = \begin{cases} U & \text{for } t \in [0,t_1] \\ u^o(t) & \text{for } t \in (t_1,t_2) \\ 0 & \text{for } t \in [t_2,T]. \end{cases} \qquad (8.24)$$

This can be sketched in the following figure.

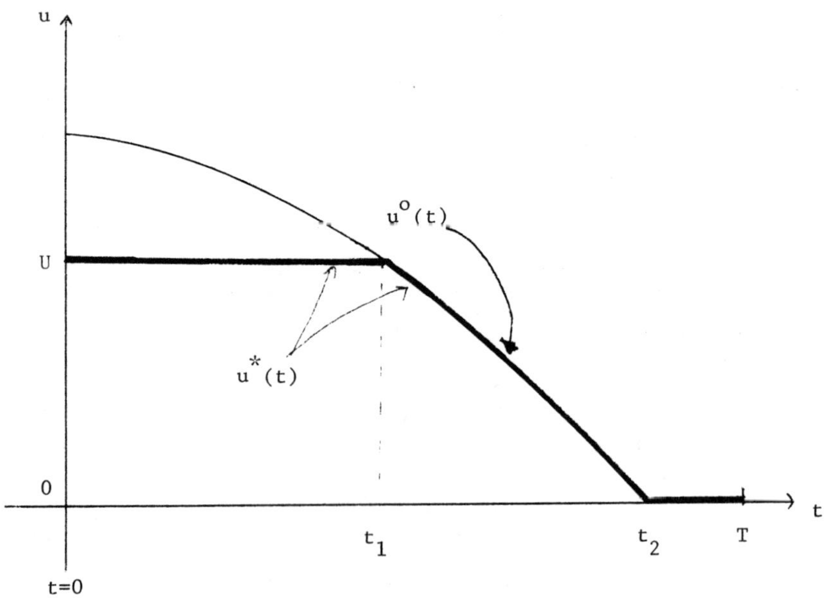

Figure 8.2 Sat Function Optimal Control

Some remarks are in order at this point. First, while the interior control is decreasing, the mode of its decrease will depend on the nature of function g. Second, to compare the model of Section (8.1.2) with that of (8.1.4) we assume that (8.3) is

$$\dot{x}(t) = -d(t) + a(t)u(t)$$

where $a(t) = g_u(u(t),t)$. In this case $g(u,t) \leq a(t)u(t)$, so that the maintenance for the linear model in Section (8.1.2) is more effective than the nonlinear model of Section (8.1.4). It can be shown that the switching time t^S of (8.11) is the same as t_2 of (8.24). Hence the optimal maintenance for the linear case is at least as great as that for the nonlinear case.

8.2 Maintenance and Replacement for a Machine Subject to Failure

In [Kamien and Schwartz, 1971] a related model is developed which has somewhat different assumptions. They assume that the production rate of the machine is independent of its age while its probability of failure increases with its age. Consistent with this assumption, the purpose of preventive maintenance in the Kamien-Schwartz model is to influence the failure rate of the machine rather than arrest the deterioration in the salvage value as before. Furthermore, their model also allows for sale of the machine at any time, provided it is still in running condition, and for its disposal as junk if it breaks down for good. The optimal control problem, therefore, is to find an optimal maintenance policy for the period of ownership and an optimal sale date at which the machine should be sold provided it has not yet failed. Other references to related models are: [Alam, Lynn and Sarma, 1976], [Alam and Sarma, 1974, 1977] and [Sarma and Alam, 1975].

8.2.1 The Model

In order to define the Kamien-Schwartz model, we use the following notation:

T = the sale date of machine to be determined,

$u(t)$ = the preventive maintenance rate at time t,

R = the constant positive rate of revenue produced by a functioning machine independent of its age at any time, net of all costs except preventive maintenance,

ρ = the continuous rate of discount,

L = the constant positive junk value independent of its age at failure,

$B(t)$ = the (exogeneously specified) resale value of the machine at time t, if it is still functioning; $\dot{B}(t) \leq 0$;

$F(t)$ = the probability that the machine fails by time t; $F(0) = 0$,

$h(t)$ = the "natural" failure rate (also termed the hazard rate in reliability theory [2a]); $h(t) \geq 0$, $\dot{h}(t) \geq 0$;

$C(u)$ = cost function for control.

For economic sense, an operable machine must at least be worth an inoperative machine and its resale value should not exceed the present value of the potential revenue generated by machine if it were functioning forever. Thus

$$0 \leq L \leq B(t) \leq R/\rho, \quad t \geq 0. \tag{8.25}$$

We also assume $B(t)$ to be decreasing and $h(t) \geq 0$ to be nondecreasing functions with machine age t. Also, for all $t \geq 0$,

$$u(t) \in \Omega = [0,1]. \tag{8.26}$$

Finally, the cost of reducing the failure rate is assumed to be proportional to the failure rate. Specifically, we assume that $C(u)h$ denotes the cost of preventive maintenance $u(t)$ at time t when the failure rate is $h(t)$. In other words, when the failure rate is $h(t)$ and a reduction in this rate of $u(t)$ is sought, it will cost $C(u)h$ dollars. It is assumed that

$$C(0) = 0, \quad C_u > 0, \quad C_{uu} > 0 \quad \text{for} \quad u \epsilon [0,1]. \tag{8.27}$$

Thus the cost of reducing the failure rate increases more than proportionately as the fractional reduction increases. But the cost of a given fractional reduction increases linearly with the natural failure rate. Hence, these conditions imply that a given absolute reduction becomes increasingly more costly as the machine gets older.

To derive the state equation for $F(t)$, we note that $\dot{F}/(1 - F)$ denotes the conditional probability density for the failure of machine at time t given that it has survived to time t. This is supposed to depend on the natural failure rate that governs the machine in the absence of preventive maintenance and on the current rate of preventive maintenance. Thus

$$\frac{\dot{F}}{1-F} = h(t)[1 - u(t)], \tag{8.28}$$

or, the state equation

$$\dot{F} = h(1 - u)(1 - F), \quad F(0) = 0. \tag{8.29}$$

Note that the modified failure rate or simply the failure rate is $(1 - u)h$. If $u = 0$, the failure rate assumes its natural value h. As u increases, the failure rate decreases and drops to zero when $u = 1$.

The expected present value of the machine is the sum of the expected present values of the total revenue less the cost of maintenance, of the junk value, and of the salvage value at the time of resale, that is,

$$J = \int_0^T \{[R - C(u)h](1 - F) + L\dot{F}\}e^{-\rho t}dt$$
$$+ e^{-\rho T}B(T)[1 - F(T)]. \qquad (8.30)$$

The optimal control problem is to maximize J subject to (8.29) and (8.26).

8.2.2 Optimal Policy

The problem is similar to model type (g) in Table 3.3 and boundary conditions (1) in Table 3.1, therefore we follow the steps for solution by the maximum principle stated in Chapter 3. The Hamiltonian is

$$H = e^{-\rho t}\{[R - C(u)h](1 - F) + L(1 - u)h(1 - F)\}$$
$$+ \lambda[(1 - u)h(1 - F)], \qquad (8.31$$

and the adjoint variable satisfies

$$\dot{\lambda} = e^{-\rho t}[R - C(u)h + L(1 - u)h] + \lambda h(1 - u),$$
$$\lambda(T) = - e^{-\rho T}B(T). \qquad (8.32$$

Since T is unspecified, the additional terminal condition analogous to (8.7) or (3.17) is (see Exercise 8.4)

$$\boxed{R - C[u^*(T^*)]h(T^*) + L[1 - u^*(T^*)]h[T^*] - [\rho + \{1 - u^*(T^*)\}h(T^*)]B(T^*) = - \frac{dB(T^*)}{dT^*}} \qquad (8.$$

We will now show that the interpretation of (8.33) is as follows: it is optimal to sell the machine at the time when the excess of expected flow value of the machine over its value in trade is just equal to the loss in resale value from postponing the sale for a small interval of time. To examine this in detail, we note that the first two terms in (8.31) give the net cash inflow (revenue - cost of preventive maintenance) to which is added, the junk value L times the probability $[1- u^*(T^*)]h(T^*)$ that it fails. From this, we subtract the value in trade given by loss of interest on resale value $\rho B(T^*)$ and the loss of the entire resale value, when the machine fails, with probability $[1 - u^*(T^*)]h(T^*)$. This completes the interpretation of the left-hand side. Finally, the right-hand side term is simply the deterioration in the resale value when the sale of the machine is postponed momentarily.

Next, we analyze the problem to obtain the optimal maintenance policy for a fixed T. Since the maximum principle states that the Hamiltonian must be maximized, then for $u^* \in (0,1)$, we have

$$H_u = - e^{-\rho t}h(1 - F)[C_u + L + e^{\rho t}\lambda] = 0. \tag{8.34}$$

Clearly, in the trivial case when the natural failure rate $h(t)$ is zero or when the machine fails with certainty by time t (i.e., $F(t) = 1$), then $u^*(t) = 0$. Assume, therefore, $h > 0$, and $F < 1$. Under these conditions, we can infer from (8.27) and (8.34) that

$$\left.\begin{array}{ll} \text{i)} & C_u(0) + L + \lambda e^{\rho t} \geq 0 \Rightarrow u^*(t) = 0 \\ \text{ii)} & C_u(1) + L + \lambda e^{\rho t} \leq 0 \Rightarrow u^*(t) = 1 \\ \text{iii)} & \text{Otherwise } C_u + L + \lambda e^{\rho t} = 0 \text{ determines } u^*(t). \end{array}\right\} \tag{8.35}$$

Using the terminal condition $(\lambda(T) = -e^{-\rho t}B(T))$ from (8.32), we can derive $u^*(T)$ satisfying (8.35). Thus

$$\boxed{\begin{array}{ll} \text{i)} & C_u(0) > B(T) - L \Rightarrow u^*(T) = 0 \\ \text{ii)} & C_u(1) < B(T) - L \Rightarrow u^*(T) = 1 \\ \text{iii)} & \text{Otherwise, } C_u = B(T) - L \Rightarrow u^*(T). \end{array}} \tag{8.36}$$

The next question is to determine how u^* changes over time. [Kamien and Schwartz, 1971] have shown that $\dot{u}^* \leq 0$; see Exercise 8.5. That means there exists $T \geq t_2 \geq t_1 \geq 0$ such that

$$u^*(t) = \begin{cases} 1 & \text{for } t \in [0, t_1] \\ \text{singular}, \dot{u}^*(t) \leq 0 & \text{for } t \in (t_1, t_2) \\ 0\ , & \text{for } t \in [t_2, T]. \end{cases} \tag{8.37}$$

Of course, $u^*(T)$ is immediately known from (8.36). If $u^*(T) \in (0,1)$, it implies $t_2 = T$; and if $u^*(T) = 1$, it implies $t_1 = t_2 = 0$.

For this model the sufficiency of the maximum principle follows from Theorem 2.1; see Exercise 8.8.

8.2.3 Determination of the Sale Date

For a fixed T, we know that the terminal optimal control $u^*(T)$ is determined by (8.36). If this $u^*(T)$ also satisfies (8.33), we have determine an optimal control trajectory as well as the optimal life of the machine. This of course, is subject to the second order condition since (8.33) is only a necessary condition for an optimal T^* to satisfy. It is clear that the determination of T, in most cases, will require numerical computations. The algorithm need only be a simple search method because it requires consideration of the single variable T.

Before we go to the next section, we remark that a business is usually a continuing entity and does not end at the sale date of the machine. Normally, an existing machine will be replaced by another which, in turn will be replaced by another and so on. The technology of the newer machines will, in general, be different from that of the existing machines. In what follows, we address ourselves to these issues.

8.3 Chain of Machines

We now extend the problem of maintenance and replacement to a chain of machines. By this we mean that given a time horizon T, we find an optimal number of machines, say ℓ, and optimal times $0 = t_0 < t_1 < t_2,\ldots,t_{\ell-1} < t_\ell = T$ of their replacements such that the existing j^{th} machine will be replaced by a new $(j + 1)$th machine at t_j. Furthermore, the optimal maintenance policy for each of the machines in the chain must be found.

Two approaches to this problem have been solved in the literature. The first attempts to solve for an infinite horizon $(\ell = \infty)$ with a simplifying assumption of identical machine lives, i.e.,

$$t_j - t_{j-1} = t_{j+1} - t_j \tag{8.38}$$

for all j, see [Sethi, 1973a]. The second relaxes the assumption (8.38) of identical machine lives ([Sethi and Morton, 1972] and [Tapiero, 1973]) at the cost of making the chain of machines finite, i.e., ℓ is finite. For a planning horizon formulation of this problem see [Sethi and Chand, 1979].

In this section, we will deal with the later problem as analyzed by [Sethi and Morton, 1972]. The problem is solved by a mixed optimization technique. The subproblems dealing with the maintenance policy are solved by appealing to the discrete maximimum principle. These subproblem solutions are then incorporated into a Wagner-Whitin model formulation [Wagner and Whitin, 1958] for solution of the full problem. The procedure is illustrated by a numerical example.

8.3.1 The Model

Consider buying a machine at the beginning of period s and salvaging it at the end of period t. Let J_{st} denote the present value of all net earnings associated with the machine. To calculate J_{st} we need the following notation for $x \leq k \leq t$.

x^k = resale value of the machine at the end of period k,

P^k = production (in dollar value) during period k,

E^k = necessary expense of the ordinary maintenance (in dollars) during period k,

$R^k \triangleq P^k - E^k$,

u^k = the rate of preventive maintenance (in dollars) during period k,

C^s = cost of purchasing machine at s,

ρ = the periodic discount rate.

It is required that

$$0 \leq u^k \leq U^{sk}, \qquad k \in [s,t]. \tag{8.38}$$

We can calculate J_{st} in terms of the variables and functions defined above.

$$J_{st} = \sum_{k=s}^{t} R^k(1+\rho)^{-k} - \sum_{k=s}^{t} u^k(1+\rho)^{-k} - C^s(1+\rho)^{-s} + x^t(1+\rho)^{-t}. \tag{8.39}$$

We must also have functions which will provide us with the way states change due to the age of the machine and the amount of preventive maintenance. Also, assuming that at time s, the only machines available are the ones up to date with respect to the technology prevailing at s, we can subscript these functions by s to reflect the effect of the machine's technology on its state at time k. Let $\Psi_s\{u^k,k\}$ be these functions such that we can write the following state equations.

$$\Delta R^k = R^{k+1} - R^k = \Psi_s[u^k,k], \quad R^s = R_o^s \tag{8.40}$$

and

$$\Delta S^k = \Phi_s[u^k,k], \quad x^s = (1-\delta)C^s \tag{8.41}$$

where δ is the fractional depreciation immediately after purchase of the machine.

To convert the problem into the Mayer form, define:

$$A^k \triangleq \sum_{i=s}^{k-1} R^i(1+\rho)^{-i} \tag{8.42}$$

and

$$B^k \triangleq \sum_{i=s}^{k-1} u^i(1+\rho)^{-i}. \tag{8.43}$$

Using equations (8.42) and (8.43), we can write the optimal control problem as

$$\max_{\{u^k\}} [J_{st} = R^t(1+\rho)^{-t} + A^t - u^t(1+\rho)^{-t} - B^t - C^s(1+\rho)^{-s} + x^t(1+\rho)^{-t}] \tag{8.44}$$

subject to

$$\Delta A^k = R^k(1+\rho)^{-k}, \qquad A^s = 0 \tag{8.45}$$

$$\Delta B^k = u^k(1+\rho)^{-k}, \qquad B^s = 0. \tag{8.46}$$

(8.40), (8.41) and (8.38).

8.3.2 Solution by the Discrete Maximum Principle

We associate the adjoint variables λ_1^{k+1}, λ_2^{k+1}, λ_3^{k+1}, and λ_4^{k+1}, respectively with the state equations (8.45), (8.46), (8.40), and (8.41), and for the Hamiltonian,

$$H = \lambda_1^{k+1}R^k(1+\rho)^{-k} + \lambda_2^{k+1}u^k(1+\rho)^{-k} + \lambda_3^{k+1}\Psi_s + \lambda_4^{k+1}\Phi_s, \tag{8.47}$$

where the adjoint variables λ satisfy the following differential equations and terminal boundary conditions.

$$\Delta\lambda_1^k = \frac{-\partial H}{\partial A} = 0; \qquad \lambda_1^t = 1, \tag{8.48}$$

$$\Delta\lambda_2^k = \frac{-\partial H}{\partial B} = 0; \qquad \lambda_2^t = -1, \tag{8.49}$$

$$\Delta\lambda_3^k = \frac{-\partial H}{\partial R} = \lambda_1^{k+1}(1+\rho)^{-k}; \qquad \lambda_3^t = (1+\rho)^{-t}, \tag{8.50}$$

and

$$\Delta\lambda_4^k = \frac{-\partial H}{\partial x} = 0; \qquad \lambda_4^t = (1+\rho)^{-t}. \tag{8.51}$$

The solutions of these equations are

$$\lambda_1^k = 1, \tag{8.52}$$

$$\lambda_2^k = -1, \tag{8.53}$$

$$\lambda_3^k = \sum_{i=k}^{t} (1+\rho)^{-i}, \tag{8.54}$$

and

$$\lambda_4^k = (1+\rho)^{-t}. \tag{8.55}$$

Note that λ_1, λ_2, and λ_4 are constants for a fixed t.

To apply the maximum principle, we substitute (8.52)-(8.55) into the Hamiltonian (8.47) collect terms containing control u, rearrange and decompose H as:

$$H = H_1 + H_2(u), \tag{8.56}$$

where H_1 is that part of H which is independent of u, and $H_2(u)$ is

$$H_2(u) = -u^k(1+\rho)^{-k} + \sum_{i=k+1}^{t} (1+\rho)^{-i}\Psi_s + (1+\rho)^{-t}\Phi_s. \tag{8.57}$$

Here we apply the maximum principle to obtain the necessary condition for the optimal schedule of preventive maintenance expenditure in dollars. The condition of optimality is that H should be maximum along the optimal control path. If u^k were unconstrained, this condition is equivalent to setting the partial derivative of H, with respect to u equal to zero, that is,

$$H_u = [H_2]_u = -(1+\rho)^{-k} + (\Psi_s)_u \sum_{i=k+1}^{t} (1+\rho)^{-i} + (\Phi_s)_u(1+\rho)^{-t} = 0. \tag{8.58}$$

Equation (8.58) is a functional equation in u with the exception of the particular case (which will be treated later in this section) when Ψ_s and Φ_s are linear in u. In general, (8.58) may or may not have a unique solution. For our case we will assume Ψ_s and Φ_s to be of the form such that we have unique solution for u. One such case occurs where Ψ_s and Φ_s are quadratic in u. In this case, (8.58) is linear in u and can be solved explicitly for a unique solution for u. Whenever a unique solution does exist, let this be

$$u^k = U^k_{st}. \tag{8.59}$$

The optimal control u^{k*} as in model type is:

$$u^{k*} = \begin{cases} 0 , & \text{if } U^k_{st} \leq 0 \\ U^k_{st} , & \text{if } 0 < U^k_{st} < U^{sk} \\ U^{sk} , & \text{if } U^k_{st} \geq U^{sk}. \end{cases} \tag{8.60}$$

8.3.3 Special Case of Bang-Bang Control

This special case occurs when the problem is linear in the control. In this case, H can be maximized simply by having the control at its maximum when coefficient of u in H is positive and minimum when it is negative. This type of control is called bang-bang.

In our problem, if Ψ_s and Φ_s assume the form

$$\Psi_s(u^k,k) = u^k\Psi^k_s \tag{8.61}$$

and

$$\Phi_s(u^k,k) = u^k\Phi^k_s , \tag{8.62}$$

where Ψ^k_s and Φ^k_s are given constants, then the coefficient of u in H, denoted by W, is

$$W^k = -(1+\rho)^{-k} + \Psi^k_s \sum_{i=k+1}^{t} (1+\rho)^{-i} + \Phi^t_0(1+\rho)^{-t} \tag{8.63}$$

and the optimal control u^* is given by

$$u^{k*} = \text{bang}[0, U^{sk}; W^k]. \tag{8.64}$$

8.3.4 Incorporation into Wagner-Whitin Framework for a Complete Solution

First we note that once we have obtained u^*, we can substitute it into (8.40) and (8.41) to obtain R^*, x^*, which in turn can be used in (8.39) to obtain the optimal value of the objective function denoted by J^*_{st}. This can be done for each s and $t \geq s$.

Let g_s denote the present value of the profit (discounted to period 0) of an optimal replacement and preventive maintenance policy for periods $s, s+1, \ldots, T$. Then

$$g_s = \max_{k=s,s+1,\ldots,T} [J^*_{sk} + g_k], \quad 0 \leq s \leq T, \tag{8.65}$$

and

$$g_{T+1} = 0. \tag{8.66}$$

The value of g_o will give the required maximum.

The mixed optimization technique presented here avoids many of the shortcomings of either pure dynamic programming or pure control theory formulations. Since the solution technique used to optimize a given machine represents a submodule of the overall method, the pure dynamic programming approach may be recognized as a special case. It should be advantageous, however, to be able to use methodology for the submodule most efficient for a given particular problem. Previous control theory formulations certainly do not seem to be easily adaptable to the situation of an existing initial machine (see [Sethi and Morton, 1972]) or other similar asymmetries.

This mixed technique can also be adapted to the case of probabilistic technological breakthroughs. Here the path of technological growth is assumed to be a tree with probabilities associated with its branches. The subproblems can be solved by using the maximum principle for stochastic networks given in [Sethi and Thompson, 1977]. However, the number of subproblems to be solved increases rapidly with the number of branches, thus putting computational limitations on the general usefulness of this extension.

Another application of the mixed technique has been used by [Pekelman and Sethi, 1978] to obtain the optimal durations of advertising copies, and the optimal level of advertising expenditures for each copy.

8.3.5 A Numerical Example

To illustrate the procedure, a simple three period example will be presented and solved, for the case when there is no existing machine at time zero.

Machines may be bought at times 0, 1, 2. The cost of a machine bought at time s is assumed to be

$$C_s = 1{,}000 + 500s^2.$$

The discount rate, the fractional instantaneous depreciation at purchase, and the maximum preventive maintenance per period are assumed to be

$$\rho = 0.06, \quad \delta = 0.25, \quad \text{and} \quad U = \$100,$$

respectively.

Let R_s^k be the net return (net of necessary maintenance) of a machine purchased in period s and operated in period k.

We assume

$$R_0^0 = \$600, \quad R_1^1 = \$1{,}000, \quad R_2^2 = \$1{,}100.$$

The incremental return function is given by $\Psi_s(u^k, k)$, which we assume to be linear. Specifically,

$$\Delta R_s^k = \Psi_s(u^k, k) = -d_s + a_s u^k,$$

where,

$$d_1 = 200, \quad d_2 = 50, \quad d_3 = 100, \quad \text{and} \quad a_s = 0.5 + 0.1s^3.$$

This means that the return on a machine purchased in period s goes down by an amount d_s every period if there is no preventive maintenance. This decrease can be offset by an amount proportional to the amount of preventive maintenance.

(Note that the function Ψ_s was assumed to be stationary over time in order to simplify the example.)

Let x_s^k be the salvage value at time k of a machine purchased at s. We assume

$$x_s^s = (1-\delta)C^s = 0.75[1,000 + 500s^2].$$

The incremental salvage value function is given by

$$\Delta x_s^k = -\ell_s C^s + b_s u^k$$

where

$$\ell_s = \begin{cases} 0.1, & \text{when } s = 0,1 \\ 0.2, & \text{when } s = 2, \end{cases}$$

and

$$b_s = (0.5 - 0.05s).$$

i.e., the decrease in salvage value is a constant percentage of the purchase price if there is no preventive maintenance. With preventive maintenance, salvage value can be enhanced by a proportional amount.

Let J_{st}^* be the optimal value of the objective function associated with a machine purchased at s and sold at t. We will now solve for J_{st}^*, $s = 0, 1, 2$, and $s < t \leq 3$, t an integer.

Before we proceed, we will denote by $W(k/s,t)$, the coefficient of u in the Hamiltonian H in (8.57).

$$W(k/s,t) = -(1+\rho)^{-k} + a_s \sum_{i=k+1}^{t} (1+\rho)^{-i} + b_s(1+\rho)^{-t}.$$

The optimal control is given by (8.58).

It is noted in passing that

$$\mathrm{sgn}[W((k+1)/s,t) - W(k/s,t)] = \mathrm{sgn}[\rho - a_s], \tag{8.67}$$

implying that

$$u^{(k+1)*} - u^{k*} \begin{cases} \geq 0 & \text{if } (\rho - a_s) > 0 \\ = 0 & \text{if } (\rho - a_s) = 0 \\ \leq 0 & \text{if } (\rho - a_s) < 0. \end{cases}$$

In this example $\rho - a_s < 0$, which implies that if there is a switching in the optimal control trajectory, the switch must be from \$100 to \$0.

Solution of Sub-Problems

We now solve the subproblems for various values of s and t $(s < t)$ by using the discrete maximum principle.

$s = 0, \quad t = 1$

$$W(0) = -1 + 0.5(1.06)^{-1} + 0.5(1.06)^{-1} < 0.$$

This implies that

$$u^{0*} = 0.$$

Now,

$$R_0^0 = 600.$$

$$R_0^1 = 600 - 200 = 400.$$

$$x_0^0 = 0.75 \times 1{,}000 = 750.$$

$$x_0^1 = 750 - 0.1 \times 1{,}000 = 650.$$

$$J_{01}^* = 600 + 400 \times (1.06)^{-1} - 1{,}000 + 650 \times (1.06)^{-1} = \$590.$$

Similar calculations can be carried out for other subproblems. We will list these results.

s = 0, t = 2

$$u^{0*} = 100, \quad u^{1*} = 0$$

$$J_{02}^{*} = 681.$$

s = 0, t = 3

$$u^{0*} = 100, \quad u^{1*} = 100, \quad u^{2*} = 0$$

$$J_{03}^{*} = 644.$$

s = 1, t = 2

$$u^{1*} = 0$$

$$J_{12}^{*} = 1242.$$

s = 1, t = 3

$$u^{1*} = 100, \quad u^{2*} = 0$$

$$J_{13}^{*} = 1869.$$

s = 2, t = 3

$$u^{2*} = 100$$

$$J_{23}^{*} = 555.$$

Wagner-Whitin Solution of the Entire Problem

With reference to the dynamic programming equation in (8.65) and (8.66) we have

$$g_3 = 0,$$

$$g_2 = J_{23}^* = \$555,$$

$$g_1 = \max [J_{13}^*, J_{12}^* + g_2]$$

$$= \max [1869,\ 1242 + 555]$$

$$= \$1869,$$

$$g_0 = \max [J_{03}^*, J_{01}^* + g_1, J_{02}^* + g_2]$$

$$= \max [644,\ 590 + 1869,\ 681 + 555]$$

$$= \$2,459.$$

Now we can summarize the optimal solution. The optimal number of machines is 2, and their optimal policies are as follows:

First Machine Optimal Policy:

Purchase at $s = 0$; sale at $t = 1$; optimal preventive maintenance policy $u^{0*} = 0$.

Second Machine Optimal Policy:

Purchase at $s = 1$; sale at $t = 3$; optimal preventive maintenance policy $u^{1*} = 100$, $u^{2*} = 0$.

The value of the objective function is $J^* = \$2,459$.

Exercises for Chapter 8

8.1 Change the values of U and $d(t)$ in example 8.1.3 to the new values

$$U = 1/2 \quad \text{and} \quad d(t) = 3$$

and resolve the problem.

8.2 Show for the model in Section 8.1.1 that if it is optimal to have maximum maintenance throughout the life of the machine, then its optimal life T must satisfy $g(T) - 1 \geq 0$. In particular, for the example in Section 8.1.3, show $T \leq 3$.

8.3 Resolve the example in Section 8.1.3 with $x(0) = 40$.

8.4 Replace the maintenance effectiveness function in example 8.1.3 by

$$g(t) = 2/(16 + t)^{1/2}$$

and solve the new problem.

8.5 Let the maintenance effectiveness function in Example 8.1.4 by

$$g(u,t) = \frac{2u^{1/2}}{(1+t)^{1/2}};$$

Solve this problem with $T = 34.8$ and compare its solution to that of Example 8.1.3.

8.6 Derive the formula in (8.33) by using (3.17)

8.7* To show that the singular control in the third alternative in (8.35) can be sustained, we set $H_{ut} = 0$ for all t during which a singular control obtains. That is

$$C_{uu}\dot{u} = C_u[\rho + (1-u)h] + \rho L - R + C(u)h \tag{8.68}$$

must hold on an optimal singular arc. Since $C_{uu} > 0$, $\text{sgn}(\dot{u}) = \text{sgn}$ (R.H.S.). Show that $\dot{u}^* \leq 0$.

8.8 For the model of Section 8.2, prove that the derived Hamiltonian H^0 is concave in F for each given λ and t, so that the sufficiency Theorem 2.1 holds.

8.9 Verify the expression in (8.67).

8.10* Extend the formulation of Sethi-Morton model in Section 8.3 to account for probabilistic technological breakthroughs. (Hint: See [Sethi and Morton, 1972] and [Sethi and Thompson, 1977].

8.11* Extend the Thompson model in Section 8.1 to account for process discontinuities. An example of this type of machine is an aeroplane assigned to passenger transportation which may, after some deterioration or obsolescence, be assigned to freight transportation before its eventual retirement. Formulate and analyze the problem. (Hint: See [Tapiero, 1971].)

8.12* Let us define the state of machine to be '0' if it is working and '1' if it is being repaired. Let λ be the breakdown rate and μ be the service rate as in waiting-line theory, we have

$$\dot{P}_o = -\lambda P_o + \mu(1-P_o), \qquad P_o(0) = 1$$

and $P_1 = 1-P_o$. This equation along with (8.3) gives us two state equations. In view of the equation for $\dot{P}_o$, we modify the objective function (8.2) to

$$J = \int_o^T [PS(t)P_o(t)-u(t)-kP_1(t)]e^{-\rho t}dt + S(T)e^{-\rho T},$$

where k characterizes the additional expenditure rate when the machine is being repaired. Solve this model to obtain the optimal control. (Hint: See [Alam and Sarma, 1974].)

9 APPLICATIONS TO NATURAL RESOURCES

The rapid increase of world population is causing a corresponding increase in the demand for consumption of natural resources. As a consequence the optimal management and utilization of natural resources is becoming increasingly important. There are two main kinds of natural resource models, those involving renewable resources such as fish, food, timber, etc., and those involving non-renewable or exhaustible resources such as petroleum, minerals, etc.

In Section 9.1 we deal with a fishery resource model the sole owner of which is considered to be a regulatory agency. The management problem of the agency is to control the rate of fishing over time so that an appropriate objective function is maximized over an infinite horizon. A differential game extension known as the common property fishery resource model is discussed in Section 12.1.3. Other kinds of renewable resource models involve predator-prey relationships, see [Clark, 1976], and [Goh, Leifman, and Vincent, 1974].

Section 9.2 deals with an optimal forest thinning model, where thinning is the process of removing some trees from the forest, to improve its growth rate and quality. An extension to a chain of forests model is presented in Section 9.2.3.

The final model presented in Section 9.3 deals with an exhaustible resour such as petroleum, which must be utilized optimally over a given horizon under the assumption that a substitute will take the place of the resources when the latter's price becomes too high. Therefore the analysis of this section can also be viewed as a problem of the optimal phasing in of the expensive substitut

9.1 The Sole Owner Fishery Resource Model

With the recent establishment of 200-mile territorial zones in the ocean for most countries having coatlines, the control of fishing in these zones has become highly regulated by these countries. In this sense fishing in territorial waters can be considered as a sole owner fishery problem. On the other hand, if the citizens and commercial fisherman of a given country are permitted to fish freely in their territorial waters, the problem becomes that of an open access fishery. The solutions of these two extreme problems are quite different, as will be shown in this section.

9.1.1 The Dynamics of Fishery Models

We introduce the following notation and terminology which is due to [Clark, 1976]:

ρ = discount rate

$x(t)$ = biomass of fish population at time t

$g(x)$ = natural growth function

$u(t)$ = rate of fishing effort at time t; $0 \leq u \leq U$.

q = catchability coefficient

p = unit price of landed fish

c = unit cost of effort

Assume that the growth function g is differentiable, concave and satisfies:

$$g(0) = 0, \quad g(X) = 0, \quad g(x) > 0 \quad \text{for} \quad 0 < x < X \tag{9.1}$$

where X denotes the "carrying capacity", that is, the maximum sustainable fish biomass.

The model equation (due to[Gordon, 1954] and [Shaefer, 1957]) is:

$$\dot{x} = g(x) - qux, \qquad x(0) = x_0 \tag{9.2}$$

and the instantaneous profit rate is

$$\pi(x,u) = (pqx - c)u \tag{9.3}$$

From (9.1) and (9.2) it follows that x will stay in the closed interval $0 \leq x \leq X$ provided x_0 is in the same interval.

An <u>open access</u> fishery is one in which exploitation is completely uncontrolled. Gordon analyzed this model, also known as the Gordon-Schaefer model, and showed that the fishing effort tends to reach an equilibrium, called <u>bionomic equilibrium</u>, at the level at which total revenue equals total cost. In other words, the so-called economic rent is completely dissipated. From (9.3) and (9.2) this level is simply

$$x_b = \frac{c}{pq}, \quad u_b = \frac{g(x_b)p}{c} \tag{9.4}$$

(We assume $u_b \leq U$.) The economic basis for this result is as follows: If the fishing effort $u > u_b$ is made then total costs exceed total revenues so that at least some fishermen will lose money, and eventually some will drop out, thus reducing the level of fishing effort. On the other hand, if fishing effort $u < u_b$ is made, then total revenues exceed total costs, thereby attracting additional fisherman, and increasing the fishing effort.

The Gordon-Schaefer model is a static model which does not (in general) maximize the present value of the total profits which can be obtained from the fish resources.

9.1.2 The Sole Owner Model

The bionomic equilibrium solution obtained from open access fishery model usually implies severe biological overfishing. Suppose a fishing regulatory agency is established to improve the operation of the fishing industry. In determining the objective of the agency it is convenient to think of it as a sole owner who has complete rights to exploit the fishing resource. It is reasonable to assume that the agency model attempt to

$$\text{Maximize}\{J = \int_0^\infty e^{-\rho t}(pqx - c)u\, dt\} \tag{9.5}$$

subject to (9.2). This is the optimal control problem to be solved.

9.1.3 Solution by Green's Theorem

The solution method presented in this section generalizes the Green's theorem procedure used in Section 6.2.2. Solving (9.2) for u we obtain

$$u = \frac{g(x) - \dot{x}}{qx} \tag{9.6}$$

which we substitute into (9.3), giving

$$J = \int_0^\infty e^{-\rho t}(pqx - c)\left(\frac{g(x) - \dot{x}}{qx}\right)dt \quad . \tag{9.7}$$

Rewriting, we have

$$J = \int_0^\infty e^{-\rho t}[M(x) + N(x)\dot{x}]\,dt \tag{9.8}$$

where

$$N(x) = -p + \frac{c}{qx} \quad \text{and} \quad M(x) = \left(p - \frac{c}{qx}\right)g(x) \quad . \tag{9.9}$$

We note that we can write $\dot{x}dt = dx$ so that (9.8) becomes the following line integral

$$J_B = \int_B [e^{-\rho t}M(x)dt + e^{-\rho t}N(x)dx] \tag{9.10}$$

where B is a state trajectory in (x,t) space, $t \in [0,\infty)$.

In this section we are only interested in the infinite horizon solution. The Green's theorem method achieves such a solution by first solving a finite horizon problem as in Section 6.2.2, and then determining the infinite horizon solution for which you are asked to verify that the maximum principle holds in Exercise 9.1. See also[Sethi, 1977a].

In order to apply Green's theorem to (9.10) let Γ denote a simple closed curve in the (x,t) space surrounding a region R in the space. Then

$$J_\Gamma = \oint [e^{-\rho t}M(x)dt + e^{-\rho t}N(x)dx]$$

$$= \iint_R \{\frac{\partial}{\partial t}[e^{-\rho t}N(x)] - \frac{\partial}{\partial x}[e^{-\rho t}M(x)]\}dt\, dx$$

$$= \iint_R -e^{-\rho t}[\rho N(x) + M'(x)]dt\, dx \quad . \tag{9.11}$$

If we let

$$I(x) = -[\rho N(x) + M'(x)],$$

we can rewrite (9.11) as

$$J_\Gamma = \iint_R e^{-\rho t} I(x)dt\, dx \quad .$$

We can now conclude, as we did in Section 6.2.2 and 6.2.4, that the turnpike level $\bar{x}$ is given by setting the integrand of (9.11) to 0; that is

$$-I(x) = [g'(x) - \rho](p - \frac{c}{qx}) + \frac{cg(x)}{qx^2} = 0 \quad . \tag{9.12}$$

In addition, a second order condition must be satisfied for the solution $\bar{x}$ of (9.12) to be a turnpike solution, see Lemma 6.1 and the subsequent discussion there. The required second order condition can be stated as:

$$I(x) < 0 \text{ for } x < \bar{x} \text{ and } I(x) > 0 \text{ for } x > \bar{x} \quad . \tag{9.12a}$$

Let $\bar{x}$ be the unique solution to (9.12) satisfying the second order condition. (The procedure can be extended to the case of nonunique solutions, see [Sethi, 1977a].)

The corresponding value $\bar{u}$ of the control which maintains $\bar{x}$ is $g(\bar{x})/q\bar{x}$. In Exercise 9.2 you are asked to show that $\bar{x} \in (x_b, X)$ and also that $\bar{u} < U$. In Figure 9.1 optimal trajectories are shown for two different initial values, $x_0 < \bar{x}$ and $x_0 > \bar{x}$.

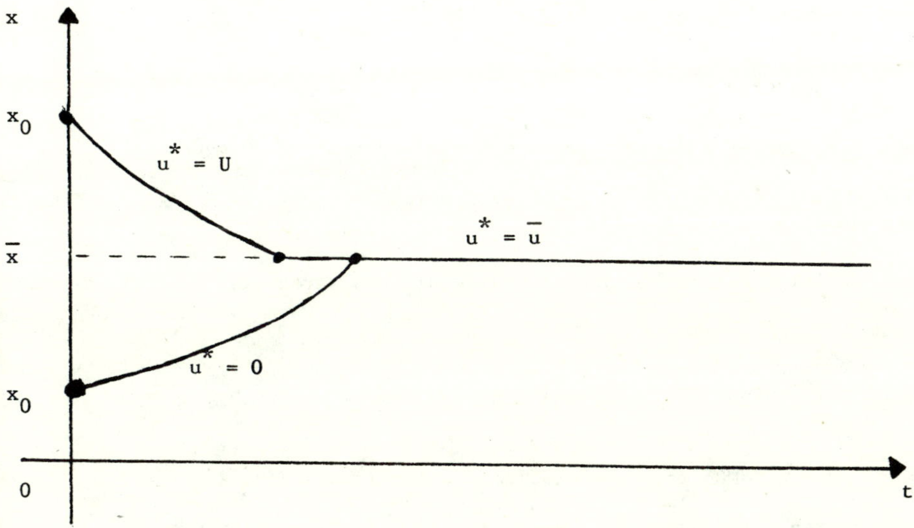

Figure 9.1

Equation (9.12) has an important economic interpretation. To see it we rewrite it as

$$\frac{d\pi(x)}{dx} = \rho\left(\frac{pqx - c}{qx}\right) \tag{9.12b}$$

where

$$\pi(x) = \frac{g(x)(pqx - c)}{qx} .$$

The interpretation of $\pi(x)$ is that it is the sustainable economic rent at fish stock level x; this can be seen by substituting $u = \frac{g(x)}{qx}$ into (9.3), which is required to maintain the fish stock at level x (see (9.2)). Suppose we have attained the equilibrium level $\bar{x}$ given by (9.12), and suppose we reduce this level to $\bar{x} - \Delta x$ by using fishing effort of $\Delta x/qx$. The immediate marginal revenue, MR, from this action is

$$MR = (pq\,x - c)\,\frac{\Delta x}{qx} \quad .$$

However, this causes a decrease in the sustained rent which equals

$$\pi'(x)\Delta x.$$

Over the infinite future the present value of this stream is, i.e. the marginal cost, MC, is

$$MC = \int_0^{\infty} e^{-\rho t}(\pi'(x)\Delta x)dt = \frac{\pi'(x)\Delta x}{\rho} \quad .$$

Equating MR and MC we obtain (9.12b) which is also (9.12).

When the discount rate is zero, equation (9.12b) reduces to

$$\pi'(x) = 0$$

thus it will give the equilibrium fish stock level $\bar{x}(\rho = 0)$ which maximizes the instantaneous profit rate $\pi(x)$. This is called in economics the golden rule level.

When $\rho = \infty$, we can assume that $\pi'(x)$ is bounded. From (9.12b) we have $pqx - c = 0$ which gives

$$\bar{x}(\rho = \infty) = x_b = \frac{c}{pq} \quad .$$

The latter is the bionomic equilibrium attained in the open access fishery solution, see (9.4).

The sole owner solution $\bar{x}$ satisfies $\bar{x} > x_b = c/pq$. If we regard a government regulatory agency as the sole owner responsible for operating the fishery at level $\bar{x}$, then it can impose restrictions, such as gear regulations, catch limitations, etc., which increase the fishing cost c. If c is increased to the level $pq\bar{x}$, then the fishery can be turned into an open access fishery subject to those regulations, and it will attain the bionomic equilibrium at level $\bar{x}$.

9.2 An Optimal Forest Thinning Model

Forests are another important kind of renewable natural resources, and their optimal management is becoming a significant current problem. In [Kilkki and Vaisanan, 1969] a model is developed for forest growth and thinning in connection with Scotch Pine forests in Finland. Thinning is the process of removing some but not all trees prior to clear cutting the forest. Besides yielding a harvest of wood, the thinning process also improves the growth rate and quality of the forest. The solution method employed by Kilkki and Vaisanan was based on dynamic programming. We shall use the maximum principle approach to solve the model, [Clark, 1976].

9.2.1 The Forestry Model

We introduce the following notation:

t_0 = initial age of the forest

ρ = discount rate

$x(t)$ = volume of usable timber in the forest at time t

$u(t)$ = rate of thinning at time t

p = constant price per unit volume of timber

c = constant cost per unit volume of thinning

$f(x)$ = growth function, which is positive, concave, and has a unique maximum at x^m; we assume $f(0) = 0$.

$g(t)$ = growth coefficient which is positive decreasing function of time .

Examples of specific function forms used in [Kilkki and Vaisanan, 1969] are as follows:

$$f(x) = xe^{-\alpha x}, \quad 0 \le x \le \frac{2}{\alpha}$$

where α is a positive constant. Note that f is concave in the relevant range and that $x_m = \frac{1}{\alpha}$;

$$g(t) = at^{-b}$$

where a and b are positive constants.

The forest growth equation is

$$\dot{x} = g(t)f(x) - u(t), \quad x(t_0) = x_0 \quad . \tag{9.13}$$

The objective function is

$$\text{Maximize}\{J = \int_{t_0}^{\infty} e^{-\rho t}(p-c)udt\} \tag{9.14}$$

subject to (9.13), the state constraint

$$x(t) \ge 0 \tag{9.15}$$

and the control constraint

$$u(t) \ge 0. \tag{9.16}$$

Constraint (9.16) implies that there is no replanting in the forest. In Section 9.2.3 we extend this model to incorporate the successive replantings of the forest each time it is clear cut.

9.2.2 Determination of Optimal Thinning

We solve the model by using the maximum principle. The Hamiltonian is

$$H = (p-c)u + \lambda[gf(x) - u] \tag{9.17}$$

with the adjoint equation

$$\dot{\lambda} = \lambda[\rho - gf'(x)] \quad . \tag{9.18}$$

The optimal control is

$$u^* = \text{bang}[0,\infty;\ p-c+\lambda] \quad . \tag{9.19}$$

The appearance of ∞ as the upper bound in (9.19) simply means that impulse control is permitted.

We did not use the Lagrangian form of the maximum principle to include (9.15) because, as we shall see, the forestry problem has a natural ending at a time T for which $x(T) = 0$.

To get the singular control solution triple $\{\bar{x},\bar{\lambda},\bar{u}\}$ we must observe that $\bar{x}$ and $\bar{u}$ will be functions of time. From (9.19)

$$\bar{\lambda} = c - p \tag{9.20}$$

which is a constant, so that $\dot{\bar{\lambda}} = 0$. From (9.18),

$$f'(\bar{x}(t)) = \frac{\rho}{g(t)} \tag{9.21}$$

which can be solved for $\bar{x}(t)$. Then from (9.13)

$$\bar{u}(t) = g(t)f(\bar{x}(t)) - \dot{\bar{x}}(t) \tag{9.22}$$

gives the singular control.

The solution of (9.21) can be illustrated as in Figure 9.2. Since $g(t)$ is decreasing function of time, it is clear from Figure 9.2 that $\bar{x}(t)$ is a

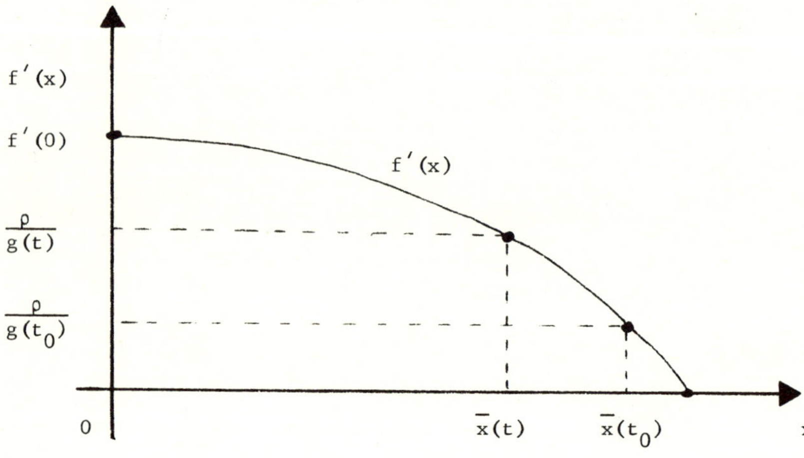

Figure 9.2

decreasing function of time, it is clear from Figure 9.2 that $\bar{x}(t)$ is a decreasing function of time, and then by (9.22) $\bar{u}(t) \geq 0$. It is also clear that $\bar{x}(\hat{T}) = 0$ at time $\hat{T}$ given by

$$\frac{\rho}{g(\hat{T})} = f'(0) \quad . \tag{9.23}$$

In Figure 9.3 we plot $\bar{x}(t)$ as a function of time t. The figure also contains an optimal control trajectory for a case in which $x_0 < \bar{x}(t_0)$. To determine the time $\hat{t}$ we first solve (9.13) with $u = 0$; let $x(t)$ be the solution. Then $\hat{t}$ is the time at which the $x(t)$ trajectory intersects the $\bar{x}(t)$ curve, see Figure 9.3. For $x_0 > \bar{x}(t_0)$, the optimal control at t_0 will be an impulse cutting to bring the level from x_0 to $\bar{x}(t_0)$ instantaneous. To complete the infinite horizon solution, $u^*(t) = 0$ for $t \geq \hat{T}$. In Exercise 9.10 you are asked to obtain $\lambda(t)$ for $t \in [0,\infty)$.

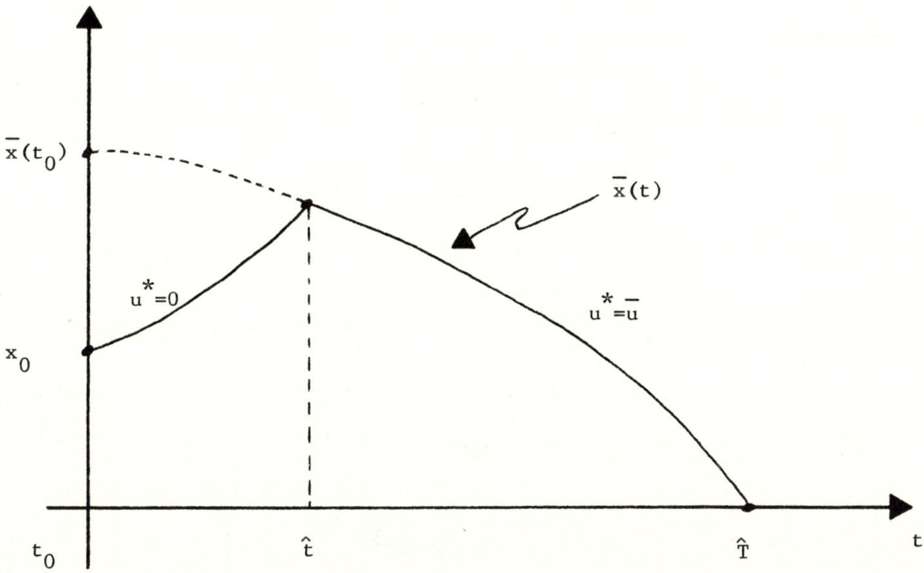

Figure 9.3

9.2.3 Chain of Forests Model

We now extend the model of Section 9.2.1 to incorporate successive replantings of the forest each time it is clearcut. This extension is similar to the chain of machines model of Section 8.3. We shall assume that successive plantings, sometimes called forest rotations, take place at equal intervals; a similar assumption was referred to in (8.38) for the machine replacement problem.

Let T be the rotation period, i.e., the time from planting to clear-cutting, which is to be determined. During the nth rotation the dynamics of the forest is given by (9.13) with $t \in [(n-1)T, T]$ and $x[(n-1)T] = 0$. The objective function is given by

$$\text{Maximize}\{J = \sum_{k=1}^{\infty} e^{-k\rho T} \int_0^T e^{-\rho t} (p-c)u\,dt$$

$$= \frac{1}{1-e^{-\rho T}} \int_0^T e^{-\rho t}(p-c)u\,dt\} \; . \tag{9.24}$$

From the solution of the model of the previous section, and the assumption that the forest is profitable, it is obvious that $0 \le T \le \hat{T}$ as shown in Figure 9.4. We have two cases to consider, depending on whether $T \ge \hat{t}$ or $T < \hat{t}$.

Case 1. $T \ge \hat{t}$. From the preceding section it is easy to conclude that the optimal trajectory is as shown in Figure 9.4. Using the turnpike terminology of Chapter 6, the trajectory from 0 to A is the entry ramp to the turnpike, the trajectory from A to B is on the turnpike, and the trajectory from B to T is the exit ramp. With this solution we can write the $J^*(T)$ of (9.24)

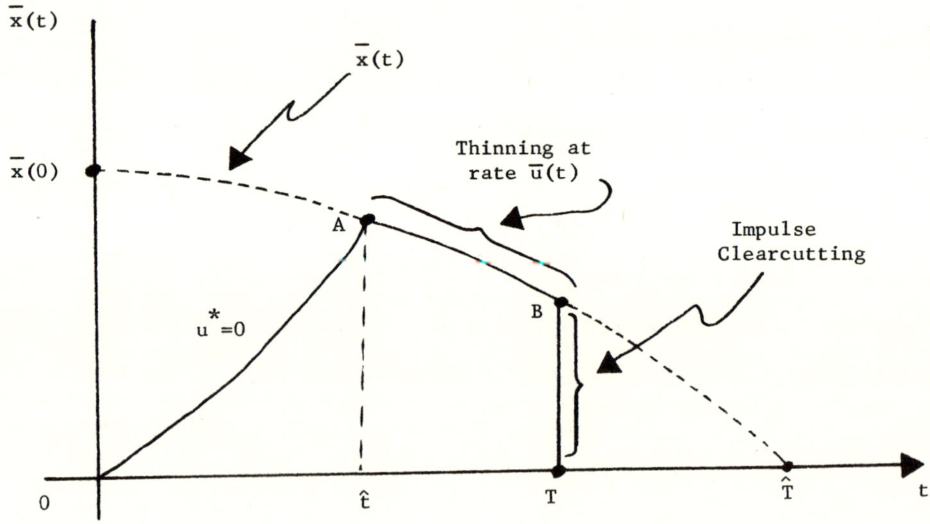

Figure 9.4

for a given T as

$$J^*(T) = \frac{1}{1-e^{-\rho T}}\left[\int_{\hat{t}}^{T-} e^{-\rho t}(p-c)\bar{u}dt + \int_{T^-}^{T} e^{-\rho t}(p-c)u^*(t)dt\right]$$

$$= \frac{1}{1-e^{-\rho T}}\left[\int_{\hat{t}}^{T} e^{-\rho t}(p-c)\bar{u}(t)dt + (p-c)\bar{x}(T)e^{-\rho T}\right] . \qquad (9.25)$$

Note that $u^* = I[\bar{x}(T),0]$ in the second integral is an impulse control bringing the forest from value $\bar{x}(t)$ to 0 by a clearcutting operation, see Exercise 9.11. To find the optimal value of T for this case we differentiate (9.25) with respect to T, equate the result to zero and simplify, obtaining (see Exercise 9.12)

$$(1-e^{-\rho T})g(T)f[\bar{x}(T)] - \rho\bar{x}(T) - \rho\int_{\hat{t}}^{T} e^{-\rho t}\bar{u}(t)dt = 0. \qquad (9.26)$$

If the solution T lies in $[\hat{t},\hat{T}]$ keep it, otherwise set $T = \hat{t}$.

<u>Case 2</u>. $T \leq \hat{t}$. The optimal trajectory in this case is as shown in Figure 9.5. In the turnpike terminology of Chapter 6, this case occurs when T is small, compare with Figure 6.10. The solution for $x(T)$ is obtained by integrating (9.13) with $u = 0$ and $x_0 = 0$. Let this solution be denoted as $x^*(t)$. Here (9.24) becomes

$$J^*(T) = \frac{e^{-\rho T}}{1-e^{-\rho T}} (p-c)x^*(T) . \qquad (9.27)$$

To find the optimal value of T for this case we differentiate (9.27), and equate to zero. We obtain

$$(e^{\rho T}-1)g(T)f[x^*(T)] - \rho e^{\rho T}X^*(T) = 0 . \qquad (9.28)$$

If the solution lies in the interval $[0,\hat{t}]$ keep it, otherwise set $T = \hat{t}$.

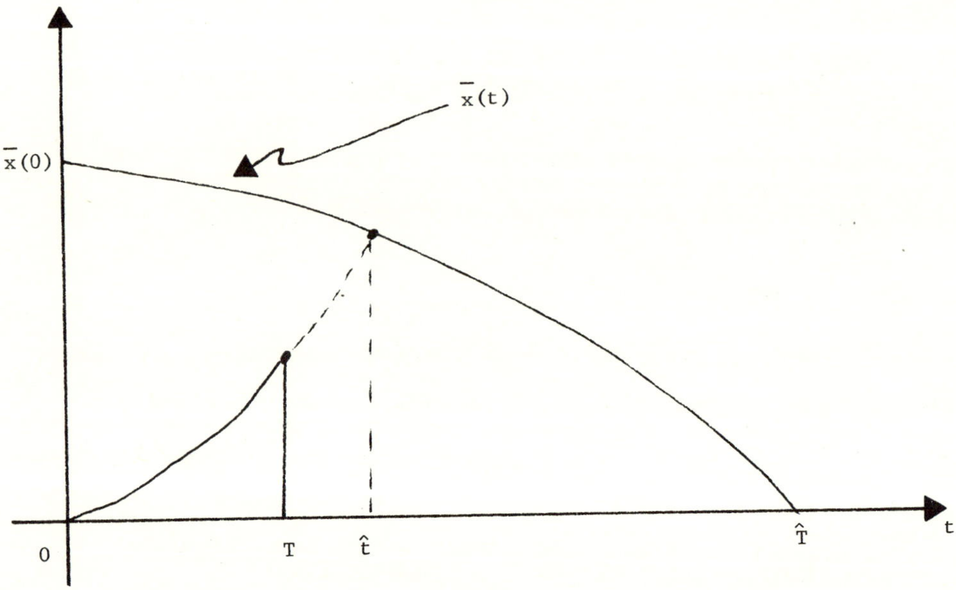

Figure 9.5

The optimal value T^* can be obtained by computing $J^*(T)$ from both cases and selecting whichever is larger, see also [Näslund, 1969] and [Sethi, 1973b].

9.3 An Exhaustible Resource Model

In the previous two sections we discussed two renewable resource models. However, many natural resources are nonrenewable or exhaustible. Examples are petroleum, mineral deposits, coal, etc. Given the growing energy shortage, the earliest optimal use of these resources is of immense importance to the world. The important work in this area is due to [Hotelling, 1931]. Since then a numb of studies have been published such as: [Dasgupta and Heald, 1974], [Solow, 19 [Weinstein and Zeckhauser, 1975], and [Manh-Hung, 1974], [Pindyek, 1978, 1978a see also other papers in the Review of Economic Studies, 1974.

In this section we discuss a simple model taken from [Sethi, 1979]. This paper analyzes optimal depletion rates by maximizing a social welfare function which involves consumer and producer surplus with various weights. Here we select a model having the equally weighted criterion function.

9.3.1 Formulation of the Model

The model will be developed under the assumption that, at a high enough price, say $\bar{p}$, a substitute, preferably renewable, will become available. For example, if the price of fossil fuel becomes sufficiently high, solar energy may become an economic substitute. In the North American context the resource under consideration could be crude oil and its expensive substitute could be coal and/or tar sands.

We introduce the following notation:

$p(t)$ = price of the natural resource at time t

q = $f(p)$ is the demand function, i.e., the quantity demanded at price p; $f' \leq 0$; $f(p) > 0$ for $p < \bar{p}$ and $f(p) = 0$ for $p \geq \bar{p}$, where $\bar{p}$ is the price at which the substitute completely replaces the natural resource. A typical graph of the demand function is shown in Figure 9.6.

c = $G(q)$ is the cost function; $G(0) = 0$, $G(q) > 0$ for $q > 0$, $G' > 0$, and $G'' \geq 0$ for $q \geq 0$, and $G'(0) < \bar{p}$. The latter assumption makes it possible for the producers to make positive profit at some prices p below $\bar{p}$.

$Q(t)$, the available stock or reserve of the natural resource at at time t, $Q(0) = Q_0 > 0$.

$\rho > 0$, the social discount rate

$T > 0$, the horizon time which is the latest time at which the substitute will become available regardless of the price of the natural resource.

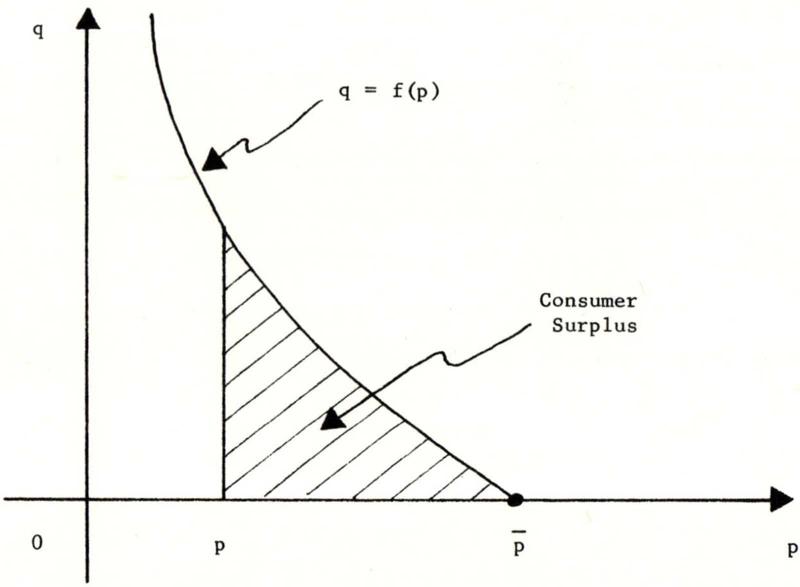

Figure 9.6 Demand Function

Before stating the optimal control problem, we need the following additional definitions and assumptions. Let

$$c = G[f(p)] = g(p) \tag{9.29}$$

for which it is obvious that $g(p) > 0$ for $p < \bar{p}$ and $g(p) = 0$ for $p \geq \bar{p}$. Let

$$\pi(p) = p\ f(p) - g(p) \tag{9.30}$$

denote the profit function of the producers, that is, the producer's surplus. Let $\underline{p}$ be the smallest price at which $\pi(p)$ is nonnegative. Assume further that $\pi(p)$ is a concave function in $[\underline{p},\bar{p}]$ and is shown in Figure 9.7. In the figure the point p^m indicates the price which maximizes $\pi(p)$.

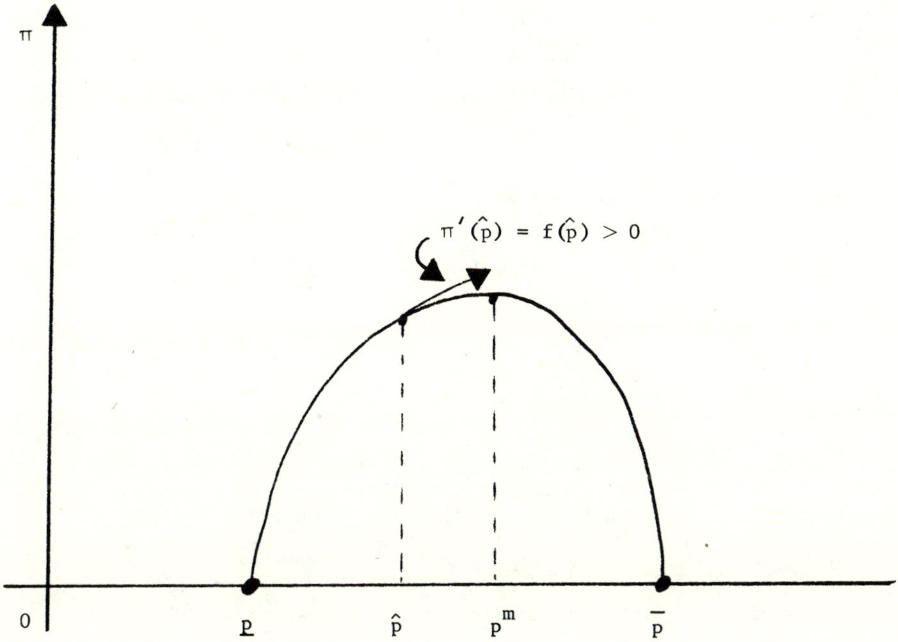

Figure 9.7 Profit Function

We also define

$$\Psi(p) = \int_{p}^{\bar{p}} f(y)dy \tag{9.31}$$

as the consumer surplus, i.e., the area shown shaded in Figure 9.6. This quantity represents the total excess amount the consumers would be willing to pay. In other words, consumers pay pq, while they would be willing to pay $pq + \Psi(p)$.

The instantaneous rate of consumer surplus and producer surplus is $\Psi(p) + \pi(p)$; let $\hat{p}$ denote the maximum of this sum, that is, $\hat{p}$ solves

$$\Psi'(\hat{p}) + \pi'(\hat{p}) = \hat{p}f'(\hat{p}) - g'(\hat{p}) = 0 \quad . \tag{9.32}$$

In Exercise 9.14 you will be asked to show that $\hat{p} < p^m$, as marked in Figure 9.7. Later we will show that the correct second order conditions hold at $\hat{p}$.

The optimal control problem is

$$\text{Maximize}\{J = \int_0^T [\Psi(p) + \pi(p)]\, e^{-\rho t}dt\} \tag{9.33}$$

Subject to

$$\dot{Q} = -f(p), \quad Q(0) = Q_0 \tag{9.34}$$

$$Q(T) \geq 0 \tag{9.35}$$

and $p \in \Omega = [\underline{p},\bar{p}]$. Note that $\Psi(p) + \pi(p)$ is a concave function.

9.3.2 Solution by the Maximum Principle

Forming the current value Hamiltonian

$$H(Q,p,\lambda) = \Psi(p) + \pi(p) + \lambda[-f(p)] \tag{9.36}$$

with λ satisfying

$$\dot{\lambda} = \rho\lambda, \quad \lambda(T) \geq 0, \quad \lambda(T)Q(T) = 0 \tag{9.37}$$

which imply

$$\lambda(t) = \begin{cases} 0 & , \text{ if } Q(T) \geq 0, \text{ not binding} \\ \lambda(T)e^{\rho(t-T)}, & \text{if } Q(T) \geq 0 \text{ is binding} \end{cases} \quad . \tag{9.38}$$

To obtain the optimal control, the Hamiltonian maximizing condition, which is both necessary and sufficient in this case (see Theorem 2.1) is

$$\frac{\partial H}{\partial p} = \Psi' + \pi' - \lambda f' = (p-\lambda)f' - g' = 0 \ . \tag{9.39}$$

To show that the solution $s(\lambda)$ to (9.39) actually maximizes the Hamiltonian, it is enough to show that the second derivative of the Hamiltonian is negative at $s(\lambda)$. Differentiating (9.39) gives

$$\frac{\partial^2 H}{\partial p^2} = f' - g'' + (p-\lambda)f'' .$$

Using (9.39) we have

$$\frac{\partial^2 H}{\partial p^2} = f' - g'' + \frac{g'}{f'} f'' . \tag{9.40}$$

From the definition of G in (9.29),

$$G'' = \frac{f'g'' - g'f''}{f'^3}$$

which when substituted into (9.40) gives

$$\frac{\partial^2 H}{\partial p^2} = f' - G'' f'^2 . \tag{9.41}$$

The right hand side of (9.41) is < 0 because $f' < 0$, and $G'' \geq 0$ by assumption. We remark that $\hat{p} = s(0)$, and hence the second order condition for $\hat{p}$ of (9.32) to give the maximum of H is verified. In Exercise 9.15 you are asked to show that $s(\lambda)$ increases from $\hat{p}$ as λ increases from 0, and that $s(\lambda) = \bar{p}$ when $\lambda = \bar{p} - G'(0)$.

Case (i). The constraint $Q(T) \geq 0$ is not binding. From (9.38), $\lambda(t) \equiv 0$ so that from (9.39) and (9.32),

$$p^* = \hat{p} . \tag{9.42}$$

With this value, the total consumption of the natural resource is $T \cdot f(\hat{p})$, which must be $\leq Q_0$ so that the constraint $Q(T) \geq 0$ is not binding.

Hence,

$$T \cdot f(\hat{p}) \leq Q_0 \tag{9.43}$$

characterizes Case (i) and its solution (9.42).

<u>Case (ii)</u>. $T \cdot F(\hat{p}) > Q_0$, so that the constraint $Q(T) \geq 0$ is binding. To find the solution requires finding a value of $\lambda^*(T)$ such that

$$\int_0^{t^*} f[s\{\lambda^*(T)e^{\rho(t-T)}\}]\,dt = Q_0 \tag{9.44}$$

where

$$t^* = \text{Minimum}\{T,\ T + \frac{1}{\rho}\ell n[\ \frac{\bar{p} - G'(0)}{\lambda^*(T)}\ ]\}\ . \tag{9.45}$$

The time t^*, if it is less than T, is the time at which $s[\lambda^*(T)e^{\rho(t^*-T)}] =$

From Exercise 9.15

$$\lambda^*(T)e^{\rho(t^*-T)} = \bar{p} - G'(0) \tag{9.46}$$

which, when solved for t^*, gives the second argument of (9.45).

One method to obtain the optimal solution is to define $\bar{T}$ as the longest time horizon during which the natural resource can be optimally used. Such a $\bar{T}$ must satisfy

$$\lambda^*(\bar{T}) = \bar{p} - G'(0)$$

and therefore

$$\int_0^{\bar{T}} f[s\{\bar{p} - G'(0)\}e^{\rho(t-\bar{T})}]\,dt = Q_0 \tag{9.47}$$

which is a trancendental equation for $\bar{T}$. We now have two subcases.

<u>Case (ii)(a)</u>. $T \geq \bar{T}$. The optimal control is

$$p^*(t) = \begin{cases} s[\{\bar{p} - G'(0)\}e^{\rho(t-\bar{T})}] & \text{for } t \le \bar{T} \\ \bar{p} & \text{for } t > \bar{T} \end{cases} \quad . \tag{9.48}$$

Clearly, in this subcase, $t^* = \bar{T}$ and

$$\lambda^*(T) = [\bar{P} - G'(0)]e^{-\rho(\bar{T}-T)} \quad .$$

A sketch of (9.48) is shown in Figure 9.8.

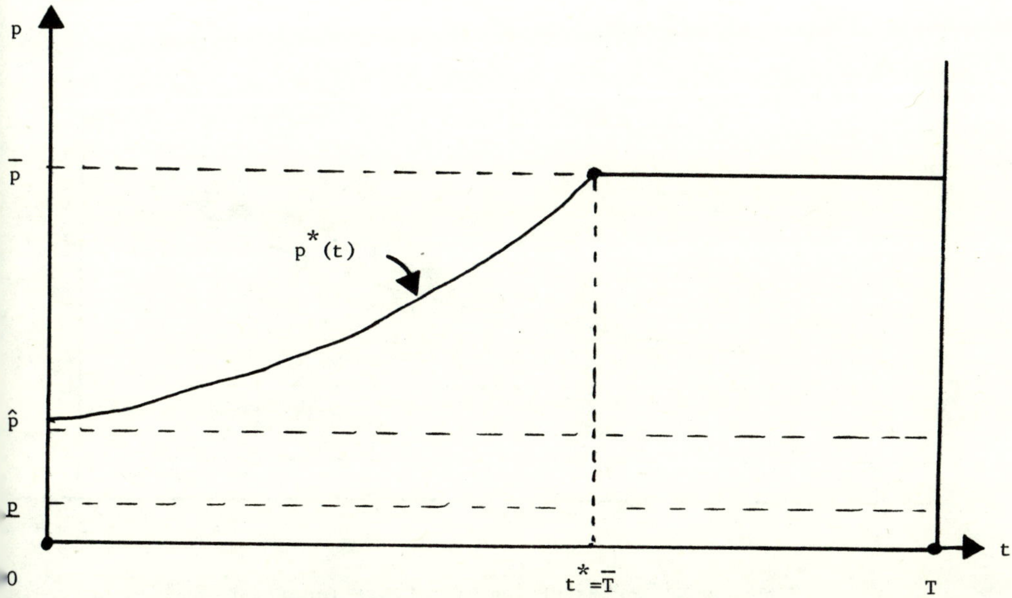

Figure 9.8 Optimal Price Trajectory for $T \ge \bar{T}$

Case (ii)(b). $T < \bar{T}$. Here the optimal price trajectory is

$$p^*(t) = s[\lambda^*(T)e^{\rho(t-T)}] \tag{9.49}$$

where $\lambda^*(T)$ is to be obtained from the trancendental equation

$$\int_0^T f[s\{\lambda^*(T)e^{\rho(t-T)}\}]dt = Q_0 \quad . \tag{9.50}$$

A sketch of (9.49) is shown in Figure 9.9.

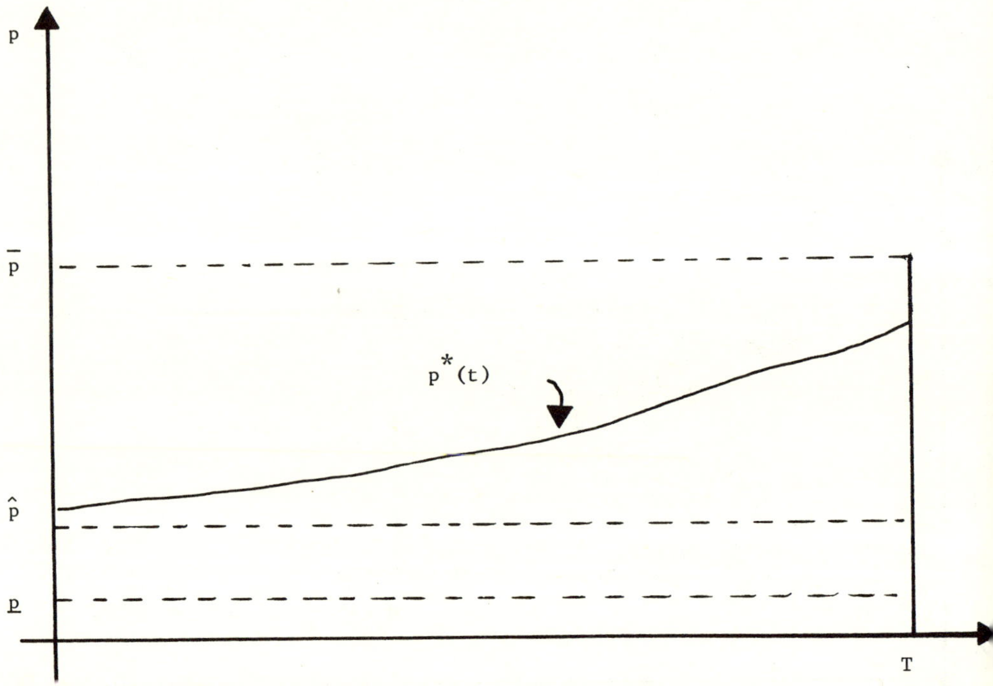

Figure 9.9 Optimal Price Trajectory for $T < \bar{T}$

In Exercise 9.16 you are given specific functions for the exhaustible resource model and asked to work out explicit optimal price trajectories for the model.

Exercises for Chapter 9

9.1 As an alternate derivation for the $\bar{x}$ of (9.12), use the maximum principle to obtain the optimal long run stationary equilibrium triple $\{\bar{x},\bar{u},\bar{\lambda}\}$.

9.2 Prove that $\bar{x} \in (x_b,X)$ and $\bar{u} < U$, where $\bar{x}$ is the solution of (9.12) and x_b is given in (9.4).

9.3 Obtain the $\bar{x}$ of (9.12) for the special case $g(x) = x(1-x)$, $p = 2$, $c = q = 1$, and $\rho = .1$.

9.4 (a) For the Schaefer model with $g(x) = rx\,(1 - \frac{x}{X})$ and $q = 1$, derive a formula for $\bar{x}$ of (9.12).

(b) [Allen, 1973] and [Clark, 1976] estimated the parameters of the Schaefer model for the Antarctic fin-whale population as: $r = .08$, $X = 400{,}000$ whales, and $x_b = 40{,}000$. Solve for $\bar{x}$ for $\rho = 0, .10$, and ∞.

9.5 Let $\pi(x,u) = [p - c(x)](qux)$ in (9.3), where $c(x)$ is a differentiable decreasing and convex. Derive an expression for $\bar{x}$ corresponding to (9.12).

9.6* Show that extinction is optimal if $p \geq c(0) < \infty$ and $\rho > 2g'(0)$ in Exercise 9.5. [Hint. Use the generalized mean value theorem.]

9.7 Let the constant price p in Exercise 9.5 be replaced by a time dependent price $p(t)$ which is differentiable. Derive an expression for $\bar{x}$ corresponding to (9.12), for this nonautonomous problem.

9.8 When $c(x) = 0$ in Exercise 9.7 show that the analogue to (9.12) reduces to

$$g'(x) = \rho - \frac{\dot{p}}{p}.$$

Give an economic interpretation of this equation.

9.9 Let $\pi(x,u)$ of Exercise 9.5 be

$$\pi(x,u) = [p - c(x)](qux) + V(x)$$

where $V(x)$, $V'(x) > 0$, is the <u>conservation value function</u> which measures the value to society of having a large fish stock. By deriving the analogue to (9.12) show that the new $\bar{x}$ is larger than the $\bar{x}$ in Exercise 9.5.

9.10 Find $\lambda(t)$, $t \in [0,\infty)$ for the infinite horizon model of Section 9.2.2.

9.11 Derive (9.25) by computing the integral from T^- to T.

9.12 Derive (9.26) by maximizing $J^*(T)$ of (9.25) with respect to T.

9.13 <u>Forest Fertilization Model</u> ([Näslund, 1969]).

Consider a forestry model in which thinning is not allowed, and the fore is to be clearcut at fixed time T. Suppose $v(t)$ is the rate of fertilization at time t, so that the growth equation is

$$\dot{x} = r(X-x) + f(v,t), \quad x(0) = x_0$$

where x is the volume of timber, r and X are positive constants, and f is an increasing, differentiable, concave function of v. The objective function is

$$\text{Maximize}\left\{J = -c\int_0^T e^{-\rho t}v(t)dt + e^{-\rho t}px(T)\right\}$$

where p is the price of a unit of timber and c is the unit cost of fertilizer.

Show that the optimal control $v^*(t)$ is given by solving the equation

$$\frac{\partial f}{\partial v} = \frac{c}{p} e^{-(\rho+\lambda)(t-T)}.$$

9.14 Show that $\hat{p}$ defined in (9.32) satisfies

$$\underline{p} \leq \hat{p} < p^m.$$

9.15 Show that $s(\lambda)$, the solution of (9.36) increases from $\hat{p}$ as λ increases from 0. Also show that $s(\lambda) = \bar{p}$ when $\lambda = \bar{p} - G'(0)$.

9.16 For the model of Section 9.3 assume

$$f(p) = \begin{cases} \bar{p}-p & \text{for } p \leq \bar{p} \\ 0 & \text{for } - > \bar{p} \end{cases}$$

$$G(q) = q^2.$$

(a) Show that $p^* = \frac{2}{3}\bar{p}$ if $T \leq \frac{3Q_0}{\bar{p}}$.

(b) Show that $\bar{T}$ satisfies $\bar{T} + \frac{e^{-\rho\bar{T}}}{\rho} = \frac{1}{\rho} + \frac{3Q_0}{\bar{p}}$ and

$$p^*(t) = \begin{cases} \frac{\bar{p}(e^{\rho(t-\bar{T})}+2)}{3} & \text{if } t \leq \bar{T} \\ \bar{p} & \text{if } t > \bar{T} \end{cases}$$

for $T \geq \bar{T}$, and

$$p^*(t) = \frac{2\bar{p}}{3} + \frac{\rho[\bar{p}T - 3Q_0]}{3e^{-\rho t}(e^{\rho T}-1)}$$

for $T > \bar{T}$.

10 SOME ECONOMIC APPLICATIONS

Optimal control theory has been extensively applied to the solution of economic problems since the papers in [Shell, 1967] and [Arrow, 1968] and [Shell 1969]. The field is too vast to be surveyed in detail here, however, several books in the area are: [Arrow and Kurz, 1970], [Hadley and Kemp, 1971], [Lesourne, 1973], and [Kamien and Schwartz, 1979]. We content ourselves with th discussion of three simple kinds of models.

Section 10.1 considers two capital accumulation models, i.e., economic growth models. In Section 10.2 we formulate and solve an epidemic control model. Finally we discuss in Section 10.3 a pollution control model.

10.1 Models of Optimal Economic Growth

In this section we develop two simple models of economic growth or capital accumulation. The earliest such model was developed by [Ramsey, 1928] for an economy having stationary population; see Exercise 10.8 for one of his models.

The first model treated in Section 10.1.1 is a finite horizon fixed end point model with stationary population. The problem is that of maximizing the present value of utility of consumption for the society and also accumulate a specified capital stock by the end of the horizon.

The second model incorporates an exogenously and exponentially growing population in the infinite horizon setting. The method of the phase diagram is used to analyze the model.

For related discussions and extensions of these models see [Arrow and Kurz, 1970], [Burmeister and Dobell, 1970], and [Intrilligator, 1971].

10.1.1 An Optimal Capital Accumulation Model

Consider a simple economy in which the stock of capital, denoted by $K(t)$, is the only factor of production. Let $F(K)$ be the output rate of the economy when K is the capital stock. Assume $F(0) = 0$, $F(K) > 0$, $F'(K) > 0$, and $F''(K) < 0$ for $K > 0$. The latter implies the diminishing marginal productivity of capital. This output can either be consumed or be reinvested for further accumulation of capital stock. Let $C(t)$ be the amount of output allocated to consumption, and let $I(t) = F[K(t)] - C(t)$ be the amount invested. Let δ be the constant rate of depreciation of capital; then the capital stock state equation is:

$$\dot{K} = F(K) - C - \delta K, \quad K(0) = K_0 \tag{10.1}$$

Let $U(C)$ be society's utility of consumption, where we assume $U'(0) = \infty$, $U'(C) > 0$, and $U''(C) < 0$ for $C \geq 0$. Let ρ denote the social discount rate and T denote the finite horizon. Then a government which is elected for a term of T years could consider the objective as

$$\text{Maximize}\{J = \int_0^T e^{-\rho t} U[C(t)]\, dt \tag{10.2}$$

subject to the further condition

$$K(T) = K_T \tag{10.3}$$

where K_T is a given positive constant. It may be noted that replacing (10.3) by $K(T) \geq K_T$ would give the same solution.

10.1.2 Solution by the Maximum Principle

Forming the current value Hamiltonian gives

$$H = U(C) + \lambda[F(K) - C - \delta K] \,. \tag{10.4}$$

The adjoint equation is

$$\dot{\lambda} = \rho\lambda - \frac{\partial H}{\partial K} = (\rho + \delta)\lambda - \lambda\frac{\partial F}{\partial K}, \quad \lambda(T) = \text{constant (to be determined)} \tag{10.5}$$

The optimal control is given by

$$\frac{\partial H}{\partial C} = U'(C) - \lambda = 0 \quad . \tag{10.6}$$

Since $U'(0) = \infty$ the solution of this condition always gives $C > 0$.

The economic interpretation of the Hamiltonian is straightforward: it consists of two terms, the first gives the utility of current consumption, and the second gives the net investment evaluated by prices λ, which, from (10.6) reflect marginal utility of consumption.

For our economic system to be run optimally the solution must satisfy the following three conditions:

(a) The static efficiency condition (10.6) which maximizes the value of the Hamiltonian at each instant of time myopically, provided $\lambda(t)$ is known.

(b) The dynamic efficiency condition (10.5) which forces the price, λ, of capital to change over time in such a way that the capital stock always yields a net rate of return which is equal to the social discount rate ρ; i.e. $d\lambda + \frac{\partial H}{\partial K} dt = \rho \lambda dt$.

(c) The long run foresight condition which establishes the terminal price of capital, $\lambda(T)$ in such a way that exactly the terminal capital stock K_T is obtained at T.

Equations (10.1), (10.3), (10.5), and (10.6) form a two-point boundary value problem which can be solved numerically. A qualitative analysis of this system can also be carried out by the phase diagram method of Chapter 6; see also [Bermeister and Dobell, 1970]. We do not give details here since a similar analysis will be given for the infinite horizon version of this model treated in Sections 10.1.3 and 10.1.4. However, in Exercise 10.1 you are asked to solve a simple version of the model in which the TPBV problem can be solved analytically.

10.1.3 A One-Sector Model with a Growing Labor Force

In the preceeding sections of this chapter we studied the simplest capital accumulation model in which the population was assumed to be fixed. We now want to introduce a new factor, labor (which for simplicity we treat the same as the population) which is growing exponentially at a fixed rate $g > 0$. It is now possible to recast the new model in terms of per capita variables so that it is formally similar to the previous model. The introduction of the per capita variables makes it possible to treat the infinite horizon version of the new model.

Let $L(t)$ denote the amount of labor at time t. Since it growing exponentially at rate g we have

$$L(t) = L(0)e^{gt} \tag{10.7}$$

Let $F(K,L)$ be the production function which is assumed to be concave and homogeneous of degree one in K and L. We can define $k = K/L$ and the per capita production function $f(k)$ as

$$f(k) = \frac{F(K,L)}{L} = F(\frac{K}{L},1) = F(k,1) \quad . \tag{10.7}$$

To derive the state equation for k, we note that

$$\dot{K} = \dot{k}L + k\dot{L} = \dot{k}L + kgL \quad .$$

Substituting for $\dot{K}$ from (10.1) and defining per capita consumption $c = C/L$, we get

$$\dot{k} = f(k) - c - \gamma k, \quad k(0) = k_0 \tag{10.8}$$

where $\gamma = g + \delta$.

Let $U(c)$ be the utility of consumption of c, where U is assumed to satisfy

$$U'(c) > 0, \quad U''(c) < 0, \quad \text{for} \quad c > 0. \tag{10.9}$$

As before we assume $\lim_{c\to 0} U'(c) = \infty$, to rule out zero consumption. The objective function is

$$\text{Maximize}\{J = \int_0^\infty e^{-\rho t} U(c)dt\} . \tag{10.10}$$

10.1.4 Solution by the Maximum Principle

The current value Hamiltonian is

$$H = U(c) + \lambda[f(k) - c - \gamma k] \quad . \tag{10.11}$$

The adjoint equation is

$$\dot{\lambda} = \rho\lambda - \frac{\partial H}{\partial k}$$

$$= (\rho + \gamma)\lambda - f'(k)\lambda. \tag{10.12}$$

To obtain the optimal control we differentiate (10.11) with respect c, set it to zero, and solve:

$$U'(c) = \lambda. \tag{10.13}$$

Let $c = h(\lambda)$ be the solution of (10.13).

To show that the maximum principle is sufficient for optimality we will show that the derived Hamiltonian $H^0(k,\lambda)$ (see Exercise 10.2) is concave in k for any λ solving (10.13). But this follows immediately from the facts that $U'(c)$ is positive because of (10.9) and $f(k)$ is assumed to be concave.

Equations (10.8), (10.12) and (10.13) now constitute a complete autonomous system since time does not enter explicitly in these equations. Therefore we can use the phase diagram solution technique of Chapter 6.

In Figure 10.1 we have drawn a phase diagram for the two equations

$$\dot{k} = f(k) - h(\lambda) - h(\lambda) - \gamma k = 0 \tag{10.14}$$

$$\dot{\lambda} = (\rho + \gamma)\lambda - f'(k)\lambda = 0 \tag{10.15}$$

obtained from (10.8), (10.12), and (10.13). In Exercise 10.3 you are asked to show that the graphs of $\dot{k} = 0$ and $\dot{\lambda} = 0$ are as shown in Figure 10.1. The point of intersection of these two graphs is $(\bar{k},\bar{\lambda})$.

The two graphs divide the plane into four regions, I, II, III, and IV as marked in Figure 10.1. To the left of the vertical line $\dot{\lambda} = 0$, $k < \bar{k}$ and $\rho + \gamma < f'(k)$ so that $\dot{\lambda} < 0$ from (10.8); therefore λ is decreasing, which is indicated by the downward pointing arrows in regions I and IV. Similarly, to the right of the vertical line, that is in regions II and III the arrows are pointed upward because λ is increasing. In Exercise 10.4 you are asked to show that the horizontal arrows which indicate the direction of change in k, point the right above the $\dot{k} = 0$ curve, that is in regions I and II; and they point to the left in regions III and IV which are below the $\dot{k} = 0$ curve.

The point $(\bar{k},\bar{\lambda})$ represents the optimal long run stationary equilibrium. The values of $\bar{k}$ and $\bar{\lambda}$ were obtained in Exercise 10.3. We now want to see if there is a path satisfying the maximum principle which converges to the equilibrium.

Clearly such a path cannot start in quadrants II and IV, because the directions of the arrows in these areas point away from $(\bar{k},\bar{\lambda})$. For $k_o < \bar{k}$ the value of λ_o (if any) must be selected so that (k_o,λ_o) is in region I; and for $k_o > \bar{k}$ the point (k_o,λ_o) must be chosen to be in region III.

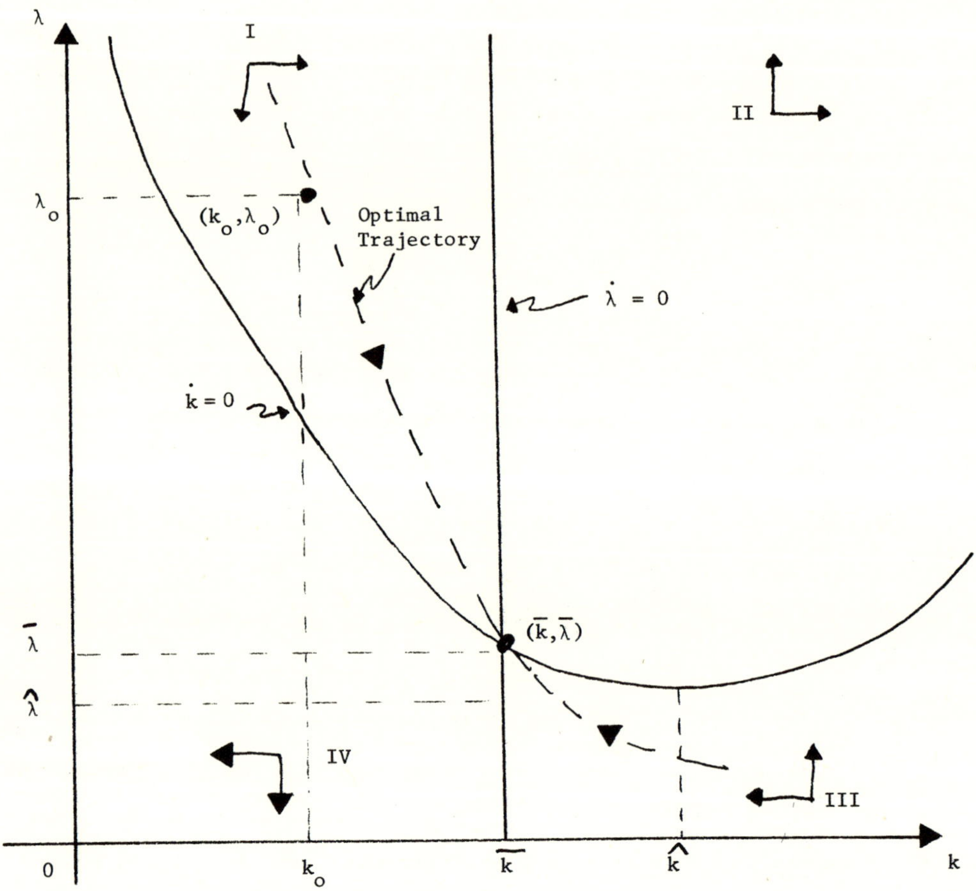

Figure 10.1 Phase Diagram for the Optimal Growth Model

We analyze the case $k_o < \bar{k}$ only, and show that there exists a unique λ_o associated with the given k_o. The locus of such (k_o, λ_o) is shown by the dotted curve in Figure 10.1.

In area I, $k(t)$ is an increasing function of t as indicated by the horizontal right-directed arrow. Therefore we can replace the independent variable t by k. Then from (10.14) and (10.15)

$$\frac{d(\ln \lambda)}{dk} = \left(\frac{1}{\lambda}\frac{d\lambda}{dt}\right) \bigg/ \frac{dk}{dt} = \frac{f'(k) - (\rho+\gamma)}{h(\lambda) + \gamma k - f(k)} . \tag{10.16}$$

For $k < \bar{k}$, the right hand side of (10.16) is negative, and since $h(\lambda)$ decreases with λ we have $d(\ln \lambda)/dk$ increasing with λ.

We show next that there can be at most one trajectory for an initial capital $k_o < \bar{k}$. Assume the contrary, that $\lambda_1^{(k)}$ and $\lambda_2^{(k)}$ are two paths leading to $(\bar{k}, \bar{\lambda})$ and are such that the selected initial values satisfy $\lambda_1(k_o) > \lambda_2(k_o)$. Since $d(\ln \lambda)/dk$ increases with λ

$$\frac{d \ln[\lambda_1(k)/\lambda_2(k)]}{dk} = \frac{d \ln \lambda_1(k)}{dk} - \frac{d \ln \lambda_2(k)}{dk} > 0 \tag{10.16}$$

whenever $\lambda_1(k) > \lambda_2(k)$. This inequality clearly holds at k_o, and by (10.16) $\lambda_1(k)/\lambda_2(k)$ increases at k_o. This in turn implies that the inequality holds at $k_o + \epsilon$, where $\epsilon > 0$ is small. Now replace k_o by $k_o + \epsilon$ and repeat the argument. Thus the ratio $\lambda_1(k)/\lambda_2(k)$ increases as k increases so that $\lambda_1(k)$ and $\lambda_2(k)$ cannot both converge to $\bar{\lambda}$ as $k \to \bar{k}$.

To show that for $k_o < \bar{k}$ there exists a λ_o such that the trajectory converges to $(\bar{k}, \bar{\lambda})$, note that for some starting values of the adjoint variable the resulting trajectory (k, λ) enters area II and then diverges, while for others it enters area IV and diverges; by continuity there exists a starting value

λ_o such that the resulting trajectory (k,λ) converges to $(\bar{k},\bar{\lambda})$.

A similar argument holds when $k > \bar{k}$.

10.2 A Model of Optimal Epidemic Control

Certain infectious epidemic diseases are seasonal in nature. Examples are the common cold, flu, and certain children's diseases. When it is beneficia to do so control measures are taken to alleviate the effects of these diseases. Here we discuss a simple control model [Sethi, 1974b] for analyzing the epidemic problem. See [Wickwire, 1977] for a good survey of optimal control theory applied to control of pests and epidemics.

10.2.1 Formulation of the Model

Let N be the total fixed population; let $x(t)$ be the number of infectives at time t so that the remaining $N-x(t)$ are the number of susceptibles. (Note that in this model no immunity is acquired so that people cured become susceptible again.) The state equation governing the dynamics of the epidemic spread in the population is

$$\dot{x} = \beta x(N-x) - vx, \quad x(0) = x_0 \tag{10.17}$$

where β is a positive constant termed <u>infectivity</u> of the disease, and v is control variable reflecting the level of medical program effort. Note that x is in $[0,N]$ if x_0 is in that interval.

The objective of the control problem is to minimize the present value of the cost stream up to a horizon time T which marks the end of the season for that disease. Let C denote the social unit cost per infective, let K denote the cost of control per unit level of program effort, and let Q be the upper bound on v which is related to the capability of the health delivery system. The optimal control problem is

$$\text{Maximize}\{J = \int_0^T -(Cx + Kv)e^{-\rho t}dt\} \tag{10.18}$$

subject to (10.17), the terminal constraint that

$$x(T) = x_T, \tag{10.19}$$

and $0 \leq v \leq Q$.

10.2.2 Solution by Green's Theorem

Rewriting (10.17) as

$$vdt = [\beta x(N-x)dt - dx]/x$$

and substituting into (10.18) yields the line integral

$$J_\Gamma = \int_\Gamma -\{[Cx + K\beta(N-x)]e^{-\rho t}dt - (\frac{K}{x})e^{-\rho t}dx\} \tag{10.20}$$

where Γ is a path from x_0 to x_T in the (x,t) space. Let Γ_1 and Γ_2 be two such paths from x_0 to x_T, and let R be the region enclosed. By Green's theorem

$$J_{\Gamma_1 - \Gamma_2} = J_{\Gamma_1} - J_{\Gamma_2} = \iint_R -[\frac{K\rho}{x} - C + K\beta]e^{-\rho t}dtdx \quad . \tag{10.21}$$

To obtain the singular control we set the integrand of (10.21) equal to zero, as we did in Chapter 6. This yields

$$x = \frac{\rho}{\frac{C}{K} - \beta} = \frac{\rho}{\theta} \tag{10.22}$$

where $\theta = C/K - \beta$. Define the singular control x^S as follows:

$$x^S = \begin{cases} \frac{\rho}{\theta} & \text{if } 0 < \frac{\rho}{\theta} < N \\ N & \text{otherwise} \end{cases} \quad . \tag{10.23}$$

The corresponding singular control level

$$v^S = \begin{cases} \beta(N - \frac{\rho}{\theta}) & \text{if } 0 < \frac{\rho}{\theta} < N \\ 0 & \text{otherwise} \end{cases} \quad . \tag{10.24}$$

We will show that x^S is the turnpike level of infectives. It is instructive to interpret (10.23) for the various cases, assuming ρ and β are constants. If $\frac{\rho}{\theta} > 0$ then $\theta > 0$ so that $C/K > \beta$. Here the smaller the ratio C/K the larger the turnpike level, i.e., the smaller the medical program effort should be. In other words the smaller the social cost per infective and/or the larger the treatment cost per infective, the smaller the medical program effort should be.

When $\frac{\rho}{\theta} < 0$, you are asked to show in Exercise 10.6 that $x^S = N$. In this case $C/K < \beta$ which means the ratio of the social cost to treatment cost is smaller than the infectivity coefficient. In case there is no terminal constraint, the optimal trajectory involves no treatment effort. An example of this case is the common cold where the social cost is low and treatment cost is high.

The optimal control for the fortituous case when $x_T = x^S$ is

$$v^*[x(t)] = \begin{cases} Q & \text{if } x(t) > x^S \\ v^S & \text{if } x(t) = x^S \\ 0 & \text{if } x(t) < x^S \end{cases} \quad . \tag{10.25}$$

When $x_T \neq x^S$ there are two cases to consider. For simplicity of exposition we assume $x_0 > x^S$ and T and Q are large.

Case (i) $x_T > x^S$. The optimal trajectory is shown in Figure 10.2. In Exercise 10.5 you are asked to show its optimality by using Green's theorem.

Case (ii) $x_T < x^S$. The optimal trajectory is shown in Figure 10.3. It can be shown that x goes asymptotically to $N - (Q/\beta)$ if $v = Q$. The level is marked in Figure 10.3.

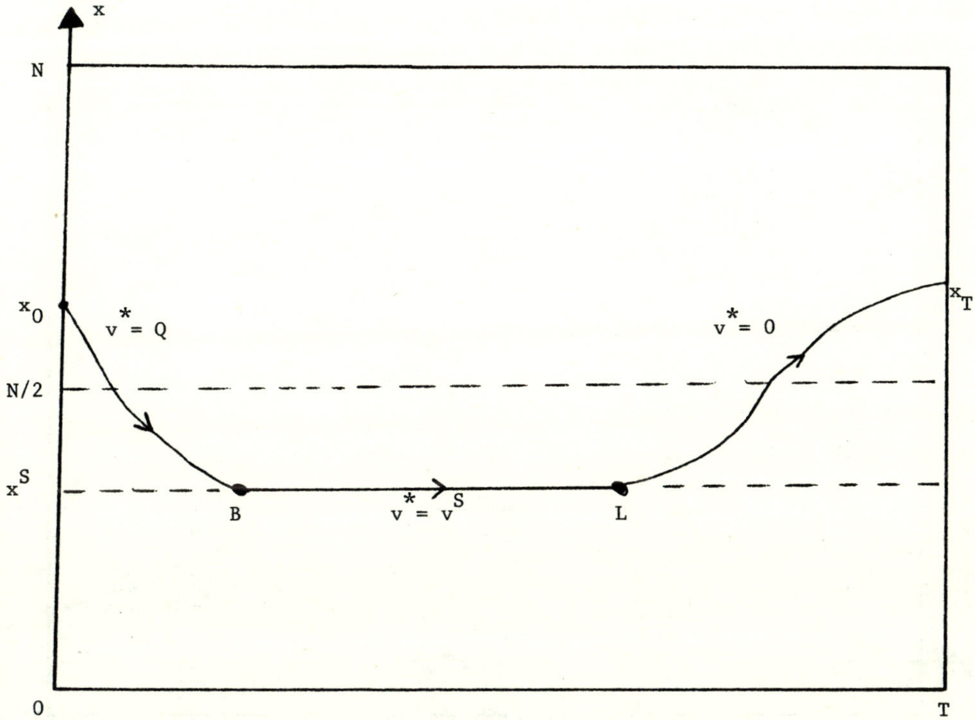

Figure 10.2 Optimal Trajectory for $x_T > x^S$

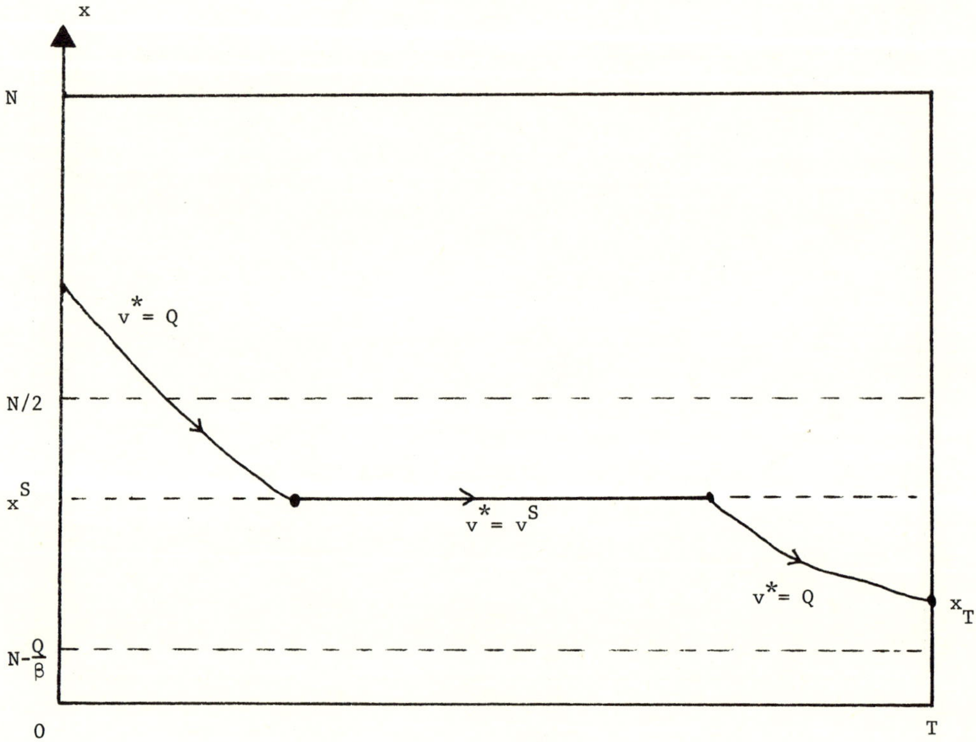

Figure 10.3 Optimal Trajectory for $x_T < x^S$

The optimal control shown in Figures 10.2 and 10.3 assumes $0 < x^S < N$. It also assumes that T is large so that the trajectory will spend some time on the turnpike and Q is large so that $x^S \geq N-(Q/\beta)$. The graphs are drawn for $x_0 > x^S$ and $x^S < N/2$. For all other cases see [Sethi, 1974b].

10.3 A Pollution Control Model

In this section we shall describe a simple pollution control model due to [Keeler, Spence, and Zeckhauser, 1971]. We shall describe this model in terms of an economic system in which labor is the only primary factor of production which is allocated between food production and DDT production. Once produced (and used) DDT is a pollutant which can only be reduced by natural decay. However, DDT is a secondary factor of production which, along with labor, determines the food output. The objective of the society is to maximize the total present value of the utility of food less the disutility of pollution due to the DDT use.

10.3.1 Model Formulation

We introduce the following notation.

L = total labor force, assumed to be constant for simplicity

v = amount of labor used for DDT production

L-v = Amount of labor used for food production

P = stock of pollution at time t

$a(v)$ = the rate of DDT output; $a(0) = 0$, $a'(v) > 0$, $a''(v) < 0$.

δ = natural exponential decay rate of DDT pollution

$C(v) = f[L-v, a(v)]$ = rate of food output; $C(v)$ is concave, $C(0) > 0$, $C(L) = 0$; $C(v)$ attains a unique maximum at $v = V > 0$, see Figure 10. Note that a sufficient condition for $C(v)$ to be strictly concave is $f_{12} \geq 0$.

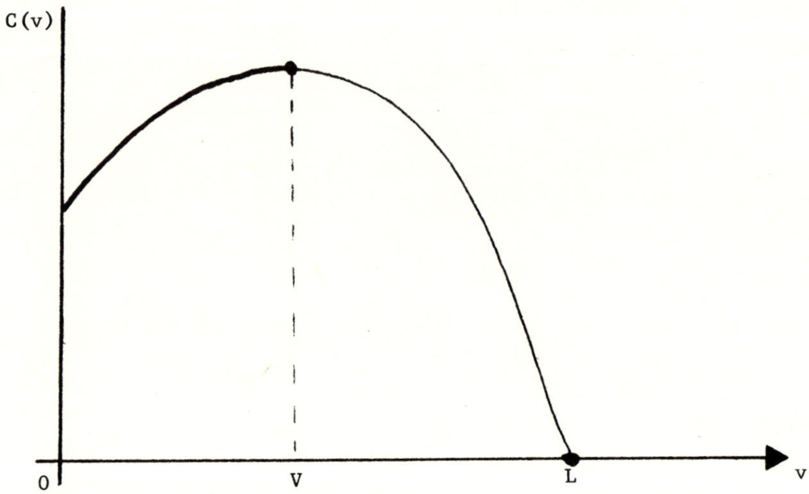

Figure 10.4 Food Output Function

$g(C)$ = utility of consumption; $g'(0) = \infty$, $g' \geq 0$, $g'' < 0$

$h(P)$ = Disutility of pollution; $h'(0) = 0$, $h' \geq 0$, $h'' > 0$.

The optimal control problem is

$$\text{Maximize}\left\{J = \int_0^T e^{-\rho t}[g[C(v)] - h(P)]dt\right\} \tag{10.26}$$

subject to

$$\dot{P} = a(v) - \delta P, \quad P(0) = P_0 \tag{10.27}$$

$$0 \leq v \leq L \quad . \tag{10.28}$$

From Figure 10.4 it is obvious that v is at most V, since the production of DDT beyond that level decreases food production as well as increases DDT pollution. Hence (10.28) can be reduced to simply

$$v \geq 0 \quad . \tag{10.29}$$

10.3.2 Solution by the Maximum Principle

Forming the current value Hamiltonian from (10.26), (10.27) and (10.29)

$$H = g[c(v)] - h(P) + \lambda[a(v) - \delta P] + \mu v \tag{10.30}$$

where

$$\dot{\lambda} = (\rho + \delta)\lambda + h'(P) \tag{10.31}$$

$$\mu \geq 0 \quad \text{and} \quad \mu v = 0 \ . \tag{10.32}$$

The optimal solution is given by

$$\frac{\partial H}{\partial v} = g'[C(v)]C'(v) + \lambda a'(v) + \mu = 0 \ . \tag{10.33}$$

Since the derived Hamiltonian is concave, Conditions (10.30)-(10.33) together with

$$\lim_{t\to\infty} \lambda(t) = \bar{\lambda} = \text{constant} \tag{10.34}$$

are sufficient for optimality; see Theorem 2.1 and Section 3.4. The phase diagram analysis presented below gives $\lambda(t)$ satisfying (10.34).

10.3.3 Phase Diagram Analysis

Since $h'(0) = 0$, $g'(0) = \infty$, and $V > 0$ it pays to produce some positive amount of DDT in equilibrium. Therefore the equilibrium value of the Lagrange multiplier is zero, i.e., $\bar{\mu} = 0$. From (10.27), (10.31) and (10.33) we get the equilibrium values $\bar{P}$, $\bar{\lambda}$, and $\bar{v}$ as follows:

$$\bar{P} = \frac{a(\bar{v})}{\delta} \tag{10.35}$$

$$\bar{\lambda} = -\frac{h'(\bar{P})}{\rho + \delta} = -\frac{g'[C(\bar{v})]\ C'(\bar{v})}{a'(\bar{v})} \ . \tag{10.36}$$

From (10.36) and the assumptions on the derivatives of g, C and a, we know that $\bar{\lambda} < 0$. From this and (10.31) we conclude that $\lambda(t)$ is always negative. The economic interpretation of λ is that $-\lambda$ is the imputed cost of pollution. Let the $v = \varphi(\lambda)$ denote the solution of (10.33) with $\mu = 0$. On account of (10.29), define

$$v^* = \text{Max}[0, \varphi(\lambda)] . \tag{10.37}$$

Because of the interpretation of λ it is reasonable to assume that

$$\frac{d\varphi}{d\lambda} > 0$$

and we will make this assumption. It follows that there exists a unique λ^c such that $\varphi(\lambda^c) = 0$, $\varphi(\lambda) < 0$ for $\lambda < \lambda^c$ and $\varphi(\lambda) > 0$ for $\lambda > \lambda^c$.

To construct the phase diagram we must plot $\dot{P} = 0$ and $\dot{\lambda} = 0$. These are

$$P = \frac{a(v^*)}{\delta} = \frac{a[\max\{0, \varphi(\lambda)\}]}{\delta} \tag{10.38}$$

$$h'(P) = -(\rho + \delta)\lambda \quad . \tag{10.39}$$

Observe that the assumption $h'(0) = 0$ implies the graph of (10.39) passes through the origin. Differentiating these equations with respect to P and using (10.37) we obtain

$$\frac{dP}{d\lambda} = \frac{a'(v)}{\delta} \frac{dv}{d\lambda} \geq 0 \tag{10.40}$$

as the slope of (10.38), and

$$\frac{dP}{d\lambda} = -\frac{(\rho + \delta)}{h''(P)} < 0 \tag{10.41}$$

as the slope of (10.39).

Using (10.35), (10.36), (10.40) and (10.41) we can draw (10.38) and (10.39) in the (P,λ) space as shown in Figure 10.5.

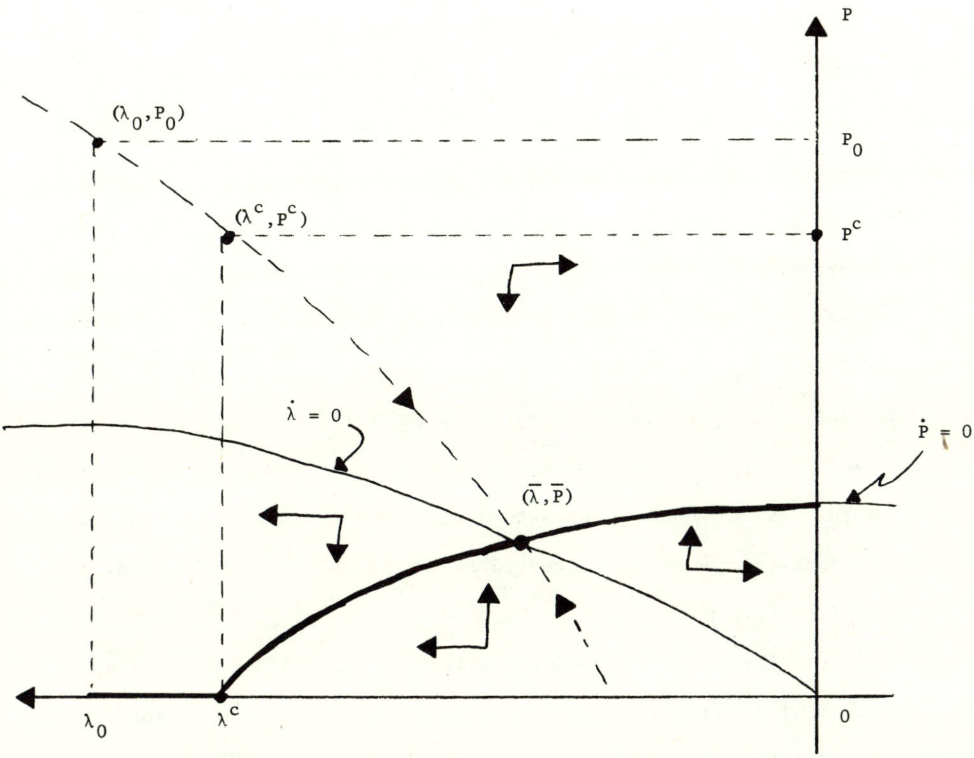

Figure 10.5 Phase Diagram for Pollution Model

The intersection point $(\bar{\lambda},\bar{P})$ of these curves denotes the equilibrium levels for the adjoint variable and pollution stock, respectively. From arguments similar to those in Section 10.1.4 it can be shown that there exists an optimal path (shown dotted in the figure) converging to the equilibrium $(\bar{\lambda},\bar{P})$.

Given λ^c as the intersection of the $\dot{P} = 0$ curve and the horizontal axis, the corresponding ordinate P^c on the optimal trajectory is the related pollution stock level. The significance of P^c is that if the existing pollution stock is larger than P^c then the optimal control is $v^* = 0$, meaning no DDT is produced.

Given an initial level of pollution P_0, the optimal trajectory curve in Figure 10.5 provides the initial value λ_0 of the adjoint variable. With these initial values the optimal trajectory is determined by (10.27), (10.31) and (10.37). If $P_0 > P^c$, as shown in Figure 10.5, then $v^* = 0$ until such time that the natural decay of pollution stock has reduced it to P^c; at this time the adjoint variable has increased to the value λ^c. The optimal control $v^* = \varphi(\lambda)$ from this point on, and the path converges to $(\bar{\lambda},\bar{P})$.

At equilibrium $\bar{v} = \varphi(\bar{\lambda}) > 0$, which implies that it is optimal to produ some DDT forever in the long run. The only time when its production is not optimal is at the beginning when the pollution stock is higher than P^c.

It is important to examine the effects of changes in the parameters on the optimal path. In particular, you are asked in Exercise 10.7 to show that an increase in the natural rate of decay of pollution, δ, will increase P^c. That is, the higher the rate of decay the higher is the level of pollution stock at which the pollutant's production is banned. For DDT, δ is small so that its complete ban, which has actually occurred, may not be far from the optimal policy.

Here we have presented a very simple model of pollution in which the problem was to choose an optimal production process. Other models in which the control variable is to determine the optimal amount to spend in reducing the pollution output of an existing dirty process have been formulated, see [Wright, 1974] and [Sethi, 1977c].

10.4 Miscellaneous Applications

The number of papers which apply control theory to problems in economics and management science is now so large that it is impossible to cover them in detail within the confines of a single book. We content ourselves by listing selected references with a brief indication of their contents.

For control theory applications to economics see: [Tu, 1969], and [Southwick and Zionts, 1974] for optimal educational investments; [Kamien and Schwartz, 1971a] for limit pricing and uncertain entry; [Treadway, 1970] for adjustment costs in the theory of competitive firms; [Vousden, 1974] for international trade; [Harris, 1976] for money demand with transaction costs; [Raviv, 1979] for the design of an optimal insurance policy; [Sethi and McGuire, 1977] for optimal training of heterogeneous labor; [Arthur and McNicoll, 1977] for population policy; [Brito and Oakland, 1977] for optimal income tax; and [Thompson, 1980] for continuous expanding economies.

For control theory applications to management science and operations research see: [Nelson, 1960] for labor assignments; [Fan and Wang, 1964], [Charnes and Kortanek, 1966], [Tapiero and Soliman, 1972], and [Bookbinder and Sethi, 1980], for distribution and transportation applications; [Nepomiastchy, 1970] and [Zimin and Ivanilov, 1971] for scheduling and network planning problems; [Lucas, 1971] for research and development; [Legey, Ripper, and Varaiya, 1973] for city congestion problems; [Taylor, 1974] for warfare models; [Mehra, 1975] for national settlement planning; and [Sethi, 1979] for optimal pilfering policies for thieves.

Exercises for Chapter 10

10.1 For the model of Section 10.1.1 and 10.1.2 assume $F(K) = bK$, and $U(C) = \frac{1}{1-\theta}(C-\bar{C})^{1-\theta}$ where $0 < \theta < 1$, $\bar{C} > 0$ constant, and $b - \delta > 0$ a constant satisfying $(b-\delta)(1-\theta) < \rho < b - \delta$. Replace $b - \delta = \beta$ and assume $\theta = \frac{1}{2}$ for simplification. We also assume that $K_0 e^{\beta T} + \frac{\bar{C}}{\beta}(1-e^{\beta t}) > K_T$ for the problem to be well-posed (note that the left hand side of thie inequality is the amount of capital at T associated with consumption rate $\bar{C}$.) Solve the new model.

10.2 Obtain the expression for the derived Hamiltonian $H^0(k,\lambda)$ from (10.11)

10.3 (a) Obtain the value of $\bar{k}$ in Figure (10.1) from equation (10.15).

(b) Show that the graph of $\dot{k} = 0$ starts from $+\infty$ when $k = 0$ and decreases to a minimum of $\hat{\lambda}$ at $\hat{k}$ and then increases. Also obtain the expression for $\bar{\lambda}$.

(c) Show that $\bar{k} < \hat{k}$.

10.4 Show that the directions of the horizontal arrows above and below the $\dot{k} = 0$ curve are as drawn in Figure 10.1.

10.5 Show that the trajectory x_0, B, L, x_T whown in Figure 10.2 is optimal for the epidemic model under the stated assumptions. Assume $0 < x^S < N$.

10.6 In (10.23) show (by using Green's theorem) that $x^S = N$ if $\frac{\rho}{\theta} < 0$.

10.7 Show that the P^c of Figure 10.5 increases as δ in equation (10.27) increases.

10.8 A variation of the optimal capital accumulation model, known as Ramsey's model, is

$$\text{Maximize}\{J = \int_0^\infty [U(c) - B]\,dt$$

subject

$$\dot{k} = f(k) - c - \delta k, \quad k(0) = k_o$$

where $B = \underset{c \geq 0}{\text{supremum}}\; U(c) > 0$ is the so-called "Bliss point", and where $\lim_{t\to\infty} U[c(t)] = B$ so that the integral in the objective function converges, and $\lim_{t\to\infty} U'[c(t)] = 0$.

(a) Show that the optimal capital stock trajectory satisfies the differential equation

$$U'[f(k) - \delta k + \dot{k}]\dot{k} = B - U[f(k) - \delta k + \dot{k}] \quad .$$

(b) From part (a), derive <u>Ramsey's rule</u>

$$\frac{d[U'(c)]}{dt} + U'(c)[f'(k) - \delta] = 0.$$

(c) Assume $U(c) = 2c - c^2/B$, $f(k) = \alpha k$ where $\gamma = \alpha - \delta > 0$, and $\gamma < B/k_o < 2\gamma$. Show that the optimal feedback consumption rule is

$$c^* = 2\gamma k - B$$

and the optimal capital k^* is given by

$$k^*(t) = \frac{1}{\gamma}[B - (B - \gamma k_o)e^{-\gamma t}].$$

11 SOME COMPUTATIONAL METHODS

11.1 Introduction

Relatively few optimal control problems can be solved in closed form. It is therefore essential to have computational methods in order to solve them. When these methods are implemented on a computer, the differential equations for the state and adjoint variables are integrated numerically and the solution is printed as tables or graphs.

There are several different methods for solving control problems, that is, solving the two point boundary value problems which arise as a consequence of applying the maximum principle. In this chapter we shall describe three of these methods, an initial value shooting methods, the Newton-Raphson method with quasilinearization, and the conjugate gradient method. We do not provide details of the proofs of these methods and their rates of convergence. We illustrate each method with simple examples. Other methods based on linear programming approaches are [Dantzig, 1966], and [Drews, Hartbergen, and Segers, 1974].

This chapter is intended to be a brief introduction to computational solution procedures for optimal control problems. Other books which treat this subject are [Polak, 1971,1973], [Moiseev, 1971], [Bellman and Kalaba, 1965], [Roberts and Shipman, 1972], [Keller, 1968].

11.2 Shooting Methods

Shooting methods are essentially guessing methods for solving TPBV problems; see Exercise 5.9 for an example. To use a shooting method, the differential equations involved must be solved using some known initial condition and guesses for the unknown initial conditions. The differential equations ar solved numerically forward in time to the terminal point (this can be thought of as shooting at the target terminal conditions.) If the computed terminal values satisfy the given terminal conditions, the problem has been solved. Otherwise, the differences between the computed and the given terminal conditions are used to revise the guesses for the missing initial conditions, and the process is repeated.

Section 11.2.1 describes the simplest of the shooting methods, termed the initial value shooting method. Section 11.2.2 gives an adaptation of this method to finding the solution of a discrete optimal control problem. This method works well for one state variable problems. For problems having more state variables more general shooting methods are needed, see [Roberts and Shipman, 1972].

11.2.1 An Initial Value Shooting Method

We shall describe an initial value shooting method for solving the continuous time optimal control problem stated in (2.4). Its solution involves solving the two point boundary value problem stated in (2.31). The procedure starts with a guess for the value of $\lambda(0)$ then, with $x(0) = x_0$ and $\lambda(0)$ as initial conditions, the two differential equations in (2.31) can be solved in forward time with the control $u(t)$ being determined by the Hamiltonian maximizing condition in (2.31). At the terminal time T, we obtain $x(T)$ and $\lambda(T)$. If $\lambda(T) = S_x[x(T)]$, then we have solved the TPBVP. If not, we must modify the initial guess for $\lambda(0)$, and reintegrate the differential equations. The most important part of the algorithm is the guessing method for $\lambda(0)$. Guessing methods for more than one state variable require mathematical ideas beyond the scope of this book.

We describe next a simple binary search guessing method for solving problems having only one state variable. For simplicity in stating the flow chart for the shooting method we shall assume that the solution of the Hamiltonian in (2.31) can be obtained as

$$u(t) = \psi(x,\lambda,t) \ .$$

With this control rule, and an initial guess $\lambda(0) = a$ we can solve the system

$$\left.\begin{aligned} \dot{x} &= f[x,\psi(x,\lambda,t),t], \quad x(0) = x_0 \\ \dot{\lambda} &= H_x[x,\psi(x,\lambda,t),t], \quad \lambda(0) = a \end{aligned}\right\} \qquad (11.1$$

to obtain the terminal values $x_a(T)$ and $\lambda_a(T)$.

The flow diagram for the initial value shooting method is shown in Figure 11.1. To start the algorithm it is necessary to find by trial and error two initial guesses $\lambda(0) = a$ and $\lambda(0) = b$ such that

$\lambda_a(T) - S_x[x_a(T)] > 0$ and $\lambda_b(T) - S_x[x_b(T)] < 0$.

The integration of the system of differential equations (11.1) would in most cases require discretization of time if a digital computer is used. Note that it is not necessary to have u solved in closed form as in (11.1), since it can also be determined by a computer subroutine which solves the static Hamiltonian maximization problem of (2.31).

11.2.2 Solution of Discrete Optimal Control Problems

The TPBV problem for the discrete control problem given in (7.49)-(7.51) is stated in (7.68). To solve (7.68) by means of the initial value shooting method is somewhat more complicated than in the previous section. This is because the optimal control u^{k*} obtained from the Hamiltonian maximizing probl in (7.68) will, in general depend on x^k, λ^{k+1}, and k. So that in the forwa time solution of the difference equations, we need u^k to solve for λ^{k+1}, b

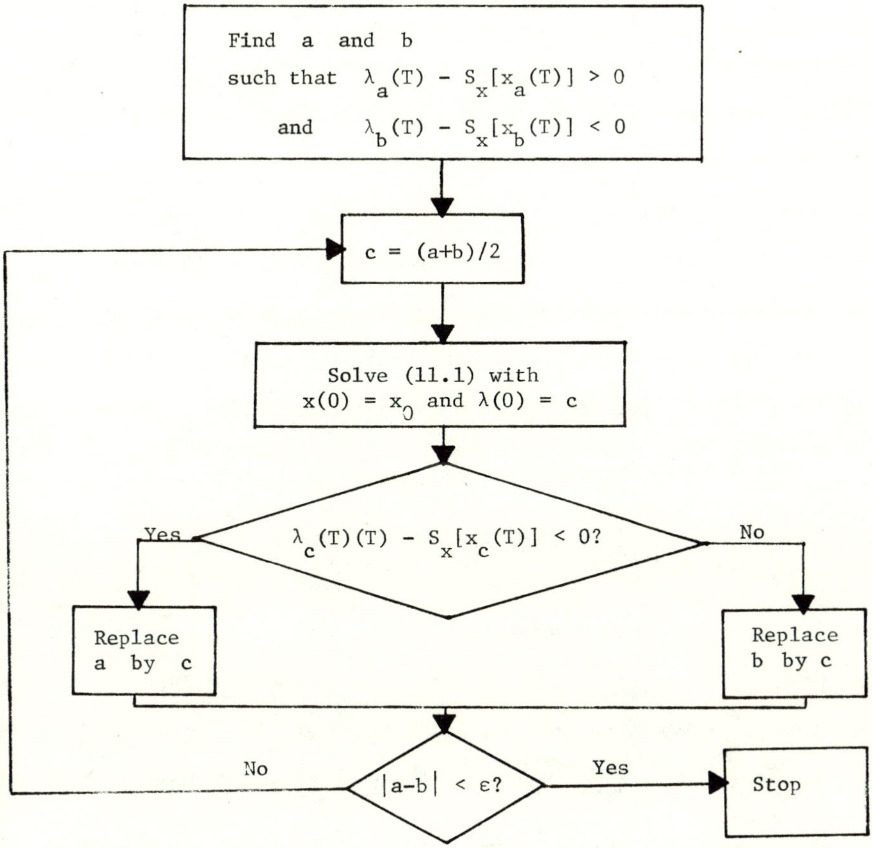

Figure 11.1 Flow diagram for the initial value shooting method

u^k depends on λ^{k+1}. Hence it is necessary to solve simultaneously the Hamiltonian maximization problem as well as the adjoint difference equation. In some cases, this difficulty can be avoided. The numerical example to be solved in the next section gives an illustration of such a case.

11.2.3 A Numerical Example

We now solve Example 7.8. The TPBV problem for this production inventory problem is stated in (7.78) (7.80) and (7.81). Since we have an explicit formula for the optimal control policy and since λ^{k+1} can be explicitly expressed in terms of I^k and λ^k, we can restate the TPBV as

$$I^{k+1} = I^k + \text{Max}[0, \hat{P} + \frac{1}{c}(\lambda^k + h(I^k - \hat{I})] - S^k, \quad I^0 = 300$$

$$\lambda^{k+1} = \lambda^k + h(I^k - \hat{I}), \quad \lambda^T = 0.$$

The computer program written to solve Exercise 11.3(b) can be easily adapted to solve this problem. We assume $\hat{P} = 30$, $\hat{I} = 15$, T = h = c = 1, and $S^k = k(k-4)(k-8) + 30$, as in Example 5.1. The computer output shown in Table 11.1. Note that P(0) = P(1) = 0 because of the large initial

k	λ(k)	P(k)	I(k)
0	-453.6148	0	300
1	-168.6148	0	131.3852
2	-52.22951	20.92623	58.15574
3	-9.073767	31.0082	25.08197
4	1.008202	27.09837	11.09017
5	-2.901627	20.28692	8.188545
6	-9.713081	18.76238	13.47546
7	-11.23762	30.00023	26.23785
8	0.000229		47.23808

Table 11.1 Solution of an Example.

The Optimal Objective Value is J = -49518.64.

inventory. The Max operation in the optimal production rule prevents negative production. In Exercise 11.5 you are asked to resolve the problem with $I(0) = 10$; for that case the production will be found to be always positive.

11.3 The Newton-Raphson Method

The Newton-Raphson method [Polak, 1971] gives a way of solving a system of nonlinear equations. Let $g\colon E^n \to E^n$ be a continuously differentiable function with a non-singular Jacobian matrix. We shall now describe the procedure which solves

$$g(z) = 0 . \tag{11.2}$$

Suppose we guess the point y to be the solution. Expanding (11.4) into the first order terms about y we obtain

$$g(z) \approx g(y) + \left.\frac{\partial g}{\partial z}\right|_y (z-y) \triangleq g(y) + J(y)(z-y) \tag{11.3}$$

where $J(y)$ is known as the Jacobian matrix. If $g(z)$ is not equal to zero we equate the righthand side of (11.3) to 0 to get a new guess for y as

$$z = y - J^{-1}(y) \cdot g(y) \quad . \tag{11.4}$$

This procedure continues until the norm of the difference between two successive approximations is sufficiently small. The procedure is given in the flow chart of Figure 11.2.

The Newton-Raphson method can be adopted to solve two-point boundary value problems. While there are many versions, we describe one which is referred to as the quasilinearization method.

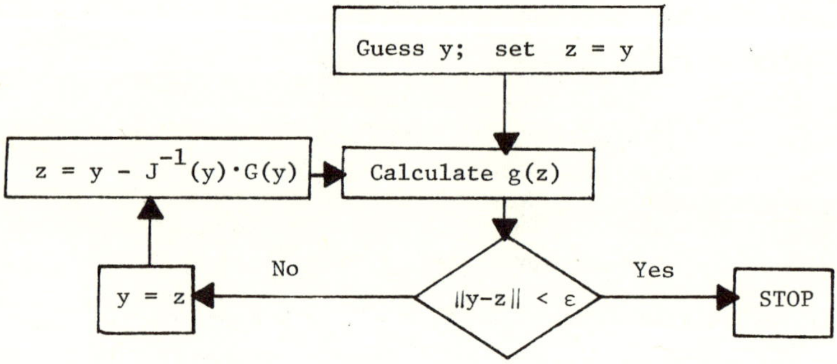

Figure 11.2 Newton-Raphson Method for Solving $g(z) = 0$

11.3.1 The Quasilinearization Method for Control Problems

We describe the quasilinearization method for solving the TPBV problem (2.31). We assume that $S \equiv 0$ and that the Hamiltonian maximization problem c be solved in closed form

$$u^* = \psi(x,\lambda,t) \quad . \tag{11.}$$

Later we will discuss how to modify the algorithm when ψ cannot be determined in closed form.

Substituting (11.5) into (2.31) the TPBV problem becomes

$$\dot{x} = g[x,\psi(x,\lambda,t),t] = g_1(x,\lambda,t), \quad x(0) = x_0 \tag{11}$$

$$\dot{\lambda} = -H_x[x,\psi(x,\lambda,t),\lambda,t] = g_2(x,\lambda,t), \quad \lambda(T) = 0 \ . \tag{11}$$

The flow chart of Figure 11.3 describes the procedure to solve (11.6) and (11.7). In the flow chart the linear differential equation box is derived fro (11.5).

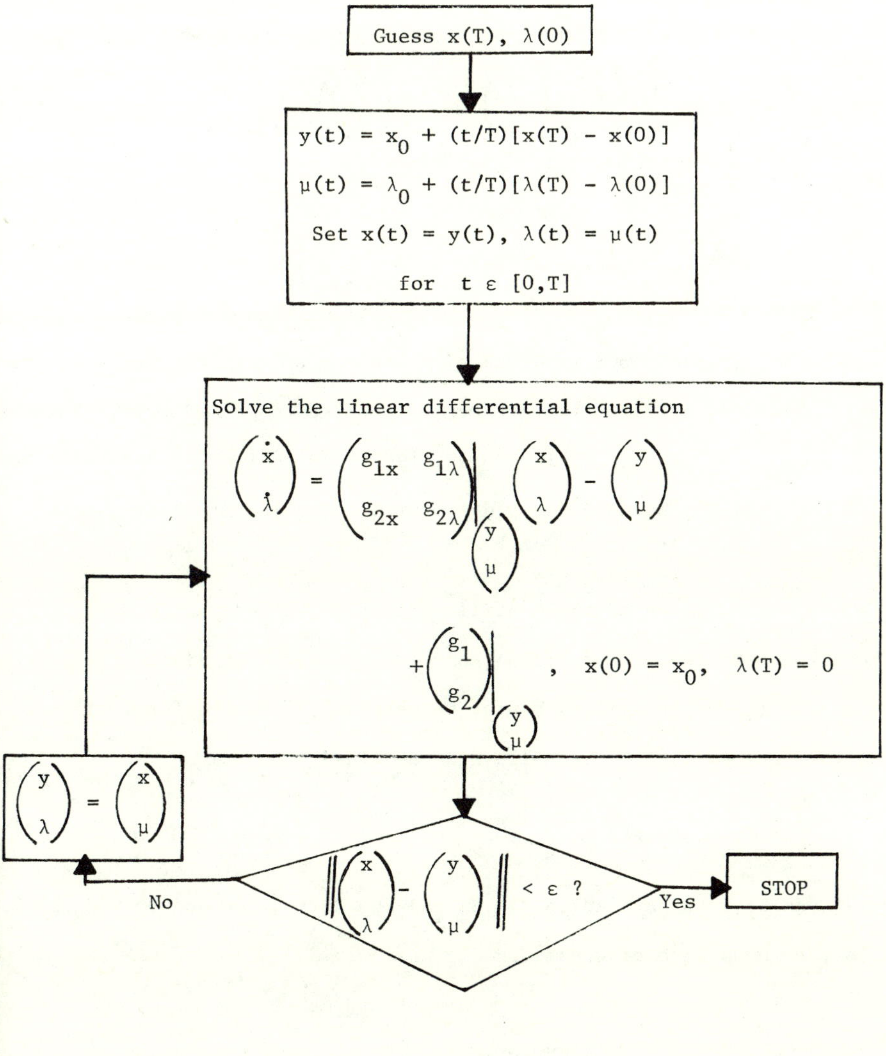

Figure 11.3 Quasilinearization Algorithm for Solving Optimal Control Problems

To see this we define

$$g\begin{pmatrix} x \\ \lambda \end{pmatrix} \triangleq \begin{pmatrix} \begin{pmatrix} x_0 \\ \lambda_0 \end{pmatrix} + \int_0^t \begin{pmatrix} g_1 \\ g_2 \end{pmatrix} dt - \begin{pmatrix} x \\ \lambda \end{pmatrix} \\ x - x_0 \\ \lambda(T) \end{pmatrix} = 0 \qquad (11.8$$

Note that g in (11.8) is a functional of state and adjoint trajectories and th quasilinearization is a Newton-Raphson scheme applied to functionals. Further details of this method are given in [Polak, 1971] and [Bellman and Kalaba, 1965] Commercial packages for solving the problem with this kind of algorithm are also available.

11.3.2 Further Discussion of the Quasilinearization Method

When the function ψ of (11.5) cannot be obtained in closed form, then g_1 and g_2 of (11.6) and (11.7) cannot be obtained in closed form. In the case that the optimal control can always be obtained by setting

$$H_u = 0 \qquad (11.$$

then (11.9) can be incorporated in (11.8) and the Newton-Raphson method can be applied to the new system of equations

$$g\begin{pmatrix} x \\ \lambda \\ u \end{pmatrix} = 0. \qquad (11.$$

This involves guessing trajectories for x, λ, and u. For details of this quasilinearization method see [Polak, 1971].

When optimal control occurs on the boundary of the admissible set Ω, then (11.9) cannot be used to determine the optimal control u. While it may be possible to extend the quasilinearization methods to this case, via penalty functions, or by adding a Hamiltonian maximizing subroutine to compute optimal control, other methods are preferable. Such methods include the already described shooting method and the generalized gradient method to be described in the Section 11.5.

11.4 The Conjugate Gradient Method

Conjugate gradient methods are another way of solving optimal control problems. Since these methods are adaptions of the conjugate method for solving an unconstrained minimization problem in E^n, we begin by first discussing the latter problem.

Consider the problem

$$\underset{x \in E^n}{\text{Maximize}} \quad f(z) \tag{11.11}$$

where $f\colon E^n \to E^1$ is a twice continuously differentiable strictly convex function. The convergence proof of the conjugate gradient method to solve (11.11) requires the assumption that the Hessian matrix $H(z) = f_{zz}$ is suitably restricted, see[Polak, 1971]. To describe the method we need to define for a given point z the set of <u>improving directions</u>

$$B(z) = \{h \mid f_z h > 0\} \quad . \tag{11.12}$$

Gradient methods for solving (11.11) are described in the flow chart of Figure 11.4.

There are several methods for selecting the improving direction h in the flow chart of Figure 11.5. If we choose

$$h = f_z \tag{11.13}$$

then the method is known as the gradient method or the method of steepest descent. For various reasons this method has rather poor convergence properties.

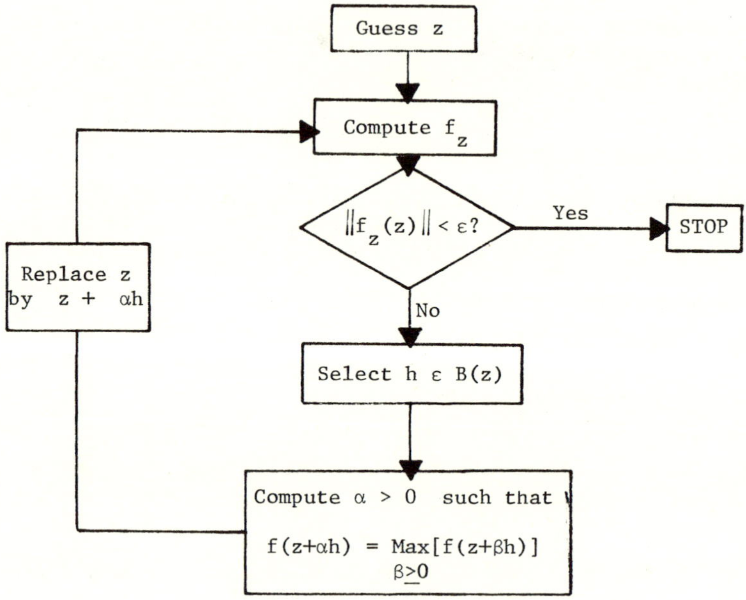

Figure 11.4 Flow Chart for Gradient Methods

The following conjugate gradient method for selecting h was proposed by [Fletcher and Reeves, 1964]. It is described in Figure 11.5. Note that if the first guess z_0 is such that $\|f_z(z_0)\| < \varepsilon$, then the algorithm stops in the first box.

To explain how the method works, the first step is to approximate the function f in the neighborhood of the point z_i by a quadratic function given by the terms up to the second degree in its Taylor's series expansion about z_i. This quadratic function is an ellipsoid which has n axes. The longest axis

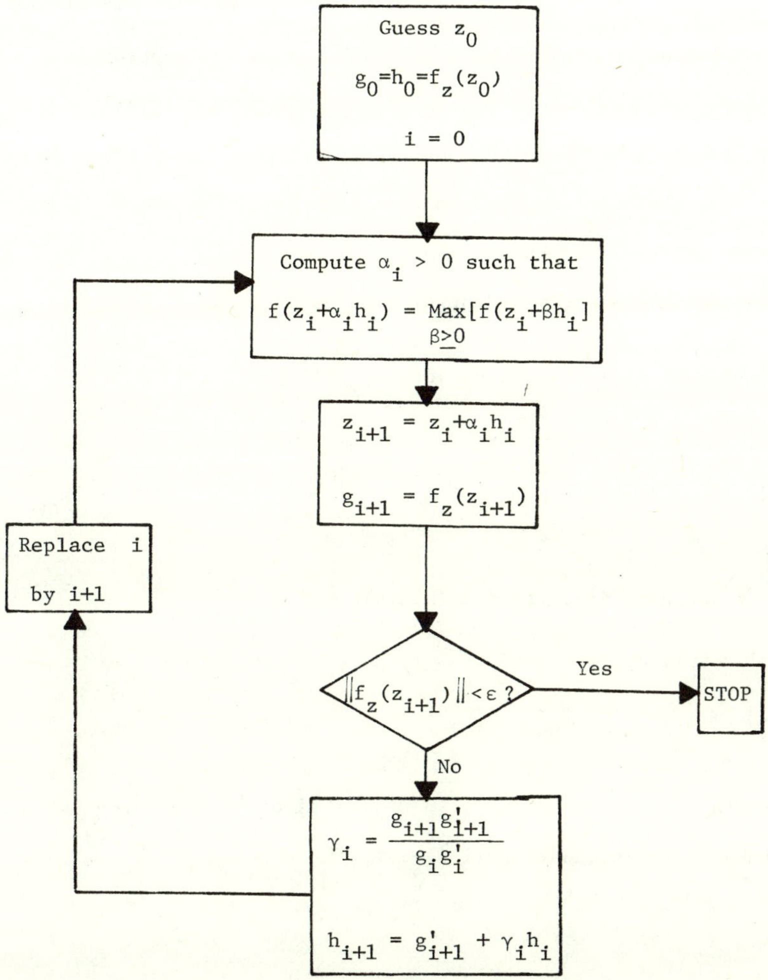

Figure 11.5 Flow Chart for the Fletcher-Reeves Conjugate Gradient Algorithm

is the principal axis. The updating formulas for g_i and h_i in Figure 11.5 cause z_{i+1} to lie on the principal axis of the ellpsoidal approximation at z_i.

11.4.1 The Conjugate Gradient Method for Optimal Control Problems

We next describe an adaptation of the Fletcher-Reeves method for solving unconstrained free end point optimal control problems. The method given here is due to [Lasdon, Mitter, and Waren, 1967].

Consider again the optimal control problem of (2.4):

$$\text{Maximize } \{J(u) = \int_0^T F(x,u,t)dt + S_x[x(T)]\} \tag{11.14}$$

Subject to

$$\dot{x} = f(x,u,t), \quad x(0) = x_0 \ . \tag{11.15}$$

The Hamiltonian is

$$H(x,u,\lambda,t) = F(x,u,t) + \lambda f(x,u,t) \tag{11.16}$$

with the adjoint variable satisfying

$$\dot{\lambda} = -H_x, \quad \lambda(T) = S_x[x(T)] \ . \tag{11.17}$$

The flow chart for the method appears in Figure 11.6. The norms used in this flow chart are defined in equation (1.17).

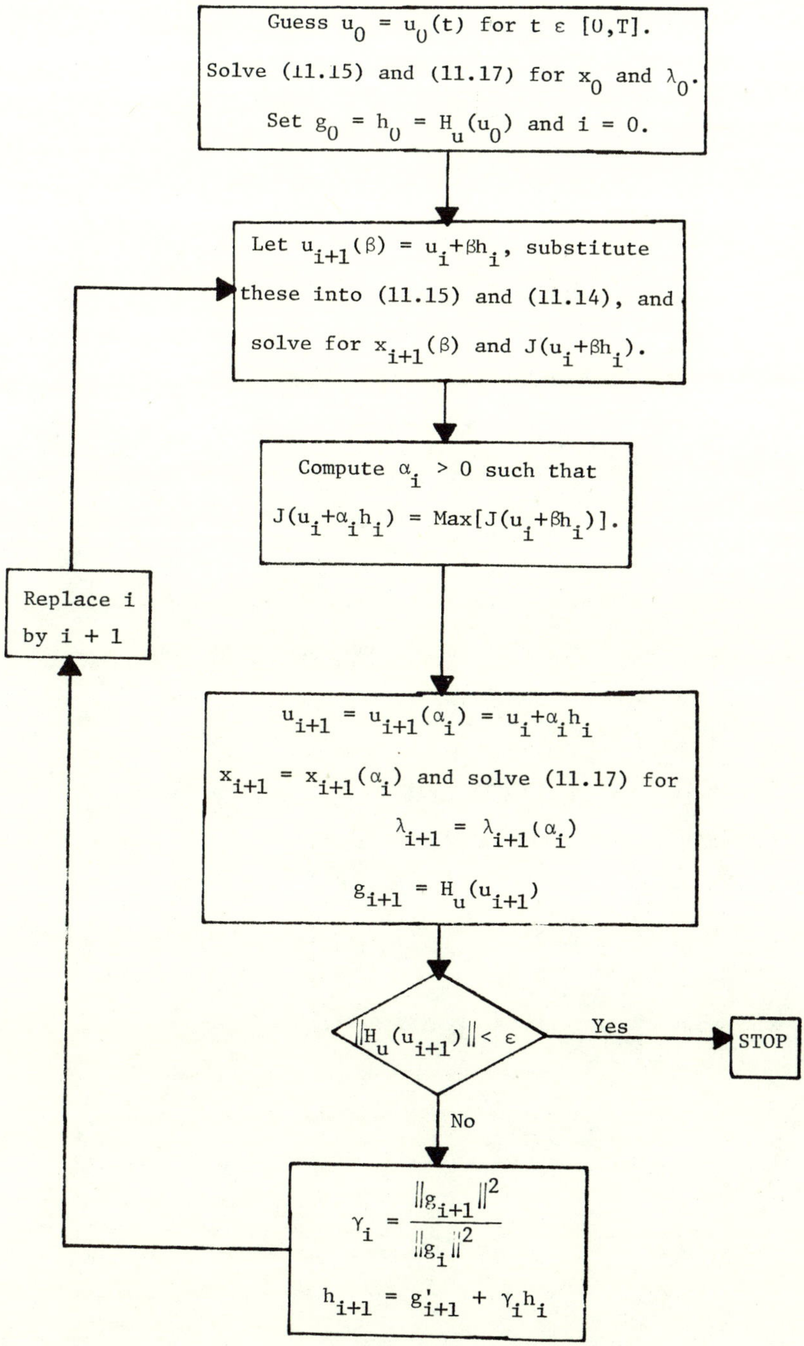

Figure 11.6 Flow Chart for Conjugate Gradient Method for Solving Optimal Control Problems

11.4.2 An Example

As an example (which is taken from [Lasdon, et al., 1967]) we consider

$$\text{Maximize } \{J = -x_2(1)\}$$

Subject to (11.18

$$\dot{x}_1 = u, \quad x_1(0) = \frac{1}{2}$$

$$\dot{x}_2 = (1+u)x_1 + u + \frac{1}{2}u^2, \quad x_2(0) = 0 \ .$$

In Exercise 11.6 you are asked to show that the optimal control for (11.18) is

$$u^*(t) = t - \frac{3}{2} \ . \qquad (11.1$$

We now illustrate the conjugate gradient method of Figure 11.6 by applying it to this example. The Hamiltonian and adjoint equations are:

$$H = \lambda_1 u + \lambda_2[(1+u)x_1 + u + \frac{1}{2}u^2] \qquad (11.2$$

$$\dot{\lambda}_1 = -\lambda_2(1+u), \quad \lambda_1(1) = 0 \qquad (11.2$$

$$\lambda_2(t) = -1 \ . \qquad (11.2$$

$$H_u = \lambda_1 - x_1 - 1 - u \ . \qquad (11.2$$

Let $u_0 \equiv 0$. Solving the state equations in (11.18) and (11.21) gives

$$x_{1,0} = \frac{1}{2}, \quad x_{2,0} = \frac{1}{2}t, \quad \lambda_{1,0}(t) = t - 1 \ . \qquad (11.2$$

Also

$$g_0 = h_0 = H_u(0) = t - \frac{5}{2} \ . \qquad (11.$$

To compute α_0, we set $u_1(\beta) = \beta(t - \frac{5}{2})$ and solve (11.18).

$$x_{1,1}(\beta) = \beta\left(\frac{t^2}{2} - \frac{5}{2}t\right) + \frac{1}{2} \tag{11.26}$$

$$x_{2,1}(\beta) = \frac{t}{2} + \beta\left[\frac{t^3}{6} - \frac{t^2}{2} - \frac{15t}{4}\right]$$

$$+ \beta^2\left[\frac{t^4}{8} - \frac{13t^3}{12} + \frac{15t^2}{8} + \frac{25t}{8}\right] \tag{11.27}$$

$$J\left[\beta\left(t-\frac{5}{2}\right)\right] = -\frac{1}{2} + \frac{49}{12}\beta - \frac{97}{24}\beta^2 \ . \tag{11.28}$$

Differentiating (11.28) with respect to β, setting the result to 0, and solving gives

$$\alpha_0 = \frac{49}{97} \ . \tag{11.29}$$

Therefore

$$u_1 = \frac{49}{97}\left(t - \frac{5}{2}\right)$$

$$\lambda_{1,1}(\alpha_0) = \frac{49}{194}t^2 - \frac{51}{194}t + \frac{1}{97}$$

$$g_1 = H_u(u_1) = \frac{48}{97}t - \frac{22}{97}$$

$$\gamma_0 = \frac{192}{194^2}$$

$$h_1 = \frac{18816}{194^2}t - \frac{9016}{194^2} \ .$$

We have completed one iteration of the flow diagram of Figure 11.6. Computing the next iteration we obtain

$$\alpha_1 = \frac{194}{196}$$

$$u_2 = \frac{49}{97}\left(t - \frac{5}{2}\right) + \alpha_1 h_1$$

$$= t - \frac{3}{2}$$

which makes $H_u(u_2) = 0$ so that the algorithm has converged to the optimal solution found in Exercise 11.7 in just two iterations.

In [Lasdon, Mitter, and Waren, 1967] several other examples are solved using the conjugate gradient method. These require the writing of computer programs.

11.4.3 Concluding Remarks

The conjugate gradient method of Figure 11.6 applies only to unconstrained problems with free end point conditions. In cases when control and state constraints are present, or when terminal conditions are imposed, the problem must be converted into an unconstrained form by means of penalty functions before applying the method, see [Mehra and Davis, 1972] and [Russell, 1965].

However, the addition of penalty functions may cause convergence difficulties. Other types of gradient methods have been suggested to deal with constrained problems. For example,[Mehra and Davis, 1972] have developed a generalized gradient method which is applicable to problems with inequality constraints and singular arcs. This algorithm selects a set of independent variables dictated by the constraints on the problem. These variables may be either control or state variables, and the selected set may be different on different parts of the trajectory. The gradient of the cost function with respec to the independent variables is called the <u>generalized gradient</u>. Further detail are given in [Mehra and Davis, 1972].

Exercises for Chapter 11

11.1 Obtain the closed form solution of a special case of the production inventory model in (5.1) - (5.2):

$$\text{Maximize } \{J = \int_0^1 -\frac{1}{2}(I^2 + P^2)dt\}$$

Subject to

$$\dot{I}(t) = P(t) - 5\sin \pi t, \quad I(0) = 1.$$

11.2 Solve Exercise 11.1 by using the initial value shooting method (Figure 11.1), and compare this numerical solution with the closed form solution found in the previous exercise.

11.3 (a) Resolve Exercise 11.1 numerically with $I(0) = 2$.

(b) Resolve (a) with the additional constraint $P(t) \geq 0$ and compare with the solution in (a).

11.4 Solve Exercise 2.17 numerically by using the initial value shooting method.

11.5 Resolve the numerical example in Section 11.2.3 with $I^0 = 10$.

11.6 Resolve Example 7.8 with the objective function (7.77) replaced by

$$J = \sum_{k=0}^{T-1} -\frac{1}{2}[h(I^k - \hat{I})^2 + c(P^k - \hat{P})^2] - \frac{h}{2}(I^T - \hat{I})^2$$

with the same values of the parameters as in Exercise 11.5.

11.7 Show that the optimal control for the problem in (11.18) is $u = t - \frac{3}{2}$.

12 OTHER CONTROL THEORY TOPICS OF INTEREST

In the last eleven chapters, we were mainly concerned with the optimal control problems formulated in Chapters 2 and 3 and their applications to various functional areas of management and to some problems in economics. These problems are described by a single objective function and a set of differential equations called the state equations defined in a deterministic framework.

Generalization of the (ordinary) optimal control problems can be made in one or more of the following directions.

There may be more than one person, each having separate objective functions which each is trying to maximize, subject to a set of differential equations. This extension of the optimal control theory is referred to as the <u>theory of differential games</u>. These will be dealt with in Section 12.1.

Another extension replaces the system of differential equations by a set of partial differential equations. These come under the classification of <u>distributed parameters systems</u> and are treated in Section 12.2.

In previous chapters we assumed that the state of the system was known with certainty. If this were not the case, one would have to measure these states or, in most cases, their functions. The measurements are usually noisy. We are thus faced with the problem of making good estimates of these state variables from the noisy measurements on their functions. This process of estimating the present values of the state variables is called optimal filtering. In Section 12.3, we will discuss one particular filter, called the Kalman filter and its continuous time analogue called the Kalman-Bucy filter.

A natural extension of the optimal filtering problem would be the stochastic optimal control problem. Here, one would like to synthesize optimal feedback controls for systems governed by stochastic differential equation in a way that maximizes the expected value of the given objective function. This topic will be covered in Section 12.4.

Finally, we treat the theory of impulsive control. This control is developed to deal with systems which, in addition to conventional controls, allows us to select instants and the amounts of discrete changes in the state variables at those instant in an optimal fashion. This theory is especially useful in dealing with inventory problems such as the dynamic lot size problems. It also allows us to charge a setup cost associated with the discrete changes in the state variables.

12.1 Differential Games

The study of differential games was initiated by [Isaacs, 1965]. After the development of Pontryagin's maximum principle, it became clear that there was a connection between differential games and optimal control theory. In fact, differential game problems represent a generalization of optimal control problems to a case in which there is more than one controller. However, differential games are conceptually far more complex than optimal control problems in the sense that it is no longer obvious what constitutes a solution, see [Case, 1979], [Ho, 1970], [Varaiya, 1970], [Friedman, 1971], [Starr and Ho, 1969], and [Westphal, 1974]. As a result, there are a number of different types of solutions such as minimax, Nash, Pareto-optimal, along with possibilities of cooperation and bargaining. We will confine ourselves to the minimax solutions for the zero-sum differential games and the Nash solutions for the nonzero-sum games.

12.1.1 Two Person Zero-Sum Differential Game

Consider the state equation

$$\dot{x} = f(x,u,v,t), \qquad x(0) = x_0 \tag{12.1}$$

where we may assume all variables to be scalar for the time being. Extension to the vector case simply requires appropriate reinterpretations of all the variables and the equations. In this equation, we let u and v denote the controls applied by controllers 1 and 2, respectively. Consider further the following objective function

$$J(u,v) = S[x(T)] + \int_0^T F(x,u,v,t)\,dt \tag{12.2}$$

which controller 1 wants to maximize and controller 2 wants to minimize. Since the gain of controller 1 represents a loss to controller 2, such games are appropriately termed zero-sum games. Clearly, we are looking for $u^*(t)$ and $v^*(t)$ such that

$$J(u^*,v) \geq J(u^*,v^*) \geq J(u,v^*) \tag{12.3}$$

subject to $u \in U$ and $v \in V$.

The necessary conditions for u^* and v^* to satisfy (12.3) are given by an extension of the maximum principle. To obtain these conditions, we form the Hamiltonian

$$H = F + \lambda f \tag{12.4}$$

with the adjoint variable λ satisfying the equation

$$\dot{\lambda} = - H_x \,, \qquad \lambda(T) = S_x[x(T)]. \tag{12.5}$$

Finally, we can state the necessary condition for u^* and v^* to be a minimax solution as

$$H(x^*,u^*,v^*,\lambda^*,t) = \min_{v \in V} \max_{u \in U} H(x^*,u,v,\lambda^*,t). \tag{12.6}$$

which can also be stated as

$$H(x^*,u^*,v,\lambda^*,t) \geq H(x^*,u^*,v^*,\lambda^*,t) \geq H(x^*,u,v^*,\lambda^*,t), \tag{12.7}$$

for $u \in U$ and $v \in V$.

Note that if u and v were unconstrained, condition (12.6) would reduce to the first order necessary conditions

$$H_u = 0 \text{ and } H_v = 0 \tag{12.8}$$

and, the second order conditions would be

$$H_{uu} \leq 0 \text{ and } H_{vv} \geq 0 . \tag{12.9}$$

We now turn to the treatment of nonzero-sum differential games.

12.1.2 Nonzero-Sum Differential Games

In this section, let us assume that we have N players where $N \geq 2$. Let u^i, $i = 1,2,\ldots,N$, represent the control variable for the ith player. Let the state equation be defined as

$$\dot{x} = f(x,u^1,u^2,\ldots u^N,t), \qquad x(0) = x_0. \tag{12.10}$$

Let J^i be defined by

$$J^i = S^i[x(T)] + \int_0^T F^i(x,u^1,u^2,\ldots u^N,t), \tag{12.11}$$

denote the objective function which the ith player wants to maximize. In this case, a Nash solution is defined by a set of N trajectories

$$\{u^{1*}, u^{2*},\ldots,u^{N*}\} \tag{12.12}$$

which have the property

$$J^i(u^{1*},u^{2*},\ldots,u^{N*}) = \max_{u^i} J^i(u^{1*},u^{2*},\ldots,u^{i-1*},u^i,$$

$$u^{i+1*},\ldots,u^{N*}), \tag{12.13}$$

for i = 1, 2,...,N.

To obtain the necessary conditions for a Nash solution for nonzero-sum differential games, we must make a distinction between open loop and closed loop controls.

Open-Loop Nash Solution. The open-loop Nash solution is defined when (12.12) is given as functions of time satisfying (12.13). To obtain maximum principle type conditions for such solutions to be a Nash solution, let us define the Hamiltonian functions,

$$H^i = F^i + \lambda^i f , \tag{12.14}$$

for $i = 1,2,\ldots,N$, with λ^i satisfying

$$\dot{\lambda}^i = -H_x^i, \quad \lambda^i(T) = S_x^i[x(T)] . \tag{12.15}$$

The Nash control u^{i*} for the ith control is obtained by maximizing the ith Hamiltonian H^i with respect to u^i, i.e., u^{i*} must satisfy

$$H^i(x^*,u^{1*},\ldots,u^{i-1*}, u^{i*},u^{i+1*},\ldots,u^{N*},\lambda^*,t)$$

$$\geq H^i(x^*,u^{1*},\ldots,u^{i-1*},u^i, u^{i+1*},\ldots,u^{N*},\lambda^*,t) \tag{12.16}$$

for all u^i t U^i, where U^i is the set of controls which the ith controller is allowed to choose from.

[Deal, Sethi, and Thompson, 1979] formulated and solved an advertising game for two players and obtained the open loop Nash solution by solving a two-point boundary value problem. In Exercise 1, you are asked to formulate this problem.

<u>Closed-loop Nash Solution</u>. A closed-loop Nash solution is defined when (12.12) is defined in terms of the state of the system. To avoid confusion, we let

$$u^{i*}(x,t) = \varphi^i(x,t), \; i = 1,2,\ldots,N. \tag{12.17}$$

For these controls to represent a Nash strategy, we must recognize the dependence of the other players' actions on the state, x. Therefore, we need to replace the adjoint equation (12.15) by

$$\dot{\lambda}^i = -H_x - \sum_{j=1,j\neq i}^{N} H^i_{u^j} \varphi^j_x . \tag{12.18}$$

The presence of the summation term in (12.18) makes the necessary condition for the closed loop solution virtually useless for deriving computational algorithms, [Starr and Ho, 1969]. It is, however, possible to use the dynamic programming approach to solve extremely simple nonzero-sum games. These require solving a partial differential equation. In Exercise 2, you are asked to formulate this partial differential equation for $N = 2$.

Note that the troublesome summation term in (12.18) is absent in optimal control problems $(N = 1)$ since $H_u u_x = 0$, in the two-person zero-sum game because $H^1 = -H^2$ so that $H^1_{u^2} u^2_x = -H^2_{u^2} u^2_x = 0$ and $H^2_{u^1} u^1_x = -H^1_{u^1} u^1_x = 0$, and in the open loop nonzero-sum game because $u^j_x = 0$. It is to be certainly expected, therefore, that the closed loop and open loop Nash solution are going to be different, in general. This can be shown for the linear-quadratic case.

We conclude this section by providing an interpretation to the adjoint variable λ^i. It is the sensitivity of profits to a perturbation in the state vector. If the other players are using closed loop (i.e., feedback) strategies, any perturbation δx in the state vector causes them to revise their controls by the amount $\varphi^j_x \, \delta x$. If the ith Hamiltonian H^i were already extremized with respect to u^j, $j \neq i$, this would not affect the ith player's profit; but since $\partial H^i / \partial u^j \neq 0$ for $i \neq j$, the reactions of the other players to the perturbation influences the ith player's profit, and the ith player must account for this effect in considering variations of the trajectory.

12.1.3 An Application to the Common-Property Fishery Resources

Consider extending the fishery model of Section 9.1 by assuming that there are two producers having unrestricted rights to exploit the fish stock in competition with each other. This gives rise to a nonzero-sum differential game analyzed by [Clark, 1976].

Equation 9.2 is modified to

$$\dot{x} = g(x) - q^1u^1x - q^2u^2x \ , \ x(0) = x_0, \tag{12.19}$$

where $u^i(t)$ represents the rate of fishing effort and q^iu^ix is the rate of catch for the ith producer, i = 1,2. The control constraints are

$$0 \leq u^i(t) \leq U^i, \qquad i = 1,2 \ , \tag{12.20}$$

the state constraints are

$$x(t) \geq 0 \tag{12.21}$$

and the objective function for the ith producer is the total present value of his profits

$$J^i = \int_0^\infty (p^iq^ix - c^i)\, u^ie^{-\rho t}dt, \ i = 1,2. \tag{12.22}$$

To find the Nash solution for this model, we let $\bar{x}^i$ denote the turnpike (or optimal biomass) level given by (9.12) on the assumption that the ith producer is the sole-owner of the fishery. Let the bionomic equilibrium x_b^i for producer i be defined by (9.4), i.e.,

$$x_b^i = \frac{c^i}{p^iq^i} \quad . \tag{12.23}$$

As shown in Exercise 9.2, $x_b^i < \bar{x}^i$. If the other producer is not fishing then producer i can maintain x_b^i by the fishing effort

$$u_b^i = \frac{g(x_b^i)p^i}{c^i} \quad . \tag{12.24}$$

(We assume $u_b^i \leq U^i$.) We assume that

$$x_b^1 < x_b^2 \tag{12.25}$$

which means that producer 1 is more efficient than producer 2, that is, producer 1 can make a positive profit at a level in the interval $(x_b^i, x_b^2]$ while producer 2 loses money in the same interval, except at x_b^2 where he breaks even. For $x > x_b^2$, both producers make positive profits.

Since, by assumption, $U^1 \geq u_b^1$, producer 1 has the capability of driving the fish stock to a level down to or below x_b^1 which, by (12.25) is less than x_b^2. This implies that producer 2 cannot operate at a sustained level $> x_b^2$; and at a sustained level below x_b^2 he cannot make a profit. Hence his optimal policy is bang-bang

$$u^{2*}(x) = \begin{cases} U^2 & \text{if } x > x_b^2 \\ 0 & \text{if } x \leq x_b^2 \end{cases} \tag{12.26}$$

As far as producer 1 is concerned, he wants to attain his turnpike level $\bar{x}^1$ if $\bar{x}^1 \geq x_b^2$. If $\bar{x}^1 < x_b^2$ and if $x_0 \geq \bar{x}^1$, then from (12.26) producer 2 will fish at his maximum rate until the fish stock is driven to x_b^2. At this level it is optimal for producer to fish at a rate which maintains the fish stock at level x_b^2 in order to keep producer 2 from fishing. Thus the optimal policy for producer 1 can be stated as:

$$u^{1*}(x) = \begin{cases} U^1 & \text{if } x > \bar{x}_1 \\ \bar{u}^1 = \dfrac{g(\bar{x}^1)}{q^1\bar{x}^1} & \text{if } x = \bar{x}_1 \\ 0 & \text{if } x < \bar{x}_1 \end{cases}, \text{ if } \bar{x}_1 < x_b^2 \quad (12.27)$$

$$u^{1*}(x) = \begin{cases} U^1 & \text{if } x > x_b^2 \\ \dfrac{g(x_b^2)}{q^1 x_b^2} & \text{if } x = x_b^2 \\ 0 & \text{if } x < x_b^2 \end{cases}, \text{ if } \bar{x}_1 \geq x_b^2 \quad (12.28)$$

The formal proof that policies (12.26)-(12.28) gives a Nash solution requires direct verification using the result of Section 9.1.2. The Nash solution for this case means that for all feasible paths u^1 and u^2

$$J^1(u^{1*},u^{2*}) \geq J^1(u^1,u^{2*}) \quad (12.29)$$

$$J^2(u^{1*},u^{2*}) \geq J^2(u^{1*},u^2) \quad (12.30)$$

The direct verification involves defining a modified growth function

$$g^1(x) = \begin{cases} g(x) - q^2U^2x & \text{if } x > x_b^2 \\ g(x) & \text{if } x \leq x_b^2 \end{cases}$$

and using the Green's theorem results of Section 9.1.2. Since $U^2 \geq u_b^2$ by assumption, we have $g^1(x) \leq 0$ for $x \geq x_b^2$. From (9.12) with g replaced by g^1 it can be shown that the new turnpike level for producer 1 is min $(\bar{x}^1;x_b^2)$, which defines the optimal policy (12.27)-(12.28) for producer 1. The optimality of (12.26) for producer 2 follows trivially.

To interpret the results of the model, suppose that Producer 1 originally has sole possession of the fishery, but anticipates a rival entry. Producer 1 will switch from his own optimal sustained yield x^{1*} to a more intensive exploitation policy <u>prior</u> to the anticipated additional entry.

We can now guess the results in situations with N producers. The fishery will see the progressive elimination of inefficient producers as the stock of fish decreases. Only the most efficient producers will survive. If, ultimately, two or more maximally efficient producers exists, the fishery will converge to an classical bioeconomic equilibrium, with zero sustained economic rent.

We have now seen that a Nash competitive solution involving $N \geq 2$ producers results in the long-run dissipation of economic rents. This conclusion depends on the assumption that producers face an infinitely elastic supply of all factors of production going into the fishing effort. But typically the methods of licensing entrants to regulated fisheries make some attempt also to control the factors of production such as requiring the license to operate only a single vessel of specific size.

In order to develop a model for this type of licensing of fishermen, we let the control variable v^i denote the capital stock of the ith producer and let the concave function $f(v^i)$, with $f(0) = 0$, denote the "fishing mortality function, for $i = 1,2,\ldots,N$. This requires the replacement of $q^i u^i$ in the previous model by $f(v^i)$. The extended model becomes non-linear in control variables. You are asked in Exercise 12.2 to formulate this new model and deve necessary conditions for a closed-loop Nash solution for this model with N producers. The reader is referred to [Clark, 1976] for further details.

12.2 Distributed Parameter Systems

Thus far our efforts have been directed to the study of controlling systems governed by systems of ordinary differential or difference equations. Such systems are often called lumped parameter systems. It is possible to generalize these to systems in which the state and control variables are defined in terms of space as well as time dimensions. These are called distributed parameter systems and are described by a set of partial differential or difference equations.

For example, in the lumped parameter advertising model we need to obtain the optimal advertising expenditures for each instant of time. However, in the analogous distributed parameter advertising model we must obtain the optimal advertising expenditure at every geographic location of interest at each instant of time. In other economic problems the spatial coordinates might be income, quality, age, etc. In Section 12.2.2 we will discuss a cattle-ranching model [Derzko, Sethi, and Thompson, 1980] in which the spatial dimension measures the age of a cow.

Let y denote a one dimensional spatial vector, let t denote time, and let $x(t,y)$ be the one dimensional state variable. Let $u(t,y)$ denote the control variable, and let the state equation be

$$\frac{\partial x}{\partial t} = f(t,y,x,\frac{\partial x}{\partial y},u) \tag{12.31}$$

for $t \in [0,T]$ and y in $[0,h]$. We denote the region $[0,T] \times [0,h]$ by D, and we let its boundary ∂D be split into two parts Γ_1 and Γ_2 as shown in Figure 12.1. The initial boundary conditions will be stated on part Γ_1 as

$$x(0,y) = x_0(y) \tag{12.32}$$

$$x(t,0) = v(t) \tag{12.33}$$

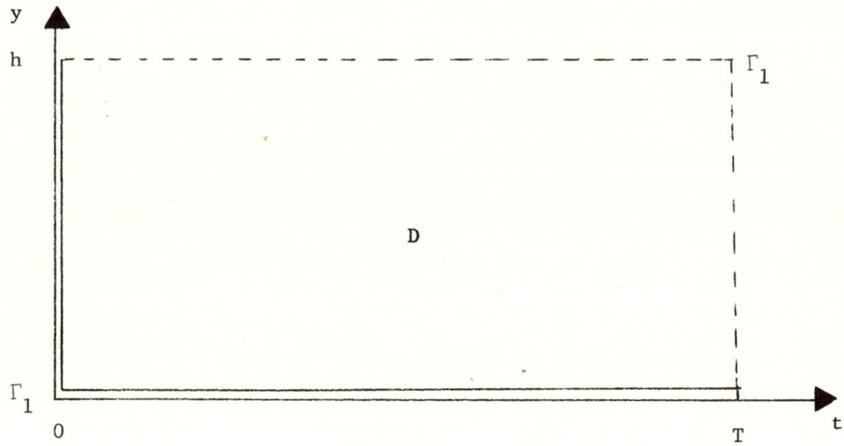

Figure 12.1 Region D with Boundaries

═ Γ_1 and ---------- Γ_2

In (12.32) the function $x_o(y)$ gives the starting distribution of x with respect to the spatial coordinate y. The function $v(t)$ in (12.33) is an exogeneous "breeding function" at time t of x when $y = 0$. In the cattle ranching example $v(t)$ measures the number of newly born calves at time t. To be consistent we make the obvious assumption that

$$x(0,0) = x_0(0) = v(0) \tag{12.34}$$

Let $F(t,y,x,u)$ denote the profit rate when $x(t,y) = x$, $u(t,y) = u$ at a point (t,y) in D; let $P(t)$ be the value of one unit of $x(t,h)$ at time t; and let $S(y)$ be the value one unit of $x(T,y)$ at time T. Then the objective function is to

$$\underset{u(t,y)\in\Omega}{\text{Maximize}}\left\{J = \int_0^T\int_0^h F\,dy\,dt + \int_0^T P(t)x(t,h)\,dt + \int_0^h S(y)x(T,y)\,dy\right\} \tag{12.35}$$

where Ω is the set of allowable controls.

12.2.1 The Distributed Parameter Maximum Principle

We will formulate, without giving proofs, a procedure for solving the problem in (12.31)-(12.35) by a distributed parameter maximum principle which is analogous to the ordinary one. A more complete treatment of this maximum principle can be found in [Sage, 1968]; [Lions, 1971], [Butkowskiy, 1969], and [Derzko, Sethi, Thompson, 1979].

In order to obtain necessary conditions for a maximum we introduce the Hamiltonian

$$H = F + \lambda f \tag{12.36}$$

where the spatial adjoint function $\lambda(t,y)$ satisfies

$$\frac{\partial \lambda}{\partial t} = -\frac{\partial H}{\partial x} + \frac{\partial}{\partial t}\left(\frac{\partial H}{\partial x_t}\right) + \frac{\partial}{\partial y}\left(\frac{\partial H}{\partial x_y}\right) \tag{12.37}$$

where $x_t = \frac{\partial x}{\partial t}$ and $x_y = \frac{\partial x}{\partial y}$. The boundaty conditions on λ are stated for the Γ_2 part of the boundary of D (see Figure 12.1) as follows

$$\lambda(t,h) = P(t) \tag{12.38}$$

$$\lambda(T,y) = S(y) \tag{12.39}$$

Once again we need a consistency requirement similar to (12.34). It is

$$\lambda(T,y) = P(T) = S(h) \tag{12.40}$$

which gives the consistency requirement in the sense that the price and the salvage value of a unit of $x(T,h)$ must agree.

We let $u^*(t,y)$ denote the optimal control function. Then the distributed parameter maximum principle requires that

$$H(t,y,x^*,x_t^*,x_y^*,u^*,\lambda^*) \geq H(t,y,x^*,x_t^*,x_y^*,u,\lambda^*) \tag{12.41}$$

for all $(t,y) \in D$ and all $u \in \Omega$.

We have stated only a simple form of the distributed parameter maximum principal which is sufficient for the cattle ranching example we deal with in the next section. More general forms of the maximum principle are available in the references cited earlier.

12.2.2 The Cattle Ranching Problem

Let t denote time, and y denote the age of an animal. Let $x(t,y)$ denote the number of cattle of age y on the ranch at time t. Let h be the age at maturity at which the cattle are slaughtered. Thus the set $[0,h]$ is the set of all possible ages of the cattle. Let $u(t,y)$ be the rate at which y-aged cattle are bought at time t, where we agree that a negative value of u denotes a sale.

To develop the dynamics of the process it is easy to see that

$$x(t + \Delta t,y) = x(t,y - \Delta t) + u(t,y)\Delta t \tag{12.42}$$

Subtracting $x(t,y)$ from both sides of (12.42), dividing by Δt, and taking the limit as $\Delta t \to 0$, yields the <u>state equation</u>

$$\frac{\partial x}{\partial t} = -\frac{\partial x}{\partial y} + u \quad . \tag{12.43}$$

The boundary conditions for x are given in (12.32)-(12.34). Here $x_0(y)$ denotes the initial distribution of cattle at various ages, and $v(t)$ is an exogeneously specified breeding rate.

To develop the objective function for the cattle rancher we let T den[ote] the horizon time. Let $P(t,y)$ be the purchase or sale price of a y-aged animal at time t. Let $P(t,h) = Q(t)$ be the slaughter value at time t and let $P(T,y) = S(y)$ be the salvage value of a y-aged animal at the horizon time T. The functions Q and S represent the proceeds of the cattle ranching business. To obtain the profit function we must subtract the costs of running the ranch from these proceeds. Let $C(y)$ be the feeding and coraling costs for a y-aged animal per unit of time. Let $\bar{u}(t,y)$ denote the "goal level" purchase rate of y-aged cattle at time t. Any deviation from this goal level is expensive, and the deviation penalty cost is given by $q[u(t,y) - \bar{u}(t,y)]^2$, where q is a constant. Thus the profit maximizing objective function is

$$J = \int_0^T \int_0^h - [q(u - \bar{u})^2 + C(y)x + Pu]\, dydt$$
$$+ \int_0^T Q(t)x(t,h)dt + \int_0^h S(y)x(T,y)dy \quad . \tag{12.44}$$

Comparing this with (12.35) we see that

$$F = -[q(u - \bar{u})^2 + C(y)x] \quad .$$

We assume $\Omega = E^1$, which means that the control variable is unconstrained.

To solve the problem we form the Hamiltonian

$$H = -[q(u - \bar{u})^2 + C(y)x + Pu] + \lambda(-\frac{\partial x}{\partial y} + u) \tag{12.45}$$

where the adjoint function λ satisfies

$$\frac{\partial \lambda}{\partial t} = - \frac{\partial \lambda}{\partial y} + C(y) \tag{12.46}$$

subject to the boundary conditions (12.38)-(12.40). In order to maximize the Hamiltonian we differentiate H with respect to u and set it to zero, giving

$$u^*(t,y) = \bar{u}(t,y) + \frac{1}{2q}[\lambda(t,y) - P(t,y)] \tag{12.47}$$

The form of this optimal control is just like that found in the production-inventory example of Chapter 5. To compute u^* we must solve for $\lambda(t,y)$. It is easy to verify that the general solution of (12.46) is of the form

$$\lambda(t,y) = -\int_y^k C(\tau)d\tau + g(t-y) \tag{12.48}$$

where g is an arbitrary one-variable function and k is a constant in the interval $[y,h]$. We will use the boundary conditions to determine g and k.

In order to state the explicit solution for λ (and later for x) we divide the region D of Figure 12.2) into the three regions D_1, D_2, and D_3 of Figure 12.2.

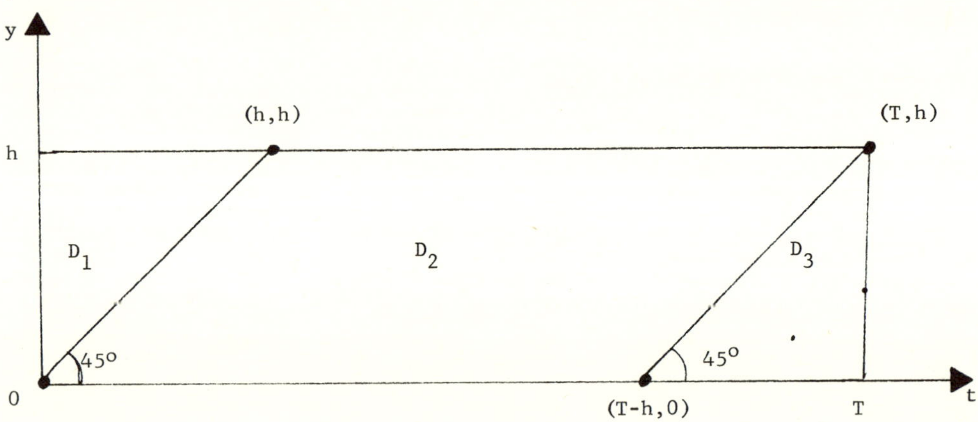

Figure 12.2

The 45° line from (0,0) to (h,h) belongs to both D_1 and D_2, and the 45° line from (T-h,0) to (T,h) belongs to both D_2 and D_3. Thus D_1, D_2, and D_3 are closed sets. The reason that these 45° lines are important in the solution for λ comes from the fact that the determination of the arbitrary function $g(t-y)$ involves the term $(t-y)$. We use (12.38) to obtain $\lambda(t,y)$ in the region $D_1 \cup D_2$. We substitute (12.38) into (12.48) and get

$$\lambda(t,h) = -\int_h^k C(\tau)d\tau + g(t-h) = Q(t).$$

We must therefore have $k = h$ and $g(t-y) = Q(t-y+h)$. For region D_3,

$$\lambda(T,y) = -\int_y^k C(\tau)d\tau + g(T-y) = S(y).$$

In order that the integral should vanish when $t = T$ we must therefore have $k = T - t + y$ and $g(t-y) = S(T-t+y)$. Note that $\lambda(T,h) = P(T,h) = Q(T) = S(h)$. We can now write the complete solution for λ as

$$\lambda(t,y) = \begin{cases} Q(t-y+h) - \int_y^h C(\tau)d\tau & \text{for } (t,y) \in D_1 \cup D_2 \\ S(T-t+y) - \int_y^{(T-t+y)} C(\tau)d\tau & \text{for } (t,y) \in D_3 \,. \end{cases} \tag{12.49}$$

The solution for x is obtained in a similar manner. We substitute (12.47) into (12.43) and use boundary conditions (12.32) - (12.34). The complete solution is given as

$$x(t,y) = \begin{cases} x_0(y-t) + \int_0^t u^*(\tau,y-t+\tau)d\tau & \text{for } (t,y) \in D_1 \\ v(t-y) + \int_0^y u^*(t-y+\tau,\tau)d\tau & \text{for } (t,y) \in D_2 \cup D_3. \end{cases} \tag{12.50}$$

We can interpret the solution (12.50) in the region D_1 as the "beginning game" which is completely characterized by the initial distribution x_0.

Also the solution (12.49) in region D_3 is the "ending game" because in this region the animals do not mature, but most be sold at what ever age they reach at terminal time T. The first expression in (12.49) and the second expression in (12.50) hold in the region D_2, which can be interpreted as the "middle game" portion of the solution.

12.2.3 Interpretation of the Adjoint Function

It is instructive to interpret the solution for $\lambda(t,y)$ in (12.49). An animal of age y at time t, where (t,y) is in $D_1 \cup D_2$, will mature at time t-y+h; its slaughter value at that time is Q(t-y+h). However, the total feeding and coral cost in keeping the animal from its age y until it matures is given by $\int_y^h C(\tau)d\tau$. Thus $\lambda(t,y)$ represents the net benefit obtained from having an animal of age y at time t. You should give a similar interpretation for λ in region D_3.

Having this interpretation for λ it is easy to interpret the optimal control u^* in (12.47) Whenever $\lambda(t,y)$ is positive we buy more than the goal level $\bar{u}(t,y)$, and when $\lambda(t,y)$ is negative we buy less than the goal level.

Another application of the distributed parameter model is to product quality deterioration [Bensoussan, Nissen, and Tapiero, 1975]. We believe tha many other applications are possible and represent fruitful areas for research

12.3 The Kalman Filter

So far in this book, we have assumed that the values of the state variable can be measured with certainty. In many cases the assumption that the value of a state variable can be directly measured and exactly determined may not be realistic.

There are two types of random disturbances present. The first kind, termed measurement noise, arises because of imprecise measuring instruments, inaccurate recording systems, etc. In many cases the measurement technique involves observations of functions of state variables, from which the values of some or all of the state variables are inferred; e.g., measuring the inventory of a natural gas reservoir involves pressure measurements together with physical laws relating pressure and volume.

The second kind can be termed system noise, in which the system itself is subjected to random disturbances. For instance, sales may follow a stochastic process, which would affect the system equation (5.1) relating inventory, production, and sales. In the cash balance example the demand for cash as well as the interest rates in (4.1) and (4.2) may be represented by stochastic processes.

In analyzing systems, in which one or both of these kinds of noises are present, it is important to be able to make good estimates of the values of the state variables. In particular, many optimal control problems require such estimates in order to determine the optimal controls; see Appendix A4.1.1. The process of estimating the current values of the state variables given the past measurements is called filtering, [Bryson and Ho, 1969], [Kalman, 1960] and [Sorensen, 1966].

Consider a dynamic stochastic system described by the following difference equation

$$x^{t+1} - x^t = A_t x^t + G_t w^t \tag{12.51}$$

or

$$x^{t+1} = \Phi_t x^t + G_t w^t = (A_t + I)x^t + G_t w^t \tag{12.52}$$

where x^t is the n-component state vector, w^t is an m-component system noise vector, and A_t is an $n \times n$ matrix and G_t is an $n \times m$ matrix. The initial state x^0 is assumed to be a gaussian (normal) random variable with mean and variance given by

$$E[x^0] = \bar{x}^0 \quad \text{and} \quad E[(x^0 - \bar{x}^0)(x^0 - \bar{x}^0)'] = M_0 \ . \tag{12.53}$$

Also w^t is a gaussian purely random sequence ([ ]) with

$$E[w^t] = \bar{w}^t \quad \text{and} \quad E[(w^t - \bar{w}^t)(w^\tau - \bar{w}^\tau)'] = Q_t \, \delta_{t\tau} \tag{12.54}$$

where

$$\delta_{t\tau} = \begin{cases} 0 & \text{if} \quad t \neq \tau \\ 1 & \text{if} \quad t = \tau \ . \end{cases} \tag{12.55}$$

Thus Q_t represents the covariance matrix of the random vector w^t, and w^t and w^τ are independent random variables for $t \neq \tau$. We also assume that the sequence w^t is independent of the initial condition x^0, that is,

$$E[(w^t - \bar{w}^t)(x^o - \bar{x}^0)] = 0. \tag{12.56}$$

The process of measurement of the state variables x^t yields a k-dimensional vector y^t which is related to x^t by the following transformation:

$$y^t = H_t \, x^t + v^t \tag{12.57}$$

where H_t is the state-to-measurement transformation matrix of dimension $k \times n$, v^t is a gaussian purely random sequence of measurement noise vectors having the following properties:

$$E[v^t] = 0, \quad E[v^t \cdot (v^\tau)'] = R_t \, \delta_{t\tau} \tag{12.58}$$

$$E[(w^t-\bar{w}^t)\cdot(v^\tau)'] = 0, \quad E[(x_0-\bar{x}_0)\cdot(\bar{v}^t)'] = 0. \tag{12.59}$$

In (12.58) the matrix T_t is the covariance matrix for the random variable v^t. The requirements in (12.59) mean that the additive measurement noise is independent of the system noise as well as the initial state.

Given a sequence of observations $y^1, y^2, \ldots, y^t$ up to time t, we would like to obtain the maximum likelihood estimate of the state x^t, or equivalently, to find the minimum weighted least squares estimate. In order to derive the estimate $\hat{x}^t$ of x^t requires the use of Bayes theorem and an application of calculus to find the unconstrained minimum of the weighted least squares function. This derivation is straightforward but lengthy. It yields the following recursive procedure for finding the estimate $\hat{x}^t$.

$$\begin{aligned}\hat{x}^t &= \bar{x}^t + (P_t H_t' R_t^{-1})(y^t - H_t \bar{x}^t) \\ &= \bar{x}^t + K_t (y^t - H_t \bar{x}^t).\end{aligned} \tag{12.60}$$

The mean $\bar{x}^t$ and the matrix P_t are calculated recursively by means of

$$\bar{x}^t = (A_{t-1} + I)\,\hat{x}^{t-1} + G_{t-1}\bar{w}^{t-1}, \quad \bar{x}^0 \text{ given} \tag{12.61}$$

$$P_t = (M_t^{-1} + H_t' R_t^{-1} H_t)^{-1} \tag{12.62}$$

where

$$M_t = (A_{t-1} + I)P_{t-1}(A_{t-1} + I)^{-1} + G_{t-1}Q_{t-1}G_{t-1}', \quad M_0 \text{ given} \tag{12.63}$$

The procedure in expressions (12.60) - (12.63) is known as the <u>Kalman filter</u> for linear discrete time processes.

The interpretation of (12.60) is that the estimate $\hat{x}^t$ is equal to the mean value $\bar{x}^t$ plus a correction term which is proportional to the difference between the actual measurement y^t and the predicted measurement $H_t\bar{x}^t$. It should be noted that

$$P_t = E[(x^t - \hat{x}^t)(x^t - \hat{x}^t)']$$

which is the variance of $(x^t - \hat{x}^t)$. It therefore is a measure of the risk of $(x^t - \hat{x}^t)$. Thus, the proportionality matrix K_t can be interpreted as the ratio between the transformed risk (P_tH_t') and the risk R_t in the measurement. Because of this property of K_t it is called the Kalman gain in the engineering literature.

To obtain an analogous filter for the continuous time case, we must define a purely random process to take the place of the gaussian purely random sequence. Such a process will be called the white noise process, which we denote by $w(t)$. This process must be gaussian at each time t, and $w(t)$ and $w(\tau)$ must be independent for $t \neq \tau$. We will now show how to obtain such a process.

Consider a gaussian random process $w(t)$ with $E[w(t)] = \bar{w}(t)$ and whose autocorrelation matrix is given by

$$E[(w(t) - \bar{w}(t))\cdot(w(\tau) - \bar{w}(\tau))'] = q(t)e^{|t-\tau|/T} \tag{12.64}$$

where the parameter T is small, and $q(t)$ is the covariance matrix,

$$q(t) = E[(w(t) - \bar{w}(t))(w(t) - \bar{w}(t))']. \tag{12.65}$$

Since T is small it is obvious from (12.64) that the correlation between $w(t)$ and $w(\tau)$ decreases rapidly as $t - \tau$ increases. The autocorrelation function is sketched in Figure 12.3 for a scalar process.

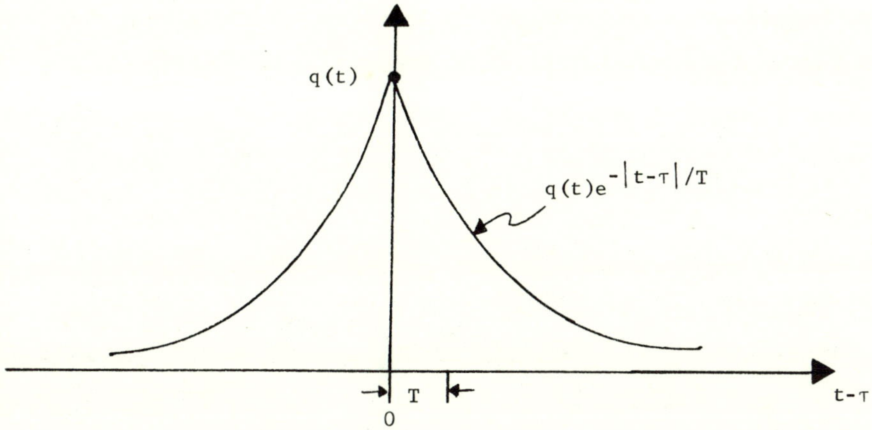

Figure 12.3 Autocorrelation Function for a Scalar Process

In order to obtain the white noise process, we let T go to 0, and define

$$Q(t) = \lim_{\substack{T\to 0 \\ q(t)\to\infty}} \{2Tq(t)\}. \tag{12.66}$$

Using the function $Q(t)$ we can approximate (12.64) by the expression

$$E[(w(t)-\bar{w}(t))(w(\tau)-\bar{w}(\tau))'] = Q(t)\delta(t-\tau) \tag{12.67}$$

where $\delta(t-\tau)$, which is called the Dirac delta function, is the limit as $\epsilon\to 0$ of

$$\delta_\epsilon(t-\tau) = \begin{cases} 0 & \text{when } |t-\tau| > \epsilon \\ \frac{1}{2\epsilon} & \text{when } |t-\tau| \le \epsilon \end{cases} . \tag{12.68}$$

It is instructive to compare (12.67) with the corresponding discrete time expression (12.54).

The above discussion permits us to formally write the continuous analogue of (12.51) as

$$dx(t) = A(t)x(t)dt + G(t)w(t)dt \; . \tag{12.69}$$

In order to interpret and solve this equation a knowledge of stochastic differential equations is required; see, e.g.,[Arnold, 1974]. Equation (12.69) can be restated as the <u>Ito stochastic differential equation</u>.

$$dx(t) = A(t)x(t)dt + G(t)dz(t) \tag{12.70}$$

where $z(t)$ denotes a Wiener process [Wiener, 1949]. Comparison of (12.69) and (12.70) suggests that $w(t)$ could be considered a "generalized derivative" of the Wiener process.

We also define the measurement process $y(t)$ in continuous time as

$$y(t) = H(t)x(t) + v(t) \tag{12.71}$$

where $v(t)$ is a white noise process with

$$E[v(t)] = 0 \quad \text{and} \quad E[v(t)v(\tau)'] = R(t)\delta(t-\tau) \; . \tag{12.72}$$

Using the theory of stochastic differential equations [Arnold, 1974] and calculus of maxima and minima, we can obtain the filter which obtains the estimate $\hat{x}(t)$ given the observations $y(s)$, $s \in [0,t]$ as follows:

$$\dot{\hat{x}} = A\hat{x} + G\bar{w} + K[y-H\hat{x}] , \; \hat{x}(0) = E[x(0)] \tag{12.73}$$

$$K = PH'R^{-1} \tag{12.74}$$

$$\dot{P} = AP + PA' - KHP + GQG' \quad (12.75)$$

$$P(0) = E[x(0)\, x'(0)] \quad . \quad (12.76)$$

This is called the Kalman-Bucy filter [Kalman and Bucy, 1961] for linear systems in continuous time. It is analogous to the Kalman filter for the discrete time case. Equation (12.75) is called the matrix Riccati equation, see Appendix A1.

Besides engineering applications, the Kalman filter and its extensions are very useful in econometric modelling [Buchanan and Norton, 1971], [Chow, 1975], and [Aoki, 1976].

12.4 Stochastic Optimal Control

In Section 12.3 on the Kalman filter, we obtained optimal state estimation for linear systems with noise and noisy measurements. We also defined the white noise process for obtaining the Kalman-Bucy filter for continuous time systems. In Appendix A4.1.1, we noted that for stochastic linear-quadratic optimal control problems, the separation principle allows us to solve the problem in two steps: to obtain the optimal estimate of the state and use it in the optimal feedback control formula for deterministic linear quadratic problems.

For more general stochastic optimal control problems, the separation principle does not hold. To find solutions for these problems requires knowledge of stochastic differential equation (see [Arnold, 1973]; [Gihman and Skorohod, 1971], [Kushner, 1977], and [Fleming and Rishel, 1975]) and partial differential equations. The solution method makes use of the Hamilton-Jacobi theory.

Let us consider the problem of maximizing

$$E\left[\int_0^T F(t, X_t, U_t)dt + S(T, X_T)\right] \quad (12.77)$$

where X is the state variable, U is the closed-loop control variable, and they are required to satisfy the Ito stochastic differential equation

$$dX_t = f(t,X_t,U_t)dt + G(t,X_t,U_t)dW_t, \quad X_0 = x_0. \tag{12.78}$$

To solve the problem, let $V(t,x)$ be the expected value of an optimal policy from t to T given $X_t = x$. Then, by the principle of optimality

$$V(t,x) = \max_u E[F(t,x,u)dt + V(t+dt,x+dX_t)]. \tag{12.79}$$

In the Ito framework

$$V(t+dt,x+dX_t) = V(t,x) + V_t dt + V_x dX_t + \frac{1}{2} V_{xx}(dX_t)^2$$

or, as shown in [1]

$$E[V(t+dt,x+dX_t)] = V(t,x) + V_t dt + V_x f\, dt + \frac{1}{2} V_{xx} G^2 dt. \tag{12.80}$$

Substituting (12.80) into (12.79) we obtain the Hamilton-Jacobi-Bellman equation

$$0 = \max_u [F + V_t + V_x f + \frac{1}{2} V_{xx} G^2] \tag{12.81}$$

for the value function $V(t,x)$ with the boundary condition

$$V(T,x) = S(x,T). \tag{12.82}$$

In the next section we shall apply the above developed theory to a simple stochastic production inventory problem, [Sethi and Thompson, 1980].

12.4.1 A Stochastic Production Planning Model

Consider a factory producing a homogeneous good and having an inventory warehouse. Define the following quantities:

X_t = inventory level at time t (state variable)

U_t = production rate at time t (control variable)

S = the constant demand rate at time t; $S > 0$

T = length of planning period

$\bar{x}$ = factory-optimal inventory level

$\bar{u}$ = factory-optimal production rate

x_o = initial inventory level

h = inventory holding cost coefficient

c = production cost coefficient

B = salvage value per unit terminal inventory

$z(t)$ = the standard Wiener process; see [Arnold, 1974]

σ = the constant diffusion coefficient; see [Arnold, 1974].

We now state the conditions of the model. The first condition is the stock-flow equation stated as an Ito stochastic differential equation:

$$dX = (U-S)dt + \sigma dW, \quad X_0 = x_0 \; . \tag{12.83}$$

We note that process dW_t can be formally expressed as $\xi(t)dt$, where $\xi(t)$ is considered to be the white noise process [Arnold, 1974]. It can be interpreted as "sales returns," "inventory spoilage," etc. which are random in nature. The second is objective function:

$$\underset{U}{\text{Minimize}} \; E\{\int_0^T [c(U-\bar{u})^2 + h(X-\bar{x})^2]dt + BX_T\}. \tag{12.84}$$

Note that we do not restrict the production rate to be nonnegative as required in Chapter 5. In other words, we permit disposal (i.e. $U < 0$). While this is done for mathematical expedience, we will state conditions under which a disposal is not required. Note further that the inventory level is allowed to be negative, i.e., we permit backlogging of demand.

The solution of the above model will be carried out via the previous development of the Hamilton-Jacobi equation satisfied by a certain 'value function.' To simplify the mathematics, we assume that

$$\bar{x} = \bar{u} = 0 \quad \text{and} \quad h = c = 1. \tag{12.85}$$

This assumption results in no loss of generality as the following analysis can be extended in a parallel manner for the case without (12.85). With (12.85), we restate (12.84) as:

$$\max E\left[\int_0^T - (U^2 + X^2)dt + BX_T\right]. \tag{12.86}$$

Let $V(t,x)$ denote the expected value of the objective function from time t to the horizon T with $X_t = x$ and using the optimal policy from t to T. The function $V(t,x)$ is referred to as the value function and it satisfies

$$0 = \underset{u}{\text{Maximum}}\left[-(u^2 + x^2) + V_t + V_x(u - S) + \frac{1}{2}\sigma^2 V_{xx}\right] \tag{12.87}$$

with the boundary condition

$$V(t,x) = Bx . \tag{12.88}$$

Note that these are (12.81) and (12.82) for the production planning problem.

It is now possible to maximize the expression inside the bracket of (12.77) with respect to u by taking its derivative with respect to u and setting it to zero. This procedure yields

$$u = \frac{V_x}{2} . \tag{12.89}$$

Substituting (12.89) into (12.87) yields the following Hamilton-Jacobi equation

$$0 = + \frac{V_x^2}{4} - x^2 + V_t - S\,V_x + \frac{1}{2}\sigma^2 V_{xx}. \tag{12.90}$$

This is a partial differential equation which must be satisfied by the value function $V(t,x)$ with the boundary condition (12.88).

It is important to remark that if production rate were restricted to be nonnegative then (12.89) would be changed to

$$u = \max\left[0, \frac{V_x}{2}\right]. \tag{12.91}$$

Substituting (12.92) in (12.88) would give us a partial differential equation which must be numerically solved for this case. We shall not consider (12.91) further here.

We now turn to solving (12.90) in the next section.

12.4.2 Solution for the Production Planning Problem

To solve equation (12.90) we let

$$V(t,x) = Q(t)x^2 + R(t)x + M(t). \tag{12.92}$$

Then

$$\begin{aligned} V_t &= \dot{Q}x^2 + \dot{R}x + \dot{M} \\ V_x &= 2Qx + R \\ V_{xx} &= 2Q \end{aligned} \tag{12.93}$$

where $\dot{Y}$ denotes dY/dt. Substituting (12.93) in (12.90) and collecting terms gives

$$x^2[\dot{Q} + Q^2 - 1] + x[\dot{R} + RQ - 2SQ] + \dot{M} + \frac{R^2}{4} - RS + \sigma^2 Q = 0. \tag{12.94}$$

Since (12.94) must hold for any value of x, we must have

$$\dot{Q} = 1 - Q^2, \qquad Q(T) = 0 \tag{12.95}$$

$$\dot{R} = 2SQ - RQ, \qquad R(T) = B \tag{12.96}$$

$$\dot{M} = RS - R^2/4 - \sigma^2 q, \qquad M(T) = 0, \tag{12.97}$$

where the boundary conditions for the system of simultaneous differential equations (12.95), (12.96) and (12.97) are obtained by comparing (12.92) with the boundary condition $B(T,x) = Bx$ of (12.88).

To solve (12.95) we expand $\dot{Q}/(1-Q^2)$ by partial fractions to obtain

$$\frac{\dot{Q}}{2}\left[\frac{1}{1-Q} + \frac{1}{1+Q}\right] = 1,$$

which can be easily integrated. The answer is

$$Q = \frac{y-1}{y+1} \tag{12.98}$$

where,

$$y = e^{2(t-T)} . \tag{12.99}$$

Since S is assumed to be constant, we can reduce (12.95) to

$$\dot{R}^o + R^o Q = 0, \quad R^o(T) = B - 2S$$

by the change of variable defined by $R^o = R - 2S$. Clearly the solution is given by

$$\log R^o\Big|_t^T = -\int_t^T Q(\tau)d\tau,$$

which can be further simplified to obtain

$$R = 2S + \frac{2(B-2S)\sqrt{y}}{y+1}. \tag{12.100}$$

Having obtained solution for R and Q, we can easily express (12.97) as

$$M(t) = -\int_t^T [RS - R^2/4 - \sigma^2 Q]dt . \tag{12.101}$$

The optimal control is defined in (12.89) and the use of (12.98) and (12.100) yields

$$u^* = V_x/2 = Qx + R/2 = S + \frac{(y-1)x + (B-2S)\sqrt{y}}{y+1} . \tag{12.102}$$

Remarks

1) The optimal production rate in (12.102) equals the demand rate plus a correction term which depends on the level of inventory and the distance from horizon. Since $(y-1) < 0$ for $t < T$, it is clear that for the lower values of x, the optimal production rate is likely to be positive. However, if x is very high, the correction term will become smaller than $-S$ and the optimal control will be negative. In other words, if inventory level is too high, the factory can save money by disposing a part of the inventory resulting in lower holding costs.

2) If the demand rate S were time-dependent, it would have changed the solution of (12.96). Having computed this new solution in place of (12.100), we can once again obtain the optimal control as $u^* = Qx + R/2$.

3) Note that when $T \to \infty$, we have $y \to 0$ and

$$u^* \to S - x. \tag{12.103}$$

But the undiscounted objective function value (12.86) in this case becomes $(-\infty)$. Clearly, any other policy will render the objective function value to be $-\infty$. In a sense, the optimal control problem becomes ill-posed. One way to get out of this difficulty is to impose a nonzero discount rate. This is carried out in [Sethi and Thompson, 1980].

4) It would help our intuition if we could draw a picture of the path of inventory levels over time. Since inventory level is a stochastic process we can only draw a typical sample path. Such a sample path is shown in Figure 12.4.

5) Recently there have been many applications of stochastic optimal control to problems in finance and natural resources, see, for example: [Merton, 1971, 1973], [Black and Scholes, 1973], [Pindyck, 1973], [Arrow and Chang, 1978], [Richard, 1979], [Vickson, 1979], [Derzko and Sethi, 1980], and [Pindyck, 1980].

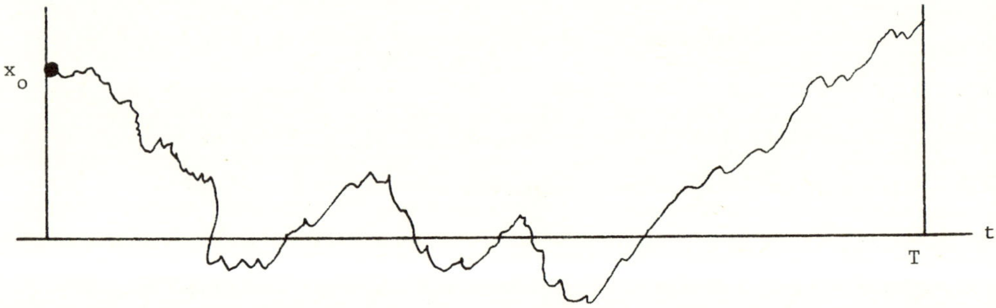

Figure 12.4 Sample Path for Positive x_o and B

If the horizon is long enough the optimal control will bring the inventory level to the goal level $\bar{x} = 0$. It will then hover around this level until t is sufficiently close to the horizon T. During ending phase, the optimal control will try to build up the inventory level in response to a positive evaluation B for ending inventory.

12.5 Impulsive Control

In Chapters 2 and 3 we studied the control of systems governed by ordinary differential equations. In these cases the state variable can only change continuously since control affects only the time derivatives of the state variables. In Chapters 4 and 6 we developed models in which the state variable did change instantaneously by a finite amount as a result of the application of an impulse control. These situations arose when an optimal control policy was bang-bang but there was no upper bound on the control variable. However, there are other kinds of situations in management science and other areas in which a finite change in the value of a state variable is explicitly permitted at certain instants of time, and these do not easily lend themselves to the impulse control methodology given in Chapters 4 and 6. For that reason we shall briefly discuss the theory of <u>impulsive control</u> which is capable of handling such problems. The state variable dynamics in these problems cannot be described by ordinary differential equations.

Consider the example of an oil producer who pumps oil from a single well. The output rate of the well is proportional to the remaining stock $x(t)$, so that the output rate is $Dx(t)$, where D is a constant. The state equation is thus

$$\dot{x} = -Dx, \quad x(0) = 1 \tag{12.104}$$

where 1 is the starting stock of a new oil well. After producing for some time, the oil remaining in the well becomes low so that the output rate is too low to be economic to pump. The oil producer then abandons the existing well and drills a new one, and repeats this procedure until the horizon time T. Let $t_1, t_2, \ldots, t_{N(T)}$ be the times at which the new wells are drilled; note that $N(T)$ is the total number of wells drilled. The graph of the stock of oil available to the producer at times in $[0,T]$ is sketched in Figure 12.5.

A convenient way of formally describing this process is to use delta functions. Let the state equation (12.104) be replaced by

$$\dot{x}(t) = Dx(t) + \sum_{i=1}^{N(T)} \delta(t-t_i)\,[1-x(t)], \; x(0) = 1 \tag{12.105}$$

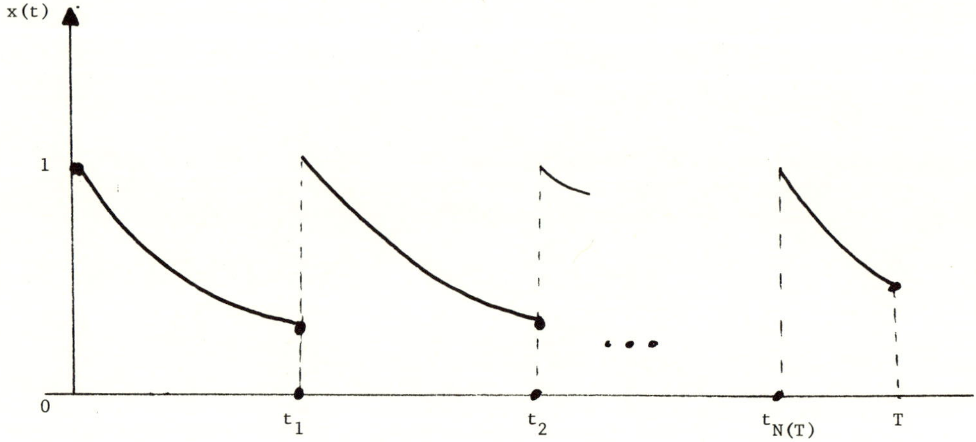

Figure 12.5

where $\delta(t-t_i)$ is the Dirac delta function explained in Section 12.3. This new equation has the same solution as that of (12.104) in the interval $[0,t_1)$. We define $x(t_1) = x(t_1^-)$. The stock of oil just after t_1, that is $x(t_1^+)$ is given by

$$x(t_1^+) = x(t_1^-) + \int_{t_1^-}^{t_1^+} \delta(t-t_1)[1-x(t)]dt = x(t_1) + [1-x(t_1)] = 1 \quad (12.106)$$

Continuation of this procedure shows that the graph of the solution of (12.105) in $[0,T]$ is the graph shown in Figure 12.5.

In equation (12,105) the times t_i at which the wells were drilled were supplied exogeneously. In an optimal impulsive control problem, the theory of which was developed by [Bensoussan and Lions, 1975], these times would have to be determined in order to maximize a given objective function. In the next section we formulate the complete version of the oil driller's problem.

12.5.1 The Oil Driller's Problem

Let $v(t)$ be the drilling control variable where $v(t) \in [0,1]$; $v(t) = 0$ will denote no drilling and $v(t) = 1$ will denote the drilling of a single well having initial stock 1. The other values of $v(t)$ in $[0,1]$ are possible but will not occur in this example because $v(t)$ apears linearly in the problem. We can now rewrite (12.105) as

$$\dot{x}(t) = Dx(t) + \sum_{i=1}^{N(T)} \delta(t-t_i)v(t)[1-x(t)], \; x(0) = 1 \; . \tag{12.107}$$

The solution of the problem involves the determination of the drilling times t_i, the magnitude $v(t_i)$, and the number $N(T)$ of drillings. Note in (12.107) that if $t \neq t_i$ for all i's then all the terms in the summation sign are zero. If $t = t_i$ then $x(t_i+) = v(t_i)$ which means that we have abandoned the old well and drilled a new well with stock equal to $v(t_i)$. A special case of this problem occurs when $N(T)$ is given. This problem reduces to an ordinary calculus problem, see Exercise 12.10.

The objective function of the oil driller is to maximize his profit, which is the difference between the revenue from the oil pumped minus the cost of drilling new wells. Since the cost of pumping the oil is fixed, we ignore it. The objective function is

$$\text{Maximize}\{J = \int_0^T P\,Dx(t)dt - \sum_{i=1}^{N(T)} Qv(t_i)\} \tag{12.108}$$

where P is the unit price of oil and Q is the drilling cost of drilling a well having initial stock 1. Therefore P is the total value of the entire stock of oil in a new well. See [Case, 1979] for the interpretation of this problem as, "The Optimal Painting of a Roadside Inn,"

In the next section we will discuss a maximum principle for impulsive optimal control problems, based on work by [Blaquiere, 1978].

The solution of the oil drillers problem appears in Section 12.5.3.

12.5.2 Maximum Principle for Impulsive Optimal Control

In (2.4) we stated the basic optimal control problem with state variable x and an ordinary control variable u. Now we will add an impulsive control variable $v \in \Omega_v$ and two associated functions. The first function is $G[x, v, t]$ which represents the cost or profit associated with impulsive control; the second function is $g[x, v, t]$ which represents the instantenous finite change in the state variable when the impulsive control is applied.

With this notation we can state the impulsive optimal control problem as:

$$\text{Maximize}\left\{J = \int_0^T F(x,u,t)dt + \sum_{i=1}^{N(T)} G[x(t_i,v(t_i),t_i] + S[x(T)]\right\}$$

Subject to

$$\dot{x}(t) = f(x(t),u(t),t) + \sum_{i=1}^{N(T)} \delta(t-t_i)g[x(t),v(t),t], \; x(0)=x_0 \tag{12.109}$$

$$u \in \Omega_u \quad \text{and} \quad v \in \Omega_v$$

As before we will have

$$x(t_i^+) = x(t_i) + g[x(t_i),v(t_i),t_i] \tag{12.110}$$

at the times t_i at which impulse control is applied, that is, $g[x(t_i),v(t_i),t] > 0$.

Blaquiere has developed [Blaiquiere, 1978] the maximum principle necessar optimal conditions to deal with the problem in (12.109). To state these conditions we first define the <u>ordinary Hamiltonian function</u>:

$$H(x,u,\lambda,t) = F(x,u,t) + \lambda f(x,u,t) \tag{12.111}$$

and the impulse Hamiltonian function:

$$H^I(x,v,t) = G(x,v,t) + \lambda(t+)g(x,v,t). \tag{12.112}$$

We now state the impulsive maximum principle. Let u^*, $v^*(t_i)$, $i = 1,\ldots,N^*(T)$, where $t_i > t_{i-1} > 0$, and associated x^* be an optimal trajectory for (12.109); then there exists an adjoint variable λ^* such that conditions (12.113) hold:

$$\begin{array}{ll}
\text{(i)} & \dot{x}^* = f(x^*,u^*,t), \quad t \in [0,T], \ t \neq t_i \\
\text{(ii)} & x^*(t_i^+) = x^*(t_i) + g[x^*(t_i),v^*(t_i),t_i] \\
\text{(iii)} & \dot{\lambda}^* = -H_x[x^*,u^*,\lambda^*,t], \ \lambda^*(T) = S_x[x^*(T)], \ t \neq t_i \\
\text{(iv)} & \lambda^*(t_i) = \lambda^*(t_i^+) + H_x^I[x^*(t_i),v^*(t_i),t_i] \\
\text{(v)} & H[x^*,u^*,\lambda^*,t] \geq H[x^*,u,\lambda^*,t] \text{ for all } u \in \Omega_u, \ t \neq t_i \\
\text{(vi)} & H^I[x^*(t_i),v^*(t_i),t_i] \geq H^I[x^*(t_i),v,t_i] \text{ for all } v \in \Omega_v \\
\text{(vii)} & H[x^*(t_i^+),u^*(t_i^+),\lambda^*(t_i^+),t_i] = \\
& \quad H[x^*(t_i),u^*(t_i),\lambda^*(t_i),t_i] + H_t^I[x^*(t_i),v^*(t_i),t_i]
\end{array} \tag{12.113}$$

If $t_1 = 0$ then the equality sign in (vii) should be replaced by a $\geq$ sign when $i = 1$. Clearly (i) and (ii) are equivalent to the state equation in (12,109) with the optimum value substituted for the variables.

12.5.3 Solution of the Oil Driller's Problem

We now give a solution to the oil driller's problem on the assumption that T is sufficiently small that no more than one drilling will be found to be optimal. We restate the problem in (12.114) for this case when t_1 is the drilling time to be determined.

$$\begin{aligned} &\text{Maximize}\{J = \int_0^T PDx(t)dt - Qv(t_1)\} \\ &\text{Subject to} \\ &\dot{x}(t) = -Dx(t) + \delta(t-t_1)v(t)[1-x(t)],\ x(0) = 1 \\ &0 \le v(t) \le 1 \end{aligned} \tag{12.114}$$

To apply the maximum principle we define the Hamiltonian functions corresponding to (12.111) and (12.112):

$$H(x,\lambda) = PDx + \lambda(-Dx) = Dx(P-\lambda) \tag{12.115}$$

$$H^I(x,v) = -Qv + \lambda(t+)v(1-x) \quad . \tag{12.116}$$

The ordinary Hamiltonian (12.115) has no ordinary control variable because there isn't any.

We now obtain the necessary conditions of (12.113) for the oil driller's problem:

$$\dot{x} = -Dx,\ t \in [0,T],\ t \neq t_1 \tag{12.117}$$

$$x[t_1^+] = x(t_1) + v(t_1)[1-x(t_1)] \tag{12.118}$$

$$\dot{\lambda} = -D(P-\lambda),\ \lambda(T) = 0,\ t \neq t_1 \tag{12.119}$$

$$\lambda(t_1) = \lambda(t_1^+) - v(t_1)\lambda(t_1^+) \tag{12.120}$$

$$[-Q + \lambda(t_1^+)(1-x)]v^*(t_i) \ge [-Q + \lambda(t_i^+)(1-x)]v \text{ for all } v \in [0,1] \tag{12.121}$$

$$Dx(t_1^+)[P-\lambda(t_1^+)] = Dx(t_1)[P-\lambda(t_1)] \quad . \tag{12.122}$$

The solution of (12.119) for $t > t_1$

$$\lambda(t) = P[1-e^{-D(T-t)}], \quad t \in (t_1,T] \quad . \tag{12.123}$$

From (12.121) the optimum impulsive control at t_1 is

$$v^*(t_1) = \text{bang}[0,1; -Q + \lambda(t_1^+)\{1-x(t_1)\}] \quad . \tag{12.124}$$

The region in the (x,t) space in which there is certain to be no drilling is given by the set of all (x,t) for which

$$-Q + \lambda(t^+)(1-x) < 0 \quad . \tag{12.125}$$

After the drilling, $\lambda(t_i^+) = P[1-e^{-D(T-t_1)}]$ which is given by (12.123). Substituting this value into (12.125) gives the condition

$$-Q + P[1-e^{-D(T-t_1)}](1-x) < 0 \quad . \tag{12.126}$$

The boundary of the no-drilling region is obtained when the $>$ sign in (12.126) is replaced by equality; this yields the curve

$$x = [1 - \frac{Q}{P(1-e^{-D(T-t_1)})}] \triangleq \varphi(t_1) \tag{12.127}$$

shown in Figure 12.6. Note that we did not label the region <u>below</u> the curve

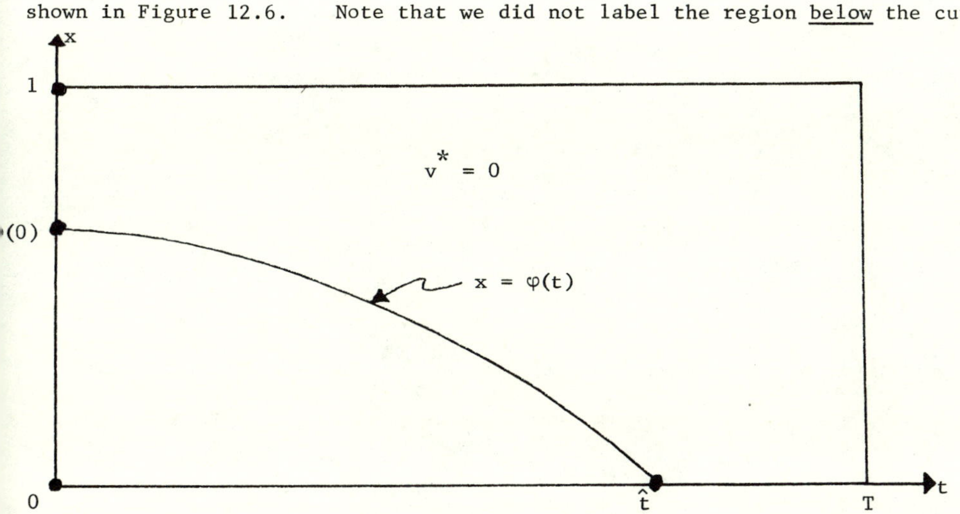

Figure 12.6

represented by the complement of (12.126) with $v^* = 1$ since it does not represent the condition $-Q + \lambda(t^+)(1-x) > 0$ prior to drilling.

Figure 12.6 is drawn under the assumption that

$$\varphi(0) = \frac{Q}{P[1-e^{-DT}]} < 1 \tag{12.128}$$

so that the solution is non trivial; see Exercises 12.7 and 12.8. In order for (12.128) to hold it is clear that Q must be relatively small, and P, D and T must be relatively large. In the contrary situation $\varphi(0) > 1$ holds and by Exercise 12.8 no drilling is optimal.

From (12.124) we have

$$v^*(t_1) = 1 .$$

Using this in (12.118) and (12.120) we have

$$x(t_1^+) = 1 \quad \text{and} \quad \lambda(t_1) = 0 \quad . \tag{12.129}$$

From (12.123)

$$\lambda(t_1^+) = P[1-e^{-D(T-t_1)}].$$

Substituting these results into (12.122) and solving we get

$$x(t_1) = e^{-D(T-t_1)} = \psi(t_1) . \tag{12.130}$$

The curve $x = \psi(t)$ is drawn in Figure 12.7. The part BC of the ψ curve lies in the no drilling region which is above the φ curve as indicated in Figure 12.6. The part AB of the ψ curve is shown darkened and represents the "drilling curve" for the problem. The optimal state trajectory starts from $x(0) = 1$ and decays exponentially at rate D until it hits the drilling

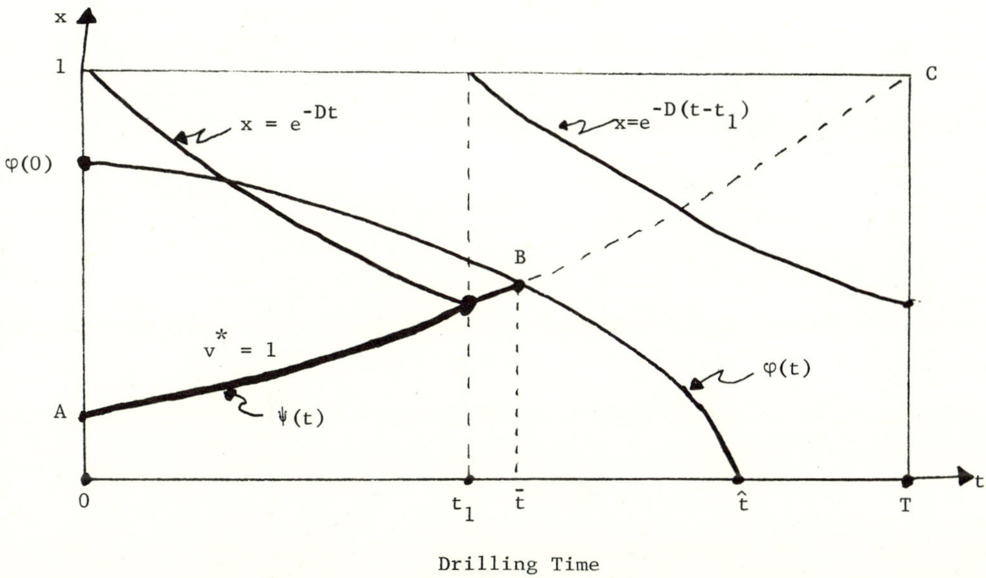

Figure 12.7

curve AB; this intersection point determines the drilling time t_1. In Exercise 12.9 you are asked to show that $t_1 = T/2$. At time t_1 the old well is abandoned, a new well is drilled making $x(t_1^+) = 1$. The stock of the new well follows a similar decay path as shown.

The complete solution for the adjoint variable $\lambda(t)$ is obtained from (12.123), (12.129) and (12.119) as

$$\lambda(t) = \begin{cases} P[1-e^{-D(T-t)}] & \text{for } t \in (t_1,T) \\ P[1-e^{-D(t_1-t)}] & \text{for } t \in [0,t_1] \end{cases} .$$

The drilling function $-Q + \lambda(t_1^+)(1-x)$ which is the argument of the function in (12.124) has a graph as shown in Figure 12.8. Note that t_1 is a point of discontinuity of the drilling function.

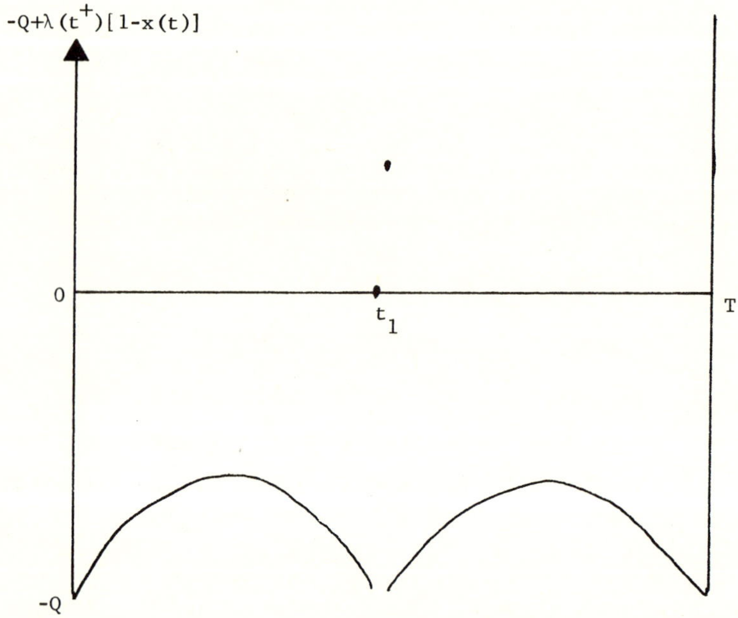

Figure 12.8

Figure 12.7 is drawn on the assumption that T is sufficiently large so that the x trajectory hit the drilling curve AB at time t_1. If T is too small the x trajectory will not hit AB and no drilling is optimal. For

$$T = \bar{T} = -\frac{2}{D}\ln\left(1 - \sqrt{\frac{Q}{P}}\right) \tag{12.131}$$

the x trajectory goes through point B of Figure 12.7, which is the point at which we are indifferent between drilling and no drilling. In Exercise 12.11 you are asked to derive (12.131) and to show that for $T > \bar{T}$ the value of $-Q + \lambda(t_1^+)(1-x)$ is positive, as shown in Figure 12.8), so that there will be drilling at t_1.

The oil driller's problem is a pure impulsive optimal control problem as it did not involve any ordinary control variable. In the next section, we formulate a machine maintenance and replacement problem containing both ordinary and impulsive control variables. The model, which generalizes Thompson's maintenance model described in Section 8.1.1 and which is similar to the Sethi-Morton model described in Section 8.3.1, is due to[Blaquiere, 1978].

12.5.4 Machine Maintenance and Replacement

In order to define the model, we use the following notation:

T = the given terminal or horizon time,

$x(t)$ = the quality of machine at time t, $0 \le x \le 1$; a higher value of x denotes a better quality,

$u(t)$ = the ordinary control variable denoting rate of maintenance at time t; $0 \le u \le U < b/g$,

b = the constant rate at which quality deteriorates in the absence of any maintenance,

g = the maintenance effectiveness coefficient,

π = the production rate per unit quality of the machine,

K = the trade-in value per unit quality, i.e., the old machine gives only a credit against the price of the new machine and that it has no terminal salvage value,

C = cost of new machine per unit quality,

t_1 = the replacement time; for simplicity, we assume at most one replacement to be optimal in the given horizon time; see Section 12.5.3, v = the replacement variable; $0 \leq v \leq 1$; v represent a fraction of the old machine replaced by the same fraction of a new machine. This interpretation will make sense because we will show that v is either 0 or 1.

With this notation, the impulsive optimal control problem is:

$$\begin{aligned} &\text{Max}\{J = \int_0^T (\pi x - u)dt + v(t_1)[Kx(t_1) - C]\} \\ &\text{subject to} \\ &\quad \dot{x} = -bx + gu + \delta(t-t_1)v(t)[1-x(t)] \\ &\quad 0 \leq u \leq U, \quad 0 \leq v \leq 1 \end{aligned} \tag{12.132}$$

In writing (12.132) we have assumed that a fraction θ of a machine with quality x has a quality θx. Furthermore, we note that the solution of the state equation will always satisfy $0 \leq x \leq 1$ on account of the assumption that $u \leq U \leq b/g$.

12.5.5 Application of the Impulsive Maximum Principle

To apply the maximum principle, we define the Hamiltonian functions:

$$H(x,u,\lambda) = \pi x - u + \lambda(-bx+gu) = (\pi-\lambda b)x + (-1+\lambda g)u \tag{12.133}$$

$$H^I(x,v) = [(Kx-C) + \lambda(t_1^+)(1-x)]v. \tag{12.134}$$

We can now state the necessary conditions of (12.113) for the machine maintenance and replacement model:

$$\dot{x} = -bx + gu, \quad t = t_1 \tag{12.135}$$

$$x(t_1^+) = x(t_1) + v(t_1)[1 - x(t_1)] \tag{12.136}$$

$$\dot{\lambda} = -\pi + \lambda b, \ \lambda(T) = 0, \quad t \neq t_1 \tag{12.137}$$

$$\lambda(t_1) = \lambda(t_1^+) + [K - \lambda(t_1^+)]v(t_1) \tag{12.138}$$

$$(\pi-\lambda b)x + (-1+\lambda g)u^* \geq (\pi-\lambda b)x + (-1+\lambda g)u, \quad \text{for all } u \in [0,U], \tag{12.139}$$

$$\{Kx(t_1) - C + \lambda(t_1^+)[1-x(t_1)]\}v^*(t_1) \geq \{Kx(t_1)-C+\lambda(t_1^+)[1-x(t_1)]\}v$$
$$\text{for all} \quad v \in [0,1], \tag{12.140}$$

$$\pi x(t_1^+) - \lambda(t_1^+)bx(t_1^+) + [-1+\lambda(t_1^+)g]u^*(t_1^+)$$
$$= \pi x(t_1) - \lambda(t_1)bx(t_1) + [-1+\lambda(t_1)g]u^*(t_1) \tag{12.141}$$

The solution of (12.137) for $t_1 < t \leq T$ is

$$\lambda(t) = \frac{\pi}{b}[1-e^{-b(T-t)}]. \tag{12.142}$$

From (12.139) the maintenance control is bang-bang:

$$u^*(t) = \text{bang}[0,U; -1 + \lambda g]. \tag{12.143}$$

The switching point t^* is given by setting $-1 + \lambda g = 0$. Thus

$$t^* = T + \frac{1}{b}\ln[1 - \frac{b}{\pi g}] \tag{12.144}$$

provided the right hand side is in the interval $[0,T]$; otherwise set $t^* = 0$. We can graph the optimal maintenance control as in Figure 12.9.

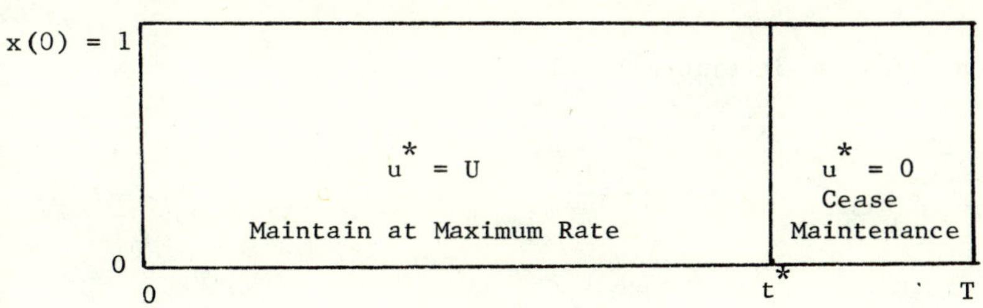

Figure 12.9

From (12.140), the optimal replacement control is also bang-bang:

$$v^*(t_1) = \text{bang}[0,1;\ Kx - C + \lambda(t_1^+)\{1 - x(t_1)\}]. \tag{12.145}$$

As in Section 12.5.3, we set the argument of (12.145) to zero to obtain $\varphi(t)$:

$$x(t_1) = \frac{C - \lambda(t_1+)}{K - \lambda(t_1+)} = \frac{C - \frac{\pi}{b}[1-e^{-b(T-t_1)}]}{K - \frac{\pi}{b}[1-e^{-b(T-t_1)}]} = \varphi(t_1). \tag{12.146}$$

To graph $\varphi(t)$, we compute

$$\varphi(0) = \frac{C - \frac{\pi}{b}[1-e^{-bT}]}{K - \frac{\pi}{b}[1-e^{-bT}]} \tag{12.147}$$

and compute the time

$$\hat{t} = T + \frac{1}{b}\ln\left(1 - \frac{bC}{\pi}\right) \tag{12.148}$$

which makes $\varphi(\hat{t}) = 0$.

For simplicity in analysis, we assume

$$gK \leq 1 < gC < g\pi/b, \quad \text{and} \quad \hat{t} > 0. \tag{12.149}$$

In Exercise 12.12, you are asked to show that,

$$0 < \hat{t} < t^* < T. \tag{12.150}$$

We can now graph $\varphi(t)$ as in Figure 12.10.

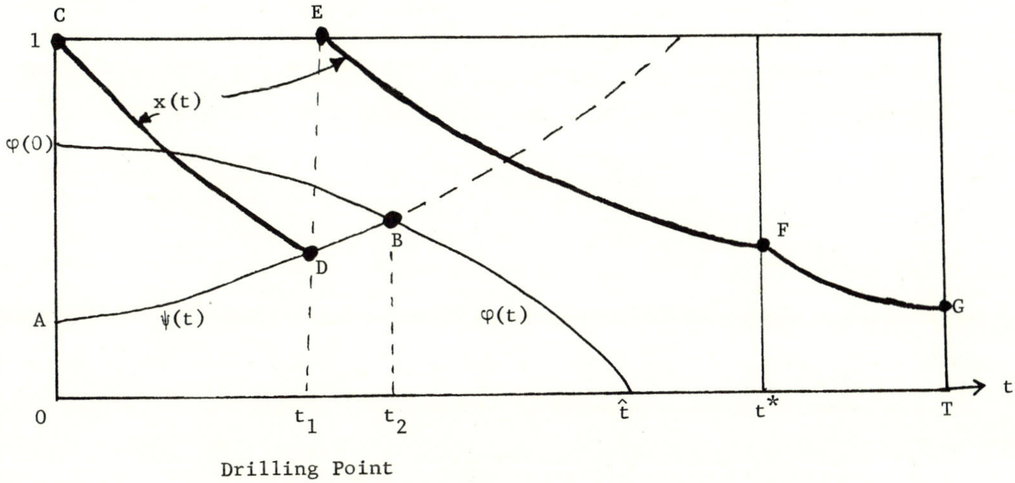

Figure 12.10

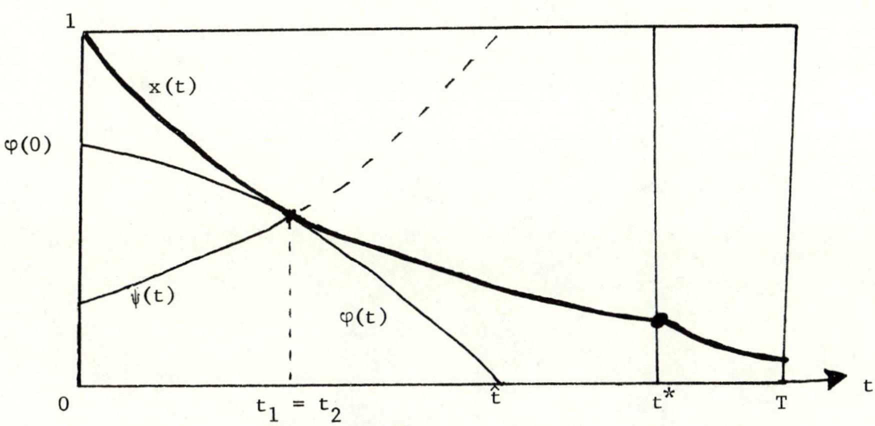

Figure 12.11

As in the oil driller's problem, we next obtain $\psi(t)$ by using (12.141). We observe that $v(t_1) = 1$ and therefore, $\lambda(t_1) = K$ from (12.138) and $x(t_1+) = 1$ from (12.136). Since $gK \leq 1$ from (12.149), we have

$$-1 + g\lambda(t_1) = -1 + gK \leq 0$$

and, thus, $u^*(t_1) = 0$ from (12.143). Since $t_1 \leq \hat{t} < t^*$ from (12.150), we have $u^*(t_1+) = U$.

Substituting all these values in (12.141), we have

$$\pi - \frac{\pi}{b}[1-e^{-b(T-t_1)}]b + \{-1 + \frac{\pi}{b}[1-e^{-b(T-t_1)}]g\} U$$

$$= \pi x(t_1) - Kbx(t_1), \tag{12.151}$$

which yields

$$x(t_1) = \frac{U(g\pi - b) + \pi(b - gU)e^{-K(T-t_1)}}{b(\pi - Kb)} \triangleq \psi(t_1). \tag{12.152}$$

The graph of $x(t) = \psi(t)$ is shown in Figure 12.10. In this figure, AB represents the replacement curve. The optimal trajectory is shown by CDEFG under the assumption that $t_1 > 0$ and $t_1 < t_2$, where t_2 is the intersection point of curves $\varphi(t)$ and $\psi(t)$, as shown in Figure 12.10.

Figure 12.11 has been drawn for a choice of the problem parameters such that $t_1 = t_2$. In this case we have the option of either replacing the machine at $t_1 = t_2$, or not replacing it as sketched in the figure. If t_1 were larger than t_2, clearly no machine replacement is needed.

To complete the solution of the problem, it is easy to compute $\lambda(t)$ corresponding to the policy CDEFG of Figure 12.10, and show that the maximum principle hold. We omit details.

Exercises for Chapter 12

12.1 The bilinear quadratic advertising model [Deal, Sethi, and Thompson, 1979]: Let x_i be the market share of firm i, and u_i be its advertising rate, for $i = 1,2$; the state equations are:

$$\dot{x}_1 = b_1u_1(1-x_1-x_2) + e_1(u_1-u_2)(x_1+x_2) - a_ix_i, \quad x_1(0) = x_{10}$$

$$\dot{x}_2 = b_2u_2(1-x_1-x_2) + e_2(u_2-u_1)(x_1+x_2) - a_2x_2, \quad x_2(0) = x_{20},$$

where b_i, e_i, a_i are given positive constants. Firm i wants to maximize

$$J_i = w_ie^{-\rho T}x_i(T) + \int_0^T (c_ix_i - u_i^2)e^{-\rho t}dt$$

where w_i, c_i, and ρ are positive constants. Derive the necessary conditions for the open loop Nash solution, and formulate the resulting boundary value problem. (In [Deal, 1979] a numerical solution to this problem with $e_i = e_2 = 0$ is given.)

12.2 Develop the nonlinear model described in the last paragraph of Section 12.1.3 by rewriting (12.19) and (12.22) for the model. Derive the adjoint equation for λ^i for the ith producer, and show that the closed-loop Nash policy for producer i is given by

$$f'(v^i) = \frac{c^i}{(p^i - \lambda^i)x}.$$

12.3 Verify that (12.48) is a solution of (12.46).

12.4 Obtain or verify that (12.50) is the solution of (12.43) with initial condition (12.32) and (12.33).

12.5 Consider the following discrete time scalar problem:

$$x^{t+1} - x^t = ax^t + w^t, \quad y^t = hx^t + v^t$$

where w^t and v^t are gaussian purely random sequences with

$$E[w^t] = E[v^t] = 0, \quad E[w^t w^\tau] = g\delta_{t\tau}$$

$$E[v^t v^\tau] = r\delta_{t\tau}$$

where h, q and r are constants. The initial condition $x(0)$ is a gaussian random variable with mean $\bar{x}^0$ and variance m_0, and let $p_0 = \dfrac{rm_0}{r + m_0 h^2}$. Show that the Kalman filter is given by:

$$\hat{x}^{t+1} - \hat{x}^t = a\hat{x}^t + \frac{p_{t+1}h}{r}(y^{t+1} - h(a+1)\hat{x}^t)$$

$$p_t = \frac{r[(a+1)^2 p_{t-1} + q]}{r + h^2[(a+1)^2 p_{t-1} + q]}, \quad p_0 \text{ as above.}$$

12.6 Consider the following continuous time scalar problem:

$$\dot{x} = w, \quad y = 2x + v$$

where w and v are white noise processes with

$$E[w(t)] = E[v(t)] = 0, \quad E[w(t)w(\tau)] = q\delta(t-\tau)$$

$$E[v(t)v(\tau)] = r\delta(t-\tau)$$

where q and r are constants. The initial condition $x(0)$ is a gaussian random variable with mean 0 and variance p_0. Show that the Kalman-Bucy filter is given by

$$\dot{\hat{x}} = \frac{2p}{r}[y - 2\hat{x}], \quad \hat{x}(0) = 0.$$

$$\dot{p} = -\frac{4p^2}{r} + q, \quad p(0) = p_0.$$

Compare these expressions with the results of Exercise 12.3.1. If you are ambitious, solve for p.

12.7 Assume $\varphi(0) < 1$ as in (12.128). Show that $Q < P$ and $0 < \hat{t} < T$.

12.8 If $\varphi(0) > 1$, show that no drilling is optimal.

12.9 Show that the point $t_1 = T/2$ in Figure 12.7.

12.10 Find the optimal drilling time for the oil drilling problem by using ordinary calculus under the assumption that there is exactly one drilling time when an oil well of unit initial stock is drilled.

12.11 Derive (12.131) and show that for $T > \bar{T}$, the drilling function

$$-Q + \lambda(t_1 +)[1 - x(t)] > 0.$$

12.12 Show that the inequalities in (12.150) follow from the assumptions in (12.149).

APPENDIX A1: SOLUTIONS OF LINEAR DIFFERENTIAL EQUATIONS

A1.1.1 Linear Differential Equations with Constant Coefficients

Linear differential equations with constant coefficients are usually written as

$$y^{(n)} + a_1 y^{(n-1)} + \dots + a_{n-1} y^{(1)} + a_n y = g \tag{A1.1.1}$$

where a_k, $k = 1,\dots,n$, are numbers, $y^{(k)} = \frac{d^k y}{dt^k}$, and $g = g(t)$ is a known function of t. We shall denote by $D = \frac{d}{dt}$ the derivative operator, so that the differential equation now becomes

$$p(D)y = (D^n + a_1 D^{n-1} + \dots + a_{n-1} D + a_n)\, y = g \quad . \tag{A1.1.2}$$

If $g(t) \equiv 0$ the equation is said to be homogeneous. If $g(t) \not\equiv 0$ then the homogeneous or reduced equation is obtained from (A1.1.2) by replacing g by 0.

If y and y^* are two different solutions to (A1.1.2) then it is easy to show that $y - y^*$ solves the reduced equation of (A1.1.2). Hence if y is any solution to (A1.1.2) it can be written as

$$y = y^* + y^h \tag{A1.1.3}$$

where y^* is any other particular solution to (A1.1.2) and y^h is a suitable solution to the homogeneous equation. Therefore solving (A1.1.2) involves

(a) Finding all solutions to the homogeneous equation, called the general solution.

(b) Finding a particular solution to the given equation.

The rest of these notes indicate how to solve these two problems.

Given (A1.1.1.2) the auxiliary equation is

$$p(m) = m^n + a_1 m^{n-1} + \ldots + a_{n-1} m + a_n = 0 \quad . \tag{A1.1.4}$$

In other words $p(m)$ is obtained from $p(D)$ by replacing D by m. The auxiliary equation is an ordinary polynomial of nth degree and has n real or complex roots, counting multiple roots according to their multiplicity. We will see that, given these roots, we can write the general solution forms of homogeneous linear differential equations.

A1.1.2 Homogeneous Equations of Order One

Here the equation is

$$(D - a)y = y' - ay = 0$$

which has $y = Ce^{at}$ as its general solution form.

A1.1.3 Homogeneous Equations of Order Two

Here the differential equation can be factored (using the quadratic formula) as

$$(D - m_1)(D - m_2)y = 0$$

where m_1 and m_2 can be real or complex. The solution forms are given in Table A1.1.1.

Examples

Differential Equation	General Solution Form
1. $y'' - 4y' + 4 = 0$	$y(t) = e^{2t}(C_1 + C_2 t)$
2. $y'' - 4y' + 3 = 0$	$y(t) = C_1 e^{3t} + C_2 e^{t}$ $= e^{2t}(C_1 \sinh t + C_2 \cosh t)$
3. $y'' - 4y + 5 = 0$	$y(t) = e^{2t}(C_1 e^{it} + C_2 e^{-it})$ $= e^{2t}(C_1 \sin t + C_2 \cos t)$

Root	General Solution Form
$m_1 \neq m_2$, real $m_1 = a + b, \quad m_2 = a - b$	$y(t) = C_1 e^{m_1 t} + C_2 e^{m_2 t} = e^{at}(C_1 e^{bt} + C_2 e^{-bt})$ or $y(t) = e^{at}(C_1 \sinh bt + C_2 \cosh bt)$
$m_1 = m_2 = m$	$y(t) = (C_1 + C_2 t)\, e^{mt}$
$m_1 \neq m_2$, complex $m_1 = a + bi, \quad m_2 = a - bi$	$y(t) = C_1 e^{m_1 t} + C_2 e^{m_2 t} = e^{at}(C_1 e^{ibt} + C_2 e^{-ibt})$ or $y(t) = e^{at}(C_1 \sin bt + C_2 \cos bt)$ or $y(t) = e^{at}[C_1 \sin (bt + C_2)]$ or $y(t) = e^{at}[C_1 \cos (bt + C_2)]$

Table A1.1.1 General Solution Forms for Second Order Linear Homogeneous Equations, Constant Coefficients

A1.1.4 Homogeneous Equations of Order n

When (A1.1.2) is of order n, the auxiliary equation $p(m) = 0$ has n roots, when multiple roots are counted according to their multiplicity. Also, complex roots occur in conjugate pairs. The general solution of the homogeneous equation is the sum of the solutions associated with each multiple root. They can be found in Table A1.1.2 for each root and should be added together to form the general solution.

Examples

Differential Equation	General Solution Form
1. $D^2(D^2-4D+4)y = 0$	$y(t) = C_1+C_2t + e^{2t}(C_3+C_4t)$
2. $(D-3)^2(D+5)^3(D-4D+5)^2y=0$	$y(t) = e^{3t}(C_1+C_2t) + e^{-5t}(C_3+C_4t+C_5t^2)$ $+ e^{2t}[(C_6+C_7t)\sin t + (C_8+C_9t)\cos t]$
3. $(D^2-2D+2)^3(D^2-2m-2)^2\ y = 0$	$y(t) = e^t[(C_1+C_2t+C_3t^2)\sin t + (C_4+C_5t+C_6x^2)\cos t$ $+(C_7+C_8t)e^{2t} + (C_9+C_{10}t)e^{-2t}]$

Root	Multiplicity	General Solution Form
m_j, real	$r_j = 1$	$y_j(t) = Ce^{m_jt}$
	$r_j > 1$	$y_j(t)=(C_1+C_2t +\ldots+ C_{r_j}t^{r_j-1})e^{m_jt}$
Complex Conjugate $a_j \pm b_ji$	$r_j = 1$	$e^{a_jt}(C_1 \sin b_jt + C_2 \cos b_jt)$
	$r_j > 1$	$e^{a_jt}[(C_1 + C_2t +\ldots+ C_{r_j}t^{r_j-1})\sin b_jt$ $+(C_{r_j+1} + C_{r_j+2}t +\ldots+ C_{2r_j}t^{r_j-1})\cos b_jt]$

Table A1.1.2. General Solution Forms for Multiple Roots of Auxiliary Equation

A1.1.5 Particular Solutions of Linear D.E. with Constant Coefficients

The particular solution to the inhomogeneous equation (A1.1.2) can usually be found by guessing the form of the answer and then verifying the guess by substitution. Table A1.1.3 shows the correct forms for guessing, for various kinds of forcing functions $g(t)$. Note that the form of the guess depends on whether certain numbers are roots of the auxiliary equation.

Examples

Differential Equation	Particular Integral
1. $y''' - 3y'' = 5$	At^2
2. $y''' - 3y'' = 1 + 3t + 5t^2$	$t^2(A_0 + A_1t + A_2t^2)$
3. $y'' - 4y' + 4 = 3 - t^2$	$A_0 + A_1t + A_2t^2$
4. $y'' - 4y' + 4 = 2 \sin t$	$A \sin t + B \cos t$
5. $y'' - 4y' + 4 = 5 \sin 2t$	$t^2(A \sin 2t + B \cos 2t)$
6. $y'' - 4y' + 4 = 10e^{5t}$	Ae^{5t}
7. $y'' - 4y' + 4 = 10e^{2t}$	$t^2(Ae^{2t})$
8. $y'' - 4y' + 5 = e^{2t} \cos t$	$t(Ae^{2t} \sin t + Be^{2t} \cos t)$
9. $y'' - 4y' + 5 = t^6 \sin t$	$(X \sin t + Y \cos t)$, X, Y, of degree 6.
10. $y'' - 4y' + 5 = t^6e^2 \cos t$	$t(X e^{2t} \sin t + Y e^{2t} \cos t)$, X, Y, of degree 6.

Forcing Function, $g(t)$	K	Particular Integral, $y(t)$
(1) c	0	A
(2) $h(t)$	0	X
(3) $c \sin qt$ or $c \cos qt$	q	$A \sin qt + B \cos qt$
(4) ce^{qt}	q	Ae^{qt}
(5) $ce^{pt}\sin qt$ or $ce^{pt}\cos qt$	$p+iq$	$Ae^{pt} \sin qt + Be^{pt} \cos qt$
(6) $h(t)e^{qt}$	q	Xe^{qt}
(7) $h(t)\sin qt$ or $h(t)\cos qt$	iq	$X \sin qt + Y \cos qt$
(8) $h(t)e^{pt}\sin qt$ or $h(t)e^{pt}\cos qt$	$p+iq$	$X e^{pt} \sin qt + Y e^{pt} \cos qt$

Notation

(a) In the forcing function column, p, q, c are given constants and $h(t)$ is a given polynomial of degree s.

(b) In the particular integral column, A and B are coefficients to be determined and

$$X = A_0 + A_1 t + \ldots + A_s t^s, \quad Y = B_0 + B_1 t + \ldots + B_s t^s$$

are s degree polynomials whose coefficients are tc be determined.

Rule

(a) If the number in the K column is not a root of the auxiliary equation, $p(m) = 0$, then the proper guess for the particular integral is as shown.

(b) If the number in the K column is a root of the auxiliary equation of degree r, then multiply the guess in the last column by t^r.

Table A1.1.3. Particular Solution Forms for Various Forcing Functions

A1.1.6 Extension of Table A1.1.3

If the forcing function $g(t)$ is the sum of several functions, $g = g_1 + g_2 + \ldots + g_k$, each having one of the forms of the table, then we solve for each g_i separately and add the results together to get the complete solution.

A1.1.7 Integrating Factor

Consider the first order linear equation

$$y' + ay = f(t) \tag{A1.1.5}$$

If we multiply both sides of the equation by the integrating factor e^{ax} we get

$$\frac{d}{dt}(ye^{at}) = y'e^{at} + aye^{at} = e^{at}f(t) .$$

Integrating from 0 to x we have

$$y(t) = y(0)e^{-at} + \int_0^t e^{a(\tau-t)}f(\tau)d\tau \tag{A1.1.6}$$

which is the complete solution (homogeneous solution plus particular solution) to the equation.

A1.1.8 Reduction of Linear Equations of Higher Order to First Order Systems

Another way of solving equation (A1.1.1) is to convert it into a system of linear first order equations. We use the transformations

$$z_1 = y, \quad z_2 = y^{(1)}, \ \ldots \ , \ z_n = y^{(n-1)} \quad . \tag{A1.1.8}$$

Then (A1.1.1) becomes

$$z' = \begin{pmatrix} z_1' \\ z_2' \\ \cdot \\ z_{n-1}' \\ z_n' \end{pmatrix} = \begin{pmatrix} 0 & 1 & 0 & \cdots & 0 \\ 0 & 0 & 1 & \cdots & 0 \\ & \cdot & & \cdots & \cdot \\ 0 & 0 & 0 & \cdots & 1 \\ -a_n & -a_{n-1} & -a_{n-2} & \cdots & -a_1 \end{pmatrix} \begin{pmatrix} z_1 \\ z_2 \\ \cdot \\ z_{n-1} \\ z_n \end{pmatrix} + \begin{pmatrix} 0 \\ 0 \\ \cdot \\ 0 \\ g \end{pmatrix} \tag{A1.1.9}$$

In vector form this equation reads

$$z' = Az + b \tag{A1.1.10}$$

with the obvious definitions obtained by comparing (A1.1.9) and (A1.1.10). We will present two ways of solving the first order system (A1.1.10).

The first method involves the matrix exponential function e^{tA} defined by the power series

$$e^{tA} = I + tA + \frac{t^2A^2}{2!} + \ldots = \sum_{k=0}^{\infty} \frac{(tA)^k}{k!} \quad . \tag{A1.1.11}$$

It can be shown that this series converges (component by component) for all values of t; also it is differentiable (component by component) for all values of t and satisfies

$$\frac{d}{dt}(e^{tA}) = Ae^{tA} = (e^{tA})A \quad . \tag{A1.1.12}$$

By analogy with Section (A1.1.7) we try e^{-At} as the integrating factor for (A1.1.10), obtaining

$$(e^{-tA})z - e^{-tA}Az = e^{-tA}b \quad .$$

(Note that the order of matrix multiplication here is important.) Using the chain rule for matrix multiplication of functions, which can be shown to be valid, the above equation becomes:

$$\frac{d}{dt}\left(e^{-tA}z\right) = e^{-At}b \quad .$$

Integrating from 0 to t gives

$$e^{-\tau A}z(\tau)\Big|_0^t = \int_0^\tau e^{-\tau A}b(\tau)d\tau \quad .$$

Evaluating and solving, we have

$$z(t) = e^{tA}z(0) + e^{tA}\int_0^t e^{-\tau A}\, b(\tau)d\tau \quad . \tag{A1.1.13}$$

The analogy between this equation and (A1.1.6) is clear.

Although (A1.1.13) represents a formal expression for the solution of (A1.1.10), it does not provide a computationally convenient way of getting explicit solutions. In order to demonstrate such a method we assume that the matrix A is diagonalizable, that is that there exists a nonsingular square matrix P such that

$$P^{-1}AP = \Lambda \quad . \tag{A1.1.14}$$

Here Λ is the diagonal matrix

$$\Lambda = \begin{pmatrix} \lambda_1 & & \bigcirc \\ & \lambda_2 & \\ & & \ddots \\ \bigcirc & & & \lambda_n \end{pmatrix} \tag{A1.1.15}$$

where the diagonal elements, $\lambda_1,\ldots,\lambda_n$ are eigenvalues of A. The ith column of P is the column eigenvector associated with the eigenvalue λ_i

(to see this multiply both sides of (A1.1.14) by P on the left.) By looking at (A1.1.11) it is easy to see that

$$P^{-1}e^{tA}P = e^{t\Lambda} \quad . \tag{A1.1.16}$$

Suppose we make the definitions

$$z = Pw, \quad z(0) = Pw(0), \quad z' = Pw' \tag{A1.1.17}$$

which in turn imply

$$w = P^{-1}z, \quad w(0) = P^{-1}z(0), \quad w' = P^{-1}w' \quad . \tag{A1.1.18}$$

Substituting (A1.1.14) into (A1.1.10) gives

$$Pw' = APw + b$$
$$w' = P^{-1}APw + P^{-1}b$$

which gives, by using (A1.1.14),

$$w' = \Lambda w + P^{-1}b \quad . \tag{A1.1.19}$$

Since Λ is a diagonal matrix it is easy to solve the homogeneous part of (A1.1.19) which is

$$w' = \Lambda w \quad . \tag{A1.1.20}$$

The solution is $w_i = w_i(0)\, e^{-t\lambda_i}$ for $i = 1,\ldots,n$.

We solve (A1.1.19) completely by multiplying through by the integrating factor $e^{-t\Lambda}$:

$$\frac{d}{dx}(e^{-t\Lambda}w) = e^{-t\Lambda}w' - e^{-t\Lambda}\Lambda w = e^{-t\Lambda}P^{-1}w \quad .$$

Integrating this equation from 0 to gives

$$w(t) = e^{t\Lambda}w(0) + e^{t\Lambda}\int_0^t e^{-\tau\Lambda} P^{-1}b(\tau)d\tau. \tag{A1.1.21}$$

Using the substitutions (A1.1.18) yields

$$z(t) = (Pe^{t\Lambda} P^{-1})z(0) + Pe^{t\Lambda}\int_0^t e^{-\tau\Lambda} P^{-1}b(\tau)d\tau \tag{A1.1.22}$$

which is the formal solution to (A1.1.10). Since well known algorithms are available for finding eigenvalues and eigenvectors of a matrix, the solution to (A1.1.22) can be found in a straightforward manner.

<u>Exercises</u>

1. If $A = \begin{pmatrix} 3 & 2 \\ 2 & 3 \end{pmatrix}$ show that $\Lambda = \begin{pmatrix} 5 & 0 \\ 0 & 2 \end{pmatrix}$ and $P = \begin{pmatrix} 1 & 1 \\ 1 & -1 \end{pmatrix}$. Use (A1.1.22) to solve (A1.1.10) for this data, given that $z(0) = \begin{pmatrix} 1 \\ 2 \end{pmatrix}$.
2. If $A = \begin{pmatrix} 3 & 3 \\ 2 & 4 \end{pmatrix}$ show that $\Lambda = \begin{pmatrix} 6 & 0 \\ 0 & 1 \end{pmatrix}$ and $P = \begin{pmatrix} 1 & 3 \\ 1 & -2 \end{pmatrix}$. Use (A1.1.22) to solve (A1.1.10) for this data, given that $z(0) = \begin{pmatrix} 0 \\ 5 \end{pmatrix}$.

A1.1.9 Solution of Linear Two Point Boundary Value Problems

In linear-quadratic control problems with linear salvage values (e.g. the production-inventory problem in Section 5.1) we require the solution of linear two-point boundary value problems of the form:

$$\begin{pmatrix} \dot{x} \\ \dot{\lambda} \end{pmatrix} = \begin{pmatrix} A_{11} & A_{12} \\ A_{21} & A_{22} \end{pmatrix}\begin{pmatrix} x \\ \lambda \end{pmatrix} + \begin{pmatrix} b_1 \\ b_2 \end{pmatrix} \tag{A1.1.23}$$

with boundary conditions

$$x(0) = x_0 \quad \text{and} \quad \lambda(T) = \lambda_T. \tag{A1.1.24}$$

The solution of this system will be of the form (A1.1.22), which can be restated as

$$\begin{pmatrix} x(t) \\ \lambda(t) \end{pmatrix} = \begin{pmatrix} Q_{11} & Q_{12} \\ Q_{21} & Q_{22} \end{pmatrix} \begin{pmatrix} x(0) \\ \lambda(0) \end{pmatrix} + \begin{pmatrix} R_1(t) \\ R_2(t) \end{pmatrix} \tag{A1.1.24}$$

where the $\lambda(0)$'s are unknowns. They can be determined by setting

$$\lambda_T = Q_{21}\, x(0) + Q_{22}\, \lambda(0) + R_2(T) \tag{A1.1.25}$$

which is a system of linear equations for the variables $\lambda(0)$.

Exercises

1. After you have read Section 5.1, redo the production-inventory example stated in equations (5.1) and (5.2), (ignoring the control constraint $P \geq 0$) by this method. The two-point linear boundary problem is stated in equations (5.6) and (5.7).

A1.2.1 Homogeneous Partial Differential Equations

A homogeneous partial differential equation is an equation containing one or more partial derivatives of an unknown function with respect to its independent variables. If the highest partial derivative appearing explicitly in the equation has order n, then the partial differential equation is said to be of order n.

As we saw in the previous sections, the general solutions of ordinary differential equations involve expressions containing arbitrary constants. Similarly, the solutions of partial differential equations are expressions containing arbitrary (differentiable) functions. Conversely, when arbitrary functions can be eliminated algebraically from a given expression, after suitable partial derivatives have been taken, then the result is a partial differential equation.

Example 1. Eliminate the arbitrary function f from the expression $u = f(ax - by)$, where a and b are non-zero constants.

Taking partial derivatives we have

$$u_x = af', \ u_y = -bf'$$

so that

$$bu_x + au_y = abf' - abf' = 0 \quad .$$

Hence $u = f(ax - by)$ is a solution for the equation $bu_x + au_y = 0$.

To show that any solution $u = g(x,y)$ can be written in this form, we set $s = ax - by$, and define

$$G(s,y) = g(\frac{s+by}{a},y) \quad .$$

Then $G_x = G_s \frac{\partial s}{\partial x} = aG_s$ and

$$G_y = G_s \frac{\partial s}{\partial y} + G_y = -bG_s + G_y \quad .$$

Since we assume G solves the equation $bu_x + au_y = 0$ we have

$$0 = bG_x + aG_y = abG_s - abG_s + aG_y = 0 \quad .$$

But this implies $G_y = 0$ so that G is a function of $s = ax - by$ only, and hence is of the required form. We conclude that $u = f(ax - by)$ is a general solution form for $bu_x + au_y = 0$.

Example 2. Eliminate the arbitrary functions f_1 and f_2 from the expression $u = f_1(x)f_2(y)$.

Taking partial derivatives we have

$$u_x = f_1'f_2, \quad u_y = f_1f_2', \quad u_{xy} = f_1'f_2'$$

so that

$$uu_{xy} - u_xu_y = f_1f_2f_1'f_2' - f_1'f_2f_1f_2' = 0.$$

As in Example 1 we conclude that $u = f_1(x)f_2(y)$ is the general solution form of the equation $uu_{xy} - u_xu_y = 0$.

The subject of partial differential equations is too vast to even survey here. However, Table A1.2.1 gives general solution forms for all the homogeneous partial differential equations we will consider in this book, as well as others.

Partial Differential Equation	General Solution Form
(1) $bu_x+au_y = 0$	$u = f(ax-by)$
(2) $xu_x+yu_y = 0,\ x \neq 0$	$u = f(y/x)$
(3) $xu_x-yu_y = 0$	$u = f(xy)$
(4) $u_x+u_y = au$	$u = f_1(x-y)e^{ax}+f_2(x-y)e^{ay}$
(5) $u_x+u_y = au^k,\ k \neq 1$	$u = [(k-1)(f_1(x-y)-ax)]^{1/(k-1)}$ $+[(k-1)(f_2(x-y)-ay)]^{1/(k-1)}$
(6) $u_{xx}-a^2u_{yy} = 0$	$u = f_1(y+ax)+f_2(y-ax)$
(7) $u_{xx}+a^2u_{yy} = 0$	$u = f_1(y+iax)+f_2(y-iax)\ ,\ i = \sqrt{-1}$
(8) $u_{xy} = 0$	$u = f_1(x)-f_2(y)$
(9) $uu_{xy}-u_xu_y = 0$	$u = f_1(x)f_2(y)$
(10) $uu_{xy}+u_xu_y = 0$	$u = f_1(x)/f_2(y)$
(11) $bcu_x+acu_y+abu_z = 0$	$u = f_1(ax-by)+f_2(by-cz)+f_3(cz-ax)$
(12) $xu_x+yu_y+zu_z = 0$	$u = f_1(x/y)+f_2(y/z)$
(13) $xu_x-yu_y+zu_z = 0$	$u = f_1(xy)+f_2(yz)$
(15) $u^2u_{xyz}-u_xu_yu_z = 0$	$u = f_1(x)f_2(y)f_3(z)$
(16) $u_{xyz} = 0$	$u = f_1(x)+f_2(y)+f_3(z)$
(17) $u_{xx}-a^2(u_{yy}+u_{zz}) = 0$	$u = f_1(y+ax)+f_2(y-ax)+f_3(z+ax)+f_4(z-ax)$
(18) $u_x+u_y+u_z = au$	$u = f_1(x-y)e^{ax}+f_2(y-z)e^{ay}+f_3(z-x)e^{az}$

Note. The functions f_i are arbitrary differentiable functions of variable; a, b, c,... stand for arbitrary (non-zero) constants.

Table A1.2.1 General Solution Forms for Some Homogeneous Partial Differential Equations.

A1.2.2 Inhomogeneous Partial Differential Equations

As in the ordinary case, an inhomogeneous partial differential equation is obtained from a homogeneous one by adding one or more forcing functions. The general case of this problem is too difficult to treat here. We consider only the case in which the forcing functions are separable, that is, can be written as a sum of functions each involving only one of the independent variables. In solving such problems we can make use of the solutions to ordinary differential equations considered earlier.

Example 1. Solve

$$u_x + u_y = 3x^2 + e^y \quad .$$

We know from the previous section that the general solution to the homogeneous equation is of the form $f(x-y)$. To get the particular solutions we solve separately the ordinary equations

$$u_x = 3x^2 \quad \text{and} \quad u_y = e^y$$

obtaining solutions x^3 and e^y. Therefore the general solution to the original equation is

$$u = f(x-y) + x^3 + e^y \quad .$$

Generally speaking, the above philosophy of finding particular solutions to separable partial differential equations (when it works) follows the same method of "dividing and conquering." Other methods involve the use of series. We will not go further here for lack of space.

APPENDIX A2:
CALCULUS OF VARIATIONS AND OPTIMAL CONTROL THEORY

In this appendix, we introduce the calculus of variations by analogy with the classical optimization of calculus. The problem of the calculus of variations is that of determining a function that maximizes a given functional, the objective function. Analogous problem in calculus is that of determining a point at which a specific function, the objective function, is maximum. This, of course, is done by taking the first derivative of the function and equating it to zero. This is what is called the first order condition for a maximum. A similar procedure will be employed to derive the first order condition for the variational problem. The analogy with classical optimization extends also to the second order condition in calculus for a maximum. Finally, we will show the relationship between the maximum principle of optimal control theory to the calculus of variations. It is noted that this relationship is similar to the one between the Kuhn-Tucker conditions in mathematical programming to classical optimization.

We start with the "simplest" variational problem.

A2.1 The Simplest Variational Problem

Assume a function $x:C[0,t]\to E^1$, where $C[0,T]$ is a class of functions defined over the interval $[0,T]$ with continuous first derivatives. (For simplicity in exposition, assume x to be a scaler function; The extension to a vector function is straightforward.) Assume further that a function in this class is termed admissible if it satisfies the terminal conditions

$$x(0) = x_0 \quad \text{and} \quad x(T) = x_T. \tag{A2.1}$$

We are thus dealing with a fixed-end-points problem. Examples of admissible functions for the problem are shown in Figure A2.1. (See Section 6 of

[Gelfand and Fomin, 1963] for problems other than the simplest problems, i.e., the problems with other kinds of specification for the end points.)

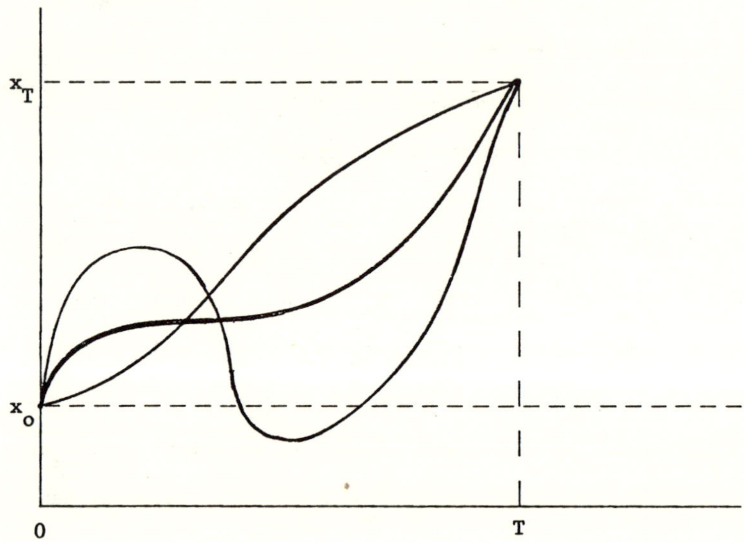

Figure A2.1 Examples of Admissible Functions for the Problem

The problem is to obtain the admissible function x^* for which the functional

$$J(x) = \int_0^T g(x,\dot{x},t)\,dt \tag{A2.2}$$

has a relative maximum. We will assume that all first and second partial derivatives of the function $g: E^1 \times E^1 \times E^1 \to E^1$ are continuous.

A2.2 The Euler Equation

The necessary first order conditions in classical optimization were obtained by considering small changes about the solution point. For the variational problem, we consider small variations about the solution function. Let $x(t)$ be the solution and let

$$y(t) = x(t) + \varepsilon\, \eta(t)$$

where $\eta(t)$: $C[0,T]\to E^1$ is an arbitrary differentiable function satisfying

$$\eta(0) = \eta(T) = 0, \tag{A2.3}$$

and $\varepsilon > 0$ is a small number. This is shown in Fig. A2.2.

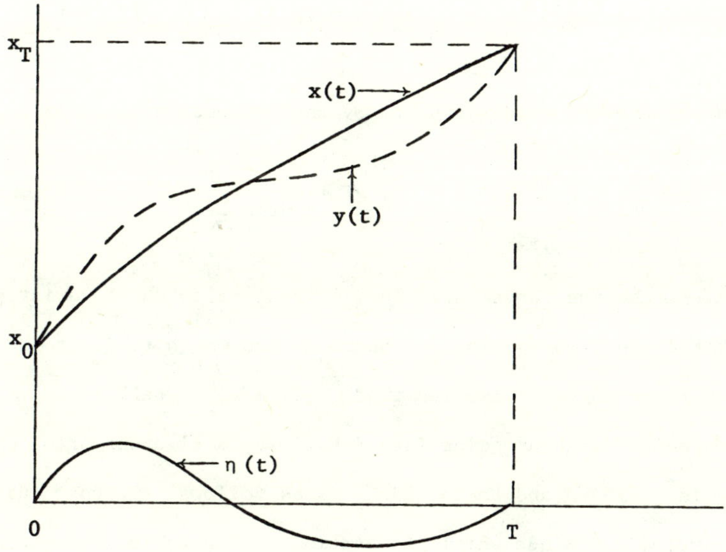

Figure A2.2 Variation about the Solution Function

The value of the objective functional associated with $y(t)$ can be considered a function of ε, i.e.,

$$V(\varepsilon) = J(y) = \int_0^T g(x+\varepsilon\eta, \dot{x}+\varepsilon\dot{\eta}, t)dt.$$

However, $x(t)$ is a solution and therefore $V(\varepsilon)$ must be a maximum at $\varepsilon = 0$. This means

$$\delta J \triangleq \left.\frac{dV}{d\varepsilon}\right|_{\varepsilon=0} = 0,$$

where δJ is known as the variation δJ in J. Differentiating $V(\varepsilon)$ with respect to ε and setting $\varepsilon = 0$ yields

$$\delta J = \left.\frac{dV}{d\varepsilon}\right|_{\varepsilon=0} = \int_0^T (g_n \eta + g_{\dot{x}} \dot{\eta})dt = 0,$$

which after integrating the second term by parts provides

$$\delta J = \left.\frac{dV}{d\varepsilon}\right|_{\varepsilon=0} = \int_0^T g_x \eta dt + [g_{\dot{x}}\eta]_0^T - \int_0^T \frac{d}{dt}(g_{\dot{x}})\eta dt = 0 \ . \qquad \text{(A2.4)}$$

On account of the end conditions on η, the expression simplifies to

$$\delta J = \left.\frac{dV}{d\varepsilon}\right|_{\varepsilon=0} = \int_0^T [g_x - \frac{d}{dt} g_{\dot{x}}]\eta dt = 0.$$

We now use the <u>fundamental lemma</u> of the calculus of variations which states that if h is a continuous function then $\int_0^T h(t)\eta(t)dt = 0$ for every continuous function $\eta(t)$ implies that $h(t) = 0, \quad \forall t \varepsilon [0,T]$. The reason that this lemma holds (without going into details of a rigorous proof which is available in [Gelfand and Fomin, 1963] is as follows. Suppose that $h(t) \neq 0$ for some $t \varepsilon [0,T]$. Since $h(t)$ is continuous, there is, therefore, an interval $(t_1,t_2) \subset [0,T]$ over which h is nonzero and has the same sign. Now selecting $\eta($ such

$$\eta(t) \ \text{ is } \begin{cases} > 0 & , \quad t \ \varepsilon \ (t_1,t_2) \\ \text{arbitrary}, & \text{otherwise}, \end{cases}$$

it is possible to make the integral $\int_0^T h(t)\eta(t) \neq 0$. Thus by contradiction, $h(t)$ must be identically zero over the entire interval $[0,T]$.

By using the fundamental lemma, we have the necessary condition

$$g_x - \frac{d}{dt} g_{\dot{x}} = 0 \tag{A2.5}$$

known as the Euler equation, which must be satisfied by a maximal solution x^*.

We note that the Euler equation is a second order ordinary differential equation. This can be seen by taking the total time derivative of $g_{\dot{x}}$ and collecting terms:

$$\ddot{x}\, g_{\dot{x}\dot{x}} + \dot{x}\, g_{x\dot{x}} + (g_{t\dot{x}} - g_x) = 0.$$

The boundary conditions for this equation are obviously the end point conditions $x(0) = x_0$ and $x(T) = x_T$.

Special case (i): When g does not depend explicitly on $\dot{x}$.

In this case, the Euler equation (A2.5) reduces to

$$g_x = 0$$

which is nothing but the first order condition of classical optimization. In this case, the dynamic problem is a succession of static classical optimization problems.

Special case (ii): When g does not depend explicitly on x.

The Euler equation reduces to

$$\frac{d}{dt} g_{\dot{x}} = 0, \tag{A2.6}$$

which we can integrate as

$$g_{\dot{x}} = \text{constant}. \tag{A2.7}$$

Special case (iii): When g does not depend explicitly on t.

Finally, we have the important special case in which g is explicitly independent of t. In this case, we write the Euler equation (A2.5) as

$$\frac{d}{dt}(g - \dot{x}g_{\dot{x}}) - g_t = 0. \tag{A2.8}$$

But $g_t = 0$ and therefore we can solve the above equation as

$$g - \dot{x}g_{\dot{x}} = C, \tag{A2.9}$$

where C is the constant of the integration.

A2.3 The Shortest Distance between Two Points on the Plane

This problem is to show that the straight line passing through two points on a plane is the shortest distance between the two points. The problem can be stated as follows:

$$\min \int_0^T \sqrt{1 + \dot{x}^2}\, dt,$$

subject to $x(0) = x_0$ and $x(T) = x_T$. Here t refers to distance rather than time. Since $g = -\sqrt{1+\dot{x}^2}$ does not depend explicitly on x, we are in the second special case and the first integral (A2.7) of the Euler equation is

$$g_{\dot{x}} = -\dot{x}(1 + \dot{x}^2)^{-1/2} = C.$$

This implies that $\dot{x}$ is a constant, which results in the solution

$$x(t) = C_1 t + C_2$$

where C_1 and C_2 are constants. These can be obtained by boundary conditions to be $C_1 = (x_T - x_0)/T$ and $C_2 = x_0$. Thus

$$x(t) = \left(\frac{x_T - x_0}{T}\right) t + x_0$$

which is the straight line passing through x_0 and x_T.

A2.4 The Brachistochrone Problem

The problem arises from the search for the shape of a wire along which a bead falls in the least time from a given point to another, under the influence of gravity.

The Brachistochrone problem is a long history. It was first studied (incorrectly) by Galilio and later associated with the names of Bernoulli, Newton, Leibniz, and L'Hospital. The problem was correctly posed by Newton and solved by Bernoulli in 1696. Note that Euler deduced the Euler's equation in 1744. We will solve the Brachistochrone problem, however, using Euler's equation. But first we must formulate the problem.

Assume the bead slides with no friction. Let m denote the mass of the bead, s denote the arc length, t denote the horizontal axiz, x denote the vertical axis (measured vertically down) and τ denote the time. Assume $t_0 = 0$, $x(t_0) = 0$, $T = 1$, $x(T) = 1$.

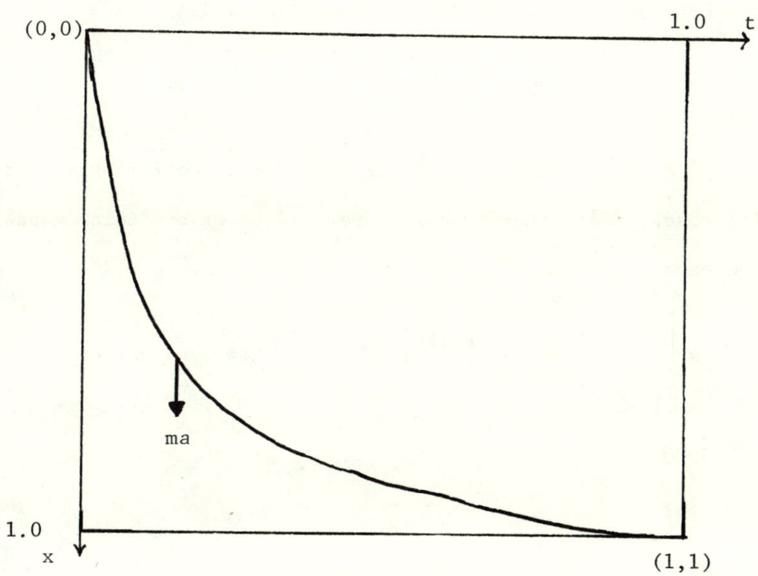

Figure A2.3 Brachistochrone Problem

We wish to minimize

$$J = \int_0^T d\tau = \int_0^{s_T} \frac{ds}{v} ,$$

where v represents velocity, and s_T is the final displacement measured on the curve. We can write

$$ds = \sqrt{1 + \dot{x}^2}\, dt$$

and, from elementary physics, it is known that if $v(t_0 = 0) = 0$ and a denotes the gravitational acceleration, then

$$v = \sqrt{2a\, x} .$$

Therefore, the variational problem can be stated as

$$\text{Minimize} \left\{ \frac{1}{\sqrt{2a}} \int_0^1 \sqrt{\frac{1+\dot{x}^2}{2}}\, dt \right\} ,$$

where $\dot{x} = dx/dt$ (note, t does not denote time here), and $x(0) = 0$ and $x(1) = 1$. Since a is a constant, we can rewrite the problem as

$$\min J(x) = \int g(x,\dot{x},t)\,dt = \int_0^1 \sqrt{\frac{1+\dot{x}^2}{x}}\, dt.$$

Since g does not depend explicitly on t, the problem belongs to the third special case. Using the first integral (A2.9) of the Euler equation for this case, we have

$$\dot{x}^2 \left[x(1+\dot{x}^2) \right]^{-1/2} - \left(\frac{1+\dot{x}^2}{x} \right)^{1/2} = C_1 \quad \text{(a constant)}.$$

We can reduce this to

$$\frac{dx}{dt} = \sqrt{\frac{1}{xC_1^2} - 1} \ .$$

To solve this equation, we separate the variables as

$$dt = \frac{\sqrt{x}\ dx}{\sqrt{1/C_1^2 - x}}$$

and substitute

$$x = \sin^2 \theta / C_1^2 = (1 - \cos 2\theta)/C_1^2. \tag{A2.10}$$

The resulting expression can be integrated to yield

$$t = [\theta - (1/2)\sin 2\theta]/C_1^2 + C_2. \tag{A2.11}$$

The condition $\theta = 0$ at $t = 0$ implies $C_2 = 0$ providing $C_1^2 > 0$. The value of C_1 can be obtained in terms of the value of θ at $t = 1$; let this be θ_1. Then since $x = 1$ at $t = 1$, we have $C_1 = \sin \theta$, where θ_1 satisfies.

$$2\theta_1 - 1 = \sin 2\theta_1 - \cos 2\theta_1.$$

This equation must be solved numerically. An iterative numerical procedure yields $\theta_1 = 1.206$ and therefore $C_1^2 = 0.873$. Defining $\phi = 2\theta$, we can rewrite (A2.10) and (A2.11) as

$$x = 0.573(1-\cos 2\phi)$$

$$t = 0.573(\phi - \sin\phi)$$

which are equations of a <u>cycloid</u> in the parametric form. The shape of the curve is shown in Fig. A2.3.

A2.5 Broken Extremals: The Weirstrass-Erdmann Corner Conditions

So far we have only considered functionals defined for smooth curves. This is, however, a restricted class of curves which qualify as solutions since it is easy to give examples of variational problems which have no solution in this class. Consider, for example, the functional

$$\min \left\{ J(x) = \int_{-1}^{1} x^2(1-\dot{x})^2 dt \right\}, \quad x(-1) = 0, \quad x(1) = 1.$$

The greatest lower bound for $J(x)$ for smooth $x = x(t)$ satisfying the boundary conditions is obviously zero. Yet there is no $x \varepsilon C[-1,1]$ with $x(-1) = 0$ and $x(1) = 1$ achieving this value of $J(x)$. In fact, the minimum is achieved for the curve

$$x(t) = \begin{cases} 0, & -1 \leq t \leq 0 \\ t, & 0 < t \leq 1, \end{cases}$$

which has a corner (i.e., a discontinuous first derivative) at $t = 0$. Such a piecewise smooth extremal with corners is called a broken extremal.

We now enlarge the class of admissible function by relaxing the requirement that they be smooth everywhere. The larger class is the class of piecewise continuous functions which are continuously differentiable almost everywhere in $[0,T]$, i.e. except at some points in $[0,T]$.

Let $x(t)$ be an extremal with a corner at $\tau \varepsilon [0,T]$. It is clear that on each of the intervals $[0,\tau)$ and $(\tau,T]$, the Euler equation must hold.

$$J(x) = \int_0^T g(x,\dot{x},t)dt = \int_0^\tau g(x,\dot{x},t)dt + \int_\tau^T g(x,\dot{x},t)dt$$

$$= J_1(x) + J_2(x).$$

To compute variations δJ_1 and δJ_2 we must recognize that the two 'pieces' of x are not fixed-end-point problems. We must require that the two pieces of x join continuously at $t = \tau$; the point $t = \tau$ can, however, move freely as shown in Fig. A2.4.

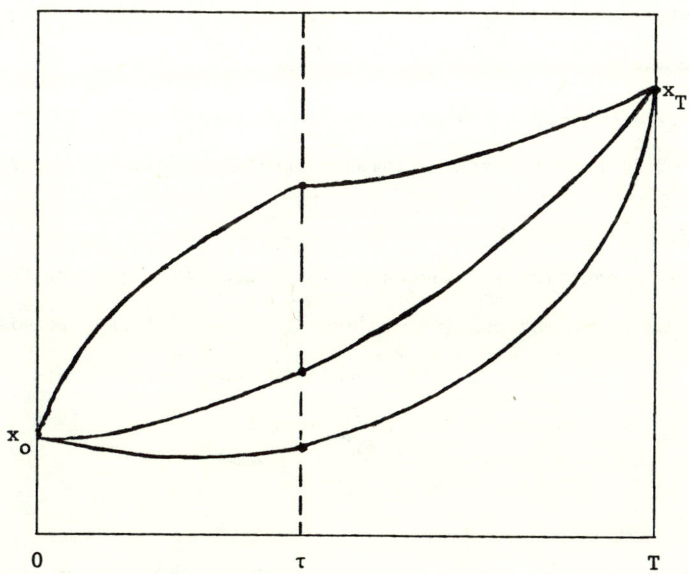

Figure A2.4 A Broken Extremal with Corner at τ

This will require a slightly modified version of formula (A2.4) for writing out the variations (see pp. 55-56 in [Gelfand and Fomin, 1963]). Equating the sum of variations

$$\delta J = \delta J_1 + \delta J_2 = 0$$

for $x(t)$ to be an extremal and using the fact that $s(t)$ must be continuous at $t = \tau$ implies

$$g_{\dot{x}}\Big|_{\tau-} = g_{\dot{x}}\Big|_{\tau+} \tag{A2.12}$$

$$\left[g-\dot{x}g_{\dot{x}}\right]_{\tau-} = \left[g-\dot{x}g_{\dot{x}}\right]_{\tau+} . \tag{A2.13}$$

These conditions are called Weirstrass-Erdmann (corner) conditions, which must hold at the point τ where the extremal has a corner.

In each of the intervals $[0,\tau)$ and $(\tau,t]$, the extremal x must satisfy the Euler equation (A2.5). Solving these two equations will provide us with four constants of integration since the Euler equations are second-order differential equations. These constants can be found from the endpoint conditions (A2.1) and Weirstrass-Erdmann conditions (A2.12) and (A2.13).

A2.6 Legendre's Condition: The Second Variation

The Euler equation is a necessary condition analogous to the first order condition for a maximum (or minimum) in the classical optimization problems of calculus. The condition analogous to the second order necessary condition for a maximum is the Legendre condition

$$g_{\dot{x}\dot{x}} \leq 0. \tag{A2.14}$$

To obtain this condition, we use the second order condition of classical optimization on function $V(\varepsilon)$ to be a maximum at $\varepsilon = 0$, i.e.,

$$\left.\frac{d^2V(\varepsilon)}{d\varepsilon^2}\right|_{\varepsilon=0} = \int_0^T (g_{xx}\eta^2+2g_{x\dot{x}}\eta\dot{\eta}+g_{\dot{x}\dot{x}}\dot{\eta}^2)dt \leq 0. \tag{A2.15}$$

Integrating the middle term by parts and using (A2.3), we can transform (A2.15) into a more convenient form

$$\int_0^T Q\eta^2 + P\dot{\eta}^2 \leq Q, \tag{A2.16}$$

where

$$Q = Q(t) = g_{x\dot{x}} - \frac{d}{dt} g_{x\dot{x}} \quad \text{and} \quad P = P(t) = g_{\dot{x}\dot{x}} .$$

While it is possible to rigorously obtain (A2.14) from (A2.16), we will only provide a qualitative argument for this. If we consider the quadratic functional (A2.16) for functions $\eta(t)$ satisfying $\eta(0) = 0$, then $\eta(t)$ will be small in [0,T] if $\dot{\eta}(t)$ is small in [0,T]. The converse is not true, however, since it is easy to construct $\eta(t)$ which is small but has a large derivative $\dot{\eta}(t)$ in [0,T]. Thus $P\dot{\eta}^2$ plays the dominant role in (A2.16); i.e., $P\dot{\eta}^2$ can be much larger than $Q\eta^2$ but it cannot be much smaller (assume $P \neq 0$). Therefore, it might be expected that the sign of the functional in (A2.6) is determined by the sign of the coefficient $P(t)$, i.e., (A2.16) implies (A2.14). For a rigorous proof, see [Gelfand and Fomin, 1963].

We note that the strengthened Legendre condition (i.e., with a strict inequality in (A2.14)) with the Euler equation and with one other condition called strengthened Jacobi condition is sufficient for a maximum. The reader can consult Ch. 5 of [Gelfand and Fomin, 1963] for details.

A2.7 Necessary Condition for a Strong Maximum

So far we have discussed necessary conditions for a weak maximum. By weak maximum we mean that the candidate extremals are smooth or piecewise smooth functions. The concept of a strong maximum on the other hand requires that the candidate extremal need only be continuous functions. Without going into details (which are available in [Gelfand and Fomin, 1963]) we state a necessary condition for a strong maximum. This is called the Weirstrass necessary condition. The condition is analogous to the one in the static case that the objective function be concave. It states that if the functional (A2.2) has a strong maximum for the extremal γ satisfying (A2.1), then

$$E(x,\dot{x},t,v) \le 0 \qquad \text{(A2.17)}$$

along γ for every finite v, where E is the <u>Weirstrass Excess Function</u> defined as

$$E(x,\dot{x},t,v) = g(x,v,t) - g(x,\dot{x},t) - g_{\dot{x}}(x,\dot{x},t)(v-\dot{x}). \qquad \text{(A2.18)}$$

Note that this condition is always met if $g(x,\dot{x},t)$ is concave in $\dot{x}$.

The proof of (A2.17) is by contradiction. Suppose there exists a $\tau \varepsilon [0,T]$ and a vector q such that

$$E(\tau,x(\tau),\dot{x}(\tau),q) > 0,$$

where $x = x(t)$ is the equation of the extremal γ. It is then possible to suitably modify γ to β which is close to γ in $C[0,T]$ such that

$$\Delta J = \int_{\beta} g(x,\dot{x},t)dt - \int_{\gamma} g(x,\dot{x},t)dt > 0$$

contradicting the hypothesis that $J(x)$ has a strong maximum for γ.

A2.8 Relation to the Optimal Control

It is possible to derive the necessary conditions of the calculus of variations from the maximum principle. This is strongly reminiscent of the relationship between classical optimization and the Kuhn-Tucker conditions.

First we note that the calculus of variations problem can be stated as an optimal control problem as follows.

$$\max \; \{J = \int_0^T g(x,u,t)dt \}$$

subject to

$$\dot{x} = u, \quad x(0) = x_0 \quad \text{and} \quad x(T) = x_T$$

and

$$u \in \Omega = E^r.$$

The Hamiltonian is

$$H(x,u,x,t) = g(x,u,t) + \lambda u \qquad \text{(A2.19)}$$

with the adjoint variable λ satisfying

$$\dot{\lambda} = -H_x = -g_x. \qquad \text{(A2.20)}$$

Maximizing the Hamiltonian with respect to u yields

$$H_u = g_{\dot{x}} + \lambda = 0 \Rightarrow \lambda = -g_{\dot{x}} \quad . \qquad \text{(A2.21)}$$

Differentiating with respect to time, we have

$$\dot{\lambda} = \frac{d}{dt} g_{\dot{x}} \quad .$$

This equation with (19) implies

$$g_x - \frac{d}{dt} g_{\dot{x}} = 0,$$

which is the Euler equation of the calculus of variations.

The second order condition for the maximization of the Hamiltonian, i.e.,

$$H_{uu} \leq 0 \Rightarrow g_{\dot{x}\dot{x}} \leq 0,$$

which is the Legendre condition,

Again, by the maximum principle, if u is an optimal control, then

$$H(x,u,\lambda,t) \geq H(x,v,\lambda,t)$$

where v is any other control. By the definition of the Hamiltonian (A2.19) and equation (A2.21), we have

$$g(x,\dot{x},t) - g_{\dot{x}}\dot{x} \geq g(x,v,t) - g_{\dot{x}}v,$$

which by transposition of terms yields the Weirstrass necessary condition

$$E(x,\dot{x},t,v) = g(x,u,t) - g(x,\dot{x},t) - g_{\dot{x}}(v-\dot{x}) \leq 0.$$

We have just proved the equivalence of the maximum principle and the Weirstrass necessary condition in the case where Ω is open. In cases when Ω is closed and when the optimal control is on the boundary of Ω, the Weirstrass necessary condition is no longer valid, in general. The maximum principle still applies, however.

Finally, according to the maximum principle, both λ and H are continuous functions of time. But

$$\lambda = -g_{\dot{x}} \quad \text{and} \quad H = g - g_{\dot{x}}\dot{x} \quad .$$

This means that the right hand sides must be continuous with respect to time, i.e., even across corners. These are precisely Weirstrass-Erdmann corner conditions.

APPENDIX A3: AN ALTERNATIVE DERIVATION OF THE MAXIMUM PRINCIPLE

Recall that in the derivation of the maximum principle in Chapter 2, we assumed the twice differentiability of the return function V. Looking at (2.27), it appears that the return function does not arise in the essence of the problem. Also since it is not an exogenously given function, there is no a priori reason to assume the twice differentiability. In many important cases, as a matter of fact, V has no derivatives at individual points, e.g., at points on switching manifolds.

In the following, we will give an alternate derivation. This proof follows the course pointed out by [Pontryagin, et. al., 1962] but with certain simplifications. It appears in [Feldbaum, 1965] and, in our opinion, it is one of the simplest proofs for the maximum principle which is not related to dynamic programming and thus permit the elimination of assumptions about the differentiability of the return function $V(x,t)$.

We select the Mayer form of the problem (2.5) for deriving the maximum principle in this section. It will be convenient to reproduce (2.5) here as (A3.1).

$$\max_{u(t)\in\Omega(t)} \left[J = cx(T) \right]$$

Subject to

$$\dot{x} = f(x,u,t), \quad x(0) = x_0 \tag{A3.1}$$

A3.1 Needle-Shaped Variation

Let $u^*(t)$ be an optimal control with corresponding state trajectory $x^*(t)$. We sketch this in Fig. A3.1 for a scalar control (i.e., $m = 1$).

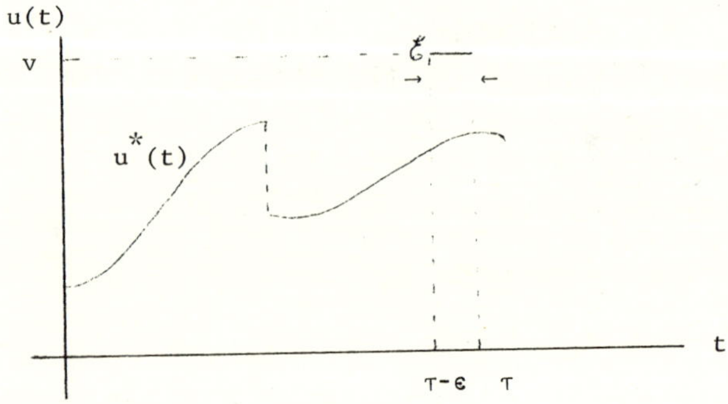

Figure A3.1 Needle-Shaped Variation

Let τ denote any time in the open interval $(0,T)$. We select a sufficiently small ϵ to insure that $\tau - \epsilon > 0$ and concentrate our attention on this small interval $(\tau - \epsilon, \tau]$. We vary the control on this interval while keeping the control on the remaining intervals $[0,\tau-\epsilon]$ and $(\tau,T]$ fixed. Specifically, the modified control is

$$u(t) = \begin{cases} v \in \Omega, & t \in (\tau-\epsilon,\tau], \\ u^*(t), & \text{otherwise}. \end{cases} \tag{A3.2}$$

This is called a 'needle-shaped' variation as shown in Fig. A3.1. It is a jump function and is different from variations in the calculus of variations (see Appendix 2). Also the difference $v - u^*$ is finite and need not be small. However, since the variation is on a small time interval, its influence on the subsequent state trajectory can be proved to be 'small.' This is done in the following.

Let the subsequent motion be denoted by $x(t) \neq x^*(t)$ for $t > \tau-\epsilon$. In Figure A3.2, we have sketched possible trajectories for the case $n = 2$.

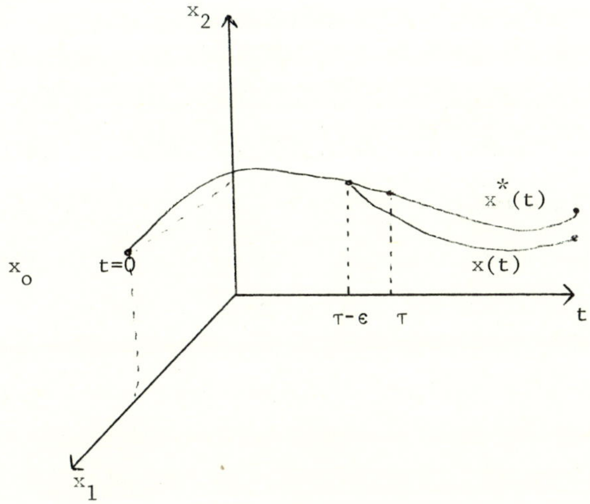

Figure A3.2

Let

$$\delta x(t) = x(t) - x^*(t), \quad t \geq \tau - \epsilon$$

denote the change in the state variables. Obviously $\delta x(\tau-\epsilon) = 0$. Clearly,

$$\delta x(\tau) \approx \epsilon[\dot{x}(s) - \dot{x}^*(s)], \tag{A3.3}$$

where s denotes some intermediate time in the interval $(\tau-\epsilon,\tau]$. In particular, we can write (A3.3) as

$$\begin{aligned}\delta x(\tau) &= \epsilon[\dot{x}(\tau) - \dot{x}^*(\tau)] + 0(\epsilon) \\ &= \epsilon[f(x(\tau),v,\tau) - f(x^*(\tau),u^*(\tau),\tau)] + 0(\epsilon).\end{aligned} \tag{A3.4}$$

But $\delta x(\tau)$ is small since f is assumed bounded. Furthermore, since f is continuous and the difference $\delta x(\tau) = x(\tau) - x^*(\tau)$ is small, we can rewrite (A3.4) as

$$\delta x(\tau) \approx \epsilon[f(x^*(\tau),v,\tau) - f(x^*(\tau),u^*(\tau),\tau)]. \tag{A3.5}$$

Since the initial difference $\delta x(\tau)$ is small and since $u^*(\tau)$ does not change from $t > \tau$ on, we may conclude that $\delta x(t)$ will be small for all $t > \tau$. Being small, the law of variation of $\delta x(t)$ can be found from linear equations for small changes in the state variables. These are called <u>variational equations</u>. From the state equation in (A3.1), we have

$$\frac{d(x^* + \delta x)}{dt} = f(x^* + \delta x, u^*, t) \tag{A3.6}$$

or,

$$\frac{dx^*}{dt} + \frac{d(\delta x)}{dt} \approx f(x^*,u^*,t) + f_x \delta x \tag{A3.7}$$

or,

$$\delta x \approx f_x(x^*,u^*,t)\delta x, \text{ for } t \geq \tau , \tag{A3.8}$$

with the initial condition $\delta x(\tau)$ given by (A3.5)

The basic idea in deriving the maximum principle is that equations (A3.8) are linear variational equations and result in an extraordinary simplification. For our derivation, first we obtain the adjoint equations.

A3.2 Derivation of the Adjoint Equation and the Maximum Principle

For this derivation, we employ two methods. The direct method (similar to [Hartberger, 1973] is the consequence of directly integrating (A3.8). The indirect method avoids this integration by a trick which is instructive.

<u>Direct Method</u>. Integrating (A3.8) we get

$$\delta x(T) = \delta x(\tau) + \int_\tau^T f_x[x^*(t),u^*(t),t]\ \delta x(t)dt, \tag{A3.9}$$

where the initial condition $\delta x(\tau)$ is given in (A3.5)

Since $\delta x(T)$ is the change in the terminal state from the optimal state $x^*(T)$, the change in the objective function δJ must be negative. Thus

$$\delta J = c\,\delta x(T) = c\,\delta x(\tau) + \int_\tau^T c f_x[x^*(t),u^*(t),t]\,\delta x(t)dt \leq 0. \tag{A3.10}$$

Furthermore, since (A3.8) in a linear homogeneous differential equation, we can write its general solution as

$$\delta x(t) = \Phi(t,\tau)\,\delta x(\tau)\ , \tag{A3.11}$$

where the fundamental solution matrix or the transition matrix $\Phi(t,\tau) \varepsilon E^{nxm}$ obeys

$$\frac{d}{dt}\Phi(t,\tau) = f_x[x^*(t),u^*(t),t]\ \Phi\ (t,\tau),\ \ \Phi(\tau,\tau) = I, \tag{A3.12}$$

where I is an nxn identity matrix.

Substituting for $\delta x(t)$ from (A3.11) into (A3.10), we have

$$\delta J = c\,\delta x(\tau) + \int_\tau^T c f_x[x^*(t),u^*(t),t]\,\Phi(t,\tau)\,\delta x(\tau)dt \leq 0. \tag{A3.13}$$

This induces the definition

$$\lambda^*(\tau) = \int_\tau^T c f_x[x^*(t),u^*(t),t]\ \Phi\ (t,\tau)dt + c, \tag{A3.14}$$

which when substituted in (A3.13), yields

$$\delta J = \lambda^*(\tau)\,\delta x(\tau) \leq 0. \tag{A3.15}$$

But $\delta x(\tau)$ is supplied in (A3.15). Noting that $\varepsilon > 0$, we can rewrite (2.45) as

$$\delta J = \lambda^*(\tau) f[x^*(\tau),v,\tau] - \lambda^*(\tau) f[x^*(\tau),u^*(\tau),\tau] \leq 0. \tag{A3.16}$$

Defining the Hamiltonian for the Mayer form as

$$H[x,u,\lambda,t] \triangleq \lambda f(x,u,t), \tag{A3.17}$$

we can rewrite (A3.16)

$$H[x^*(\tau),u^*(\tau),\lambda^*(\tau),\tau] \geq H[x^*(\tau),v,\lambda^*(\tau),\tau]. \tag{A3.18}$$

Since this can be done for almost every τ, we have the required Hamiltonian-maximizing condition.

The differential equation form of the adjoint equation (A3.14) can be obtained by taking its derivative with respect to τ. Thus

$$\frac{d\lambda^*(\tau)}{d\tau} = \int_\tau^T cf_x[x^*(t),u^*(t),t] \frac{d\Phi(t,\tau)}{d\tau} dt$$

$$- cf_x[x^*(\tau),u^*(\tau),\tau]. \tag{A3.19}$$

From Appendix 1 , we know that the transition matrix has the property:

$$\frac{d\Phi(t,\tau)}{d\tau} = - \Phi(t,\tau)f_x[x^*(\tau),u^*(\tau),\tau],$$

which can be used in (A3.19) to obtain

$$\frac{d\lambda^*(\tau)}{d\tau} = - \int_\tau^T cf_x[x^*(t),u^*(t),t] \Phi(t,\tau)f_x[x^*(\tau),u^*(\tau),\tau]dt$$

$$- cf_x[x^*(\tau),u^*(\tau),\tau]. \tag{A3.20}$$

Using the definition (A3.14) of $\lambda^*(\tau)$ in (A3.20) we have

$$\frac{d\lambda^*(\tau)}{d\tau} = - \lambda^*(\tau)f_x[x^*(\tau),u^*(\tau),\tau],$$

with $\lambda^*(T) = c$, or using (A3.17) and noting that τ is arbitrary, we have

$$\dot{\lambda}^* = - \lambda^* f_x[x^*,u^*,t] = - H_x(x^*,u^*,\lambda^*,t), \ \lambda^*(T) = c. \tag{A3.21}$$

This completes the derivation of the maximum principle with the adjoint equation using the direct method.

Indirect Method: The indirect method employs a trick which simplifies considerably the derivation. Instead of integrating (A3.8) explicitly, we now assume that the result of this integration yields $\delta x(T)$ as the change in the state at the terminal time. As in (A3.10), we have

$$\delta J = c\,\delta x(T) \leq 0. \tag{A3.22}$$

First we define

$$\lambda^*(T) \triangleq c, \tag{A3.23}$$

which makes it possible to write (A3.22) as

$$\delta J = c\,\delta x(T) = \lambda^*(T)\delta x(T) \leq 0.\ ^{1/} \tag{A3.24}$$

Now, $\lambda^*(T)\delta x(T)$ is the change in the objective function due to a change $\delta x(T)$ at the terminal time T. That is, $\lambda^*(T)$ is the marginal return or the marginal change in the objective function per unit change in the state at time T. But $\delta x(T)$ cannot be known without integrating

---

1/ Note that if the objective function $J = S(x(T))$, we must define $\lambda^*(T) \triangleq \frac{\partial S[x(T)]}{\partial x(T)}$ giving us $\delta J = \frac{\partial S[x(T)]}{\partial x(T)} \delta x(T) = \lambda^*(T)x(T)$.

(A3.8). We do know, however, the value of the change $\delta x(\tau)$ at time τ which caused the terminal change $\delta x(T)$ via (A3.8).

We would therefore like to pose the following problem [9]; we would like to obtain the change δJ in the objective function in terms of the known value $\delta x(\tau)$. Simply stated, we would like to obtain the marginal return $\lambda(\tau)$ per unit change in state at time τ. Thus

$$\lambda^*(\tau)\delta x(\tau) = \delta J = \lambda^*(T)x(T) \leq 0. \tag{A3.25}$$

Obviously, knowing $\lambda^*(\tau)$ will make it possible to make inference about δJ which is directly related to the needle-shaped variation applied in the small interval $(\tau-\epsilon,\tau]$.

However, since τ is arbitrary, our problem of finding $\lambda^*(\tau)$ can be translated to one of finding $\lambda^*(t)$, $t \in [0,T]$, such that

$$\lambda^*(t)\delta x(t) = \lambda^*(T)x(T), \quad t \in [0,T], \tag{A3.26}$$

or, in other words

$$\lambda^*(t)\delta x(t) = \text{constant}, \quad \lambda^*(T) = c. \tag{A3.27}$$

It turns out that the differential equation which $\lambda^*(t)$ must satisfy can be easily found. From (A3.27)

$$\frac{d}{dt}[\lambda^*(t)\delta x(t)] = \lambda^*\dot{\delta x} + \dot{\lambda}^*\delta x = 0, \tag{A3.28}$$

which after substituting for $\dot{\delta x}$ from (2.38) becomes

$$\lambda^* f_x \delta x + \dot{\lambda}^*\delta x = (\lambda^* f_x + \dot{\lambda}^*)\ \delta x = 0. \tag{A3.29}$$

Since (A3.29) is true for arbitrary δx, we have

$$\dot{\lambda}^* = -\lambda^* f_x = -H_x \tag{A3.20}$$

using the definition (2.47) for the Hamiltonian.

The Hamiltonian maximizing condition can be obtained by substituting for $\delta x(\tau)$ from (A3.5) into (A3.25). This is same as what we did in (A3.15) through (A3.18).

The purpose of the alternative proof was to demonstrate the validity of the maximum principle for a simple problem without knowledge of any return function. For more complex problems, one needs complicated mathematical analysis to rigorously prove the maximum principle without making use of return functions. A part of the rigor is in proving the existence of an optimal solution without which necessary conditions are meaningless in the opinion of some mathematicians [Young, 1969].

APPENDIX A4: SPECIAL TOPICS IN OPTIMAL CONTROL

In this appendix, we will discuss three specialized topics. These are linear-quadratic problems, second order variations, and singular control. These topics are referred to but not discussed in the main body of the text because of their advanced nature. While we shall not be able to go into a great detail, we will provide an adequate description of these topics and list relevant references.

A4.1 Linear-Quadratic Problems

An important problem in systems theory, especially engineering sciences, is to synthesize feedback controllers. These controllers provide optimal control as a function of the state of the system. A usual method of obtaining these controllers is to solve the Hamilton-Jacobi-Bellman partial differential equation (2.18). This equation is only rarely treatable for most nonlinear systems, making it difficult to develop exact explicit feedback control schemes.

It is however, feasible, in many cases, to obtain perturbation feedback control [Bryson and Ho, 1969], which refers to control in the vicinity of an optimal path. These perturbation schemes require approximating the problem by a linear-quadratic problem in the vicinity of an optimal path (see Section A4.2) and feedback control for the approximating problem is easy to obtain.

More specifically, a linear-quadratic control problem is a problem with linear dynamics and quadratic objective function. Thus,

$$\max\{J = \frac{1}{2} x' \Sigma x + \frac{1}{2} \int_0^T (x'Cx + u'Du)dt\} \tag{A4.1}$$

subject to

$$\dot{x} = Ax + Bu, \quad x(0) = x_0 . \tag{A4.2}$$

The matrices Σ, C, and D are assumed to be negative definite. Note that this problem is a special case of row (c) of Table 3.3.

To solve this problem for the explicit feedback controller, we write the Hamilton-Jacobi-Bellman equation (2.18) as

$$0 = \max_u [H - V_t] = \max_u \{\frac{1}{2}(x'Cx + u'Du) + V_x[Ax + Bu] + V_t\} \tag{A4.3}$$

with the terminal boundary condition

$$V(x,t) = \frac{1}{2}x'Sx. \tag{A4.4}$$

The maximization of the maximand in (A4.3) can be carried out by taking its derivative with respect to u and setting it to zero. Thus

$$\frac{\partial(H + V_t)}{\partial u} = \frac{\partial H}{\partial u} = (Du)' + V_x B = 0 \Rightarrow u' = -V_x B(D')^{-1}. \tag{A4.5}$$

Note that (A4.5) is same as the Hamiltonian maximizing condition. Substituting (A4.5) in (A4.3) and simplifying, we obtain

$$0 = \frac{1}{2}\, x'Cx + V_x A_x - \frac{1}{2}\, V_x\ BD^{-1}B'V_x\ . \tag{A4.6}$$

This is a nonlinear partial differential equation of first order and it has a solution of the form

$$V = \frac{1}{2}\, x'S(t)_x\ . \tag{A4.7}$$

Substitute this in (A4.6) yields

$$0 = \frac{1}{2}\, x'[\dot{S} + SA + A'S - SBD^{-1}B'S + C]x. \tag{A4.8}$$

Since (A4.8) must hold for all x, it implies the following matrix differential equation

$$\dot{S} + SA + A'S - SBD^{-1}B'S + C = 0, \tag{A4.9}$$

called a <u>matrix Riccati equation</u> with the boundary condition

$$S(T) = \Sigma. \tag{A4.10}$$

A solution procedure for Riccati equations appears in [Bryson and Ho, 1969]. Once we have the solution $S(t)$ of (A4.9) and (A4.10), the optimal feedback control can be written as

$$u^*(t) = D(t)^{-1}B'(t)S(t)x(t). \tag{A4.11}$$

A generalization of (A4.1), which would be useful in the next section on the second variation, is to set

$$J = \frac{1}{2}x'\Sigma x + \frac{1}{2}\int_0^T (x'u')\begin{pmatrix} C & N \\ N' & D \end{pmatrix}\begin{pmatrix} x \\ u \end{pmatrix} dt. \tag{A4.12}$$

The state equation is given by (A4.2). It is possible to derive the optimal control for this problem as

$$u^*(t) = D(t)^{-1}[N'(t) + B'(t)S(t)]x(t), \tag{A4.13}$$

where

$$\dot{S} + SA + A'S - (SB + N)D^{-1}(B'S + N') + C = 0, \quad S(T) = \Sigma. \tag{A4.14}$$

For other variations and extensions of the linear quadratic problem (A4.1) and (A4.2), for which explicit feedback controllers can be developed. The reader is referred to [Bryson and Ho, 1969].

A4.1.1 Certainty Equivalence or Separation Principle

Suppose equation (A4.2) is changed by the presence of a gaussian white noise $w(t)$ and becomes

$$\dot{x} = Ax + Bu + w$$

where

$$E[w(t)] = 0, \quad E[w(t)w(\tau)'] = Q(t)\delta(t-\tau)$$

and $x(0)$ is a normal random variable with

$$E[x(0)] = 0, \quad E[x(0)x(0)'] = P_0.$$

Because of the presence of uncertainty in the system equation, we must modify the objective function in (A4.12) as follows

$$\text{Maximize}\{J = E[\frac{1}{2} x'\Sigma x + \frac{1}{2}\int_0^T (x'u') \begin{pmatrix} C & N \\ N' & D \end{pmatrix} \begin{pmatrix} x \\ u \end{pmatrix} dt\} \cdot$$

Assume further that x cannot be directly measured and the measurement process is given by (12.3.21), that is

$$y(t) = H(t)x(t) + v(t)$$

where $v(t)$ is a white noise as defined in (12.3.22).

The optimal control $u^*(t)$ for this linear quadratic stochastic optimal control problem can be shown to be given by (A4.13) with $x(t)$ replaced by its estimate $\hat{x}(t)$, see [Bryson and Ho, 1969]. Thus

$$u^*(t) = D(t)^{-1}[N'(t) + B'(t)S(t)]\hat{x}(t)$$

where S is given in (A.14) and $\hat{x}$ is given by the Kalman-Bucy filter:

$$\dot{\hat{x}} = A\hat{x} + Bu^* + \bar{w} + K[y - H\hat{x}], \quad \hat{x}(0) = 0$$

$$K = P H' R^{-1}$$

$$\dot{P} = A P + P A' - K H P + Q, \quad P(0) = P_0.$$

The above procedure has received two different names in the literature. In economics it is called the <u>certainty equivalence principle</u>, see [Simon, 1956]. In engineering and mathematics literature it is called the <u>separation principle</u>, [Joseph and Tou, 1961].When we call it the certainty equivalence principle we are emphasizing the fact that $\hat{x}(t)$ can be used for the purposes of optimal feedback control as if it were the certain value of the state variable $x(t)$. Whereas the term separation principle emphasizes the fact that the process of determining the optimal control can be broken down into two steps: first, estimate x by using the optimal filter; second, use that estimate in the optimal feedback control formula for the deterministic problem.

A4.2 Second Order Variations

Second order variation in optimal control theory is analogous to the second order conditions in classical optimization problem of calculus. To discuss the second order variational condition is difficult when the control variable u is contained in the control set Ω. So we make the simplifying assumption that $\Omega = R^m$ and thus controls u are unconstrained. Thus, we are now dealing with

$$\max_u \{J = \int_0^T F(x,u,t)dt + S[x(T)]\} \tag{A4.15}$$

subject to

$$\dot{x} = f(x,u,t), \quad x(0) = x_0. \tag{A4.16}$$

From Chapter 2, we know that the first-order necessary conditions for this problem is given by

$$\dot{\lambda} = -H_x, \quad \lambda(T) = 0 \tag{A4.17}$$

$$H_u = 0 \tag{A4.18}$$

where the Hamiltonian H is given by

$$H = F + \lambda f . \tag{A4.19}$$

Since u is unconstrained, these conditions may be easily derived by the method of calculus of variations. To see this, we write the augment objective function

$$\bar{J} = \Phi[x(T)] + \int_0^T [H(x,u,\lambda,t) - \lambda\dot{x}]dt. \tag{A4.20}$$

Consider small perturbation from the extremal path given by (A4.16 - A4.19) as a result of small perturbations $\delta x(0)$ in the initial state. Define the resulting perturbations in state, adjoint, and control variables by $\delta x(t)$, $\delta\lambda(t)$, and $\delta u(t)$. These, of course, will be obtained by linearizing (A4.16 - A4.18) around the external path:

$$\dot{\delta x} = f_x \delta_x + f_u \delta_u, \quad \delta_x(0) \text{ specified} \tag{A4.21}$$

$$\dot{\delta\lambda} = -(H_{xx}\delta_x)' - \delta\lambda f' - (H_{xu}\delta_u)' \tag{A4.22}$$

$$\begin{aligned} \delta H_u &= (H_{ux}\delta_x)' + \delta\lambda\, H_{u\lambda}{}' + (H_{uu}\delta_u)' \\ &= (H_{ux}\delta_x)' + \delta\lambda f_u + (H_{uu}\delta_u)' = 0. \end{aligned} \tag{A4.23}$$

Alternatively, we may consider an expansion of the objective function and the state equation to second order. Since the first order terms vanish about a trajectory which satisfies (A4.15 - A4.18). From [Bryson and Ho, 1969], this may be accomplished by expanding (A4.20) to second order and all the constraints to first order. Thus we have

$$\delta^2\bar{J} = \frac{1}{2}(\delta_x{}'\,\Phi_{xx}\delta_x)_{t=T} + \frac{1}{2}\int_0^T (\delta x'\ \delta u')\begin{pmatrix} H_{xx} & H_{xu} \\ H_{ux} & H_{uu} \end{pmatrix}\begin{pmatrix} \delta_x \\ \delta_u \end{pmatrix} dt \tag{A4.24}$$

subject to

$$\delta\dot{x} = f_x\delta_x + f_u\delta_u, \quad \delta_x(0) \text{ specified.} \tag{4.25}$$

Since we are interested in a neighboring extremal path, we must determine $\delta u(t)$ so as to maximize $\delta^2\bar{J}$ subject to (A4.25). This problem is a linear-quadratic problem discussed in the previous section. For this problem, the optimal control $\delta u^*(t)$ is given by formula (A4.14), provided $H_{uu}(t)$ is nonsingular for $0 \leq t \leq T$ (see the next Section A4.3 when $H_{uu}(t)$ is singular for a finite time interval). Thus, recognizing that $\Sigma = \Phi_{xx}$, $C = H_{xx}$, $N = H_{xu}$, $D = H_{uu}$, $A = f_x$ and $B = f_u$, we have

$$\delta u^*(t) = H_{uu}^{-1}[H_{ux} + f_u' S(t)]\ \delta x(t) \tag{A4.26}$$

where

$$\dot{S} + Sf_x + f_x' S - (Sf_u + H_{xu}) H_{uu}^{-1} (f_u'S + H_{ux}) + H_{xx} = 0, \quad S(T) = \Phi_{xx}. \tag{4.27}$$

While a number of second order conditions can be obtained by proceeding further from this manner , we shall only be interested in the concavity condition (or strengthened Legendre-Clebsch condition). It is possible to show that neighboring stationary paths exist (in a weak sense; i.e. δx and δu are small) if

$$H_{uu}(t) < 0 \quad \text{for} \quad 0 \le t \le T; \tag{A4.28}$$

or in other words, $H_{uu}(t)$ is negative semidefinite. First order conditions, condition (A4.28) and the condition that $S(t)$ is finite for $0 \le t \le T$ represent sufficient condition for a trajectory to be a local minimum. We are not being specific here because we would mostly be relying on the sufficiency condition developed in Section 2.4. We are stating (A3.28) because its similarity with the second order condition for a local maximum in the classical maximization problem.

We must note that

$$H_u = 0 \quad \text{and} \quad H_{uu} \le 0 \tag{A4.29}$$

form necessary conditions for a trajectory to be a local maximum.

A4.3 Singular Control

In some optimization problems including some problems treated in this text, extremal arcs satisfying $H_u = 0$ occur on which the matrix H_{uu} is singular. Such arcs are called singular arcs. Note that these arcs satisfy (A4.29) but not the strengthened condition (A4.28). While no general sufficiency conditions are available for singular arcs, some additional necessary conditions known as the generalized Legendre Clebsch conditions have been developed. A good reference on singular control is [Bell and Jacobson, 1975].

We shall only discuss the case in which the Hamiltonian is linear in one or more of the control variables. For these systems, $H_u = 0$ implies that coefficient of the linear control term vanishes identically along a singular arc. Thus, the control is not determined in terms of x and λ by the Hamiltonian maximizing condition $H_u = 0$. Instead, the control is determined by the requirement that the coefficient of these linear terms remain zero on the singular arc. That is, the time derivatives of H_u must be zero. Having obtained the control satisfying $\frac{dH_u}{dt} = 0$ (or setting higher time derivatives equal to zero) along the singular arc, we must check an additional necessary condition analogous to the concavity condition (A4.28). For a maximum problem with a single control variable, the condition

$$(-1)^k \frac{\partial}{\partial u} [(\frac{d}{dt})^{2k} H_u] \leq 0, \quad k = 0,1,2,\ldots . \tag{A4.30}$$

The conditions (A4.30) is called the generalized Legendre Clebsch conditions.

<u>Example</u>: We present an example treated by[Johnson and Gibson, 1963].

$$\max \{J = -\frac{1}{2} \int_0^T x_1^2 \, dt \} \tag{A4.31}$$

subject to

$$\dot{x}_1 = x_2 + u, \quad x_1(0) = a \quad \text{given} \tag{A4.32}$$

$$\dot{x}_2 = -u, \qquad x_2(0) = b \quad \text{given} \tag{A4.33}$$

$$\text{and} \quad x_1(T) = x_2(T) = 0. \tag{A4.34}$$

We form the Hamiltonian

$$H = -\frac{1}{2} x_1^2 + \lambda_1(x_2 + u) + \lambda_2(-u) \tag{A4.35}$$

where the adjoint equations are

$$\dot{\lambda}_1 = +x_1, \quad \dot{\lambda}_2 = -\lambda_1 \; . \tag{A4.36}$$

The optimal control is bang bang plus singular (see Row of Table). Singular arcs must satisfy

$$H_u = \lambda_1 - \lambda_2 = 0 \quad , \tag{A4.37}$$

for a finite time interval. The optimal control can, therefore, be obtained by

$$\frac{d}{dt} H_u = \dot{\lambda}_1 - \dot{\lambda}_2 = + x_1 + \lambda_1 = 0 \quad . \tag{A4.38}$$

Differentiating once more with respect to time t, we obtain

$$\frac{d^2}{dt^2} H_u = + \dot{x}_1 + \dot{\lambda}_1 = +x_2 + u + x_1 = 0$$

which implies

$$u = -(x_1 + x_2) \tag{A4.39}$$

along the singular arc. We now verify the generalized Legendre Clebsch condition (A4.30):

$$\frac{\partial}{\partial u} \frac{d^2}{dt^2} \left(\frac{\partial H}{\partial u} \right) = -1 \leq 0. \tag{A4.40}$$

Many examples of singular control occur throughout this book.

BIBLIOGRAPHY

We list below only the references actually cited in this book. For further references on applications of optimal control theory to management science we refer the reader to the surveys in [Sethi, 1977, 1978a], [Tapiero, 1977], [Wickwire, 1977], [Kleindorfer, 1978], [Pekelman, 1978], and [Bookbinder and Sethi, 1980].

Agnew, C.E., Dynamic Modeling and Control of Congestion - Prone Systems, Operations Research 24 (1976) 400-419.

Alam, M., Lynn, J.W., and Sarma, V.V.S., Optimal Maintenance Policy for Equipment Subject to Random Deterioration and Random Failure, Int. J. of Systems Science 7 (1976) 1071-1080.

Alam, M., and Sarma, V.V.S., Optimal Maintenance Policy for an Equipment Subject to Deterioration and Random Failure, I.E.E.E. Transactions on System, Man, and Cybernetics SMC-4 (1974) 172-175.

Alam, M. and Sarma, V.V.S., An Application of Optimal Control Theory to Repairman Problem with Machine Interference, IEEE Transactions on Reliability R-26 (1977) 121-124.

Allen, K.R., Analysis of the Stock-Recruitment Relation in Antarctic Fin Whales, in B. Parrish (ed.) Fish Stock and Recruitment, J. du Conseil International pour l'Ex loration de la Mer, Rapports et Procés-Verbaux de Réunions 164 (1973) 132-137.

Aoki, M., Dynamic Economics: A System Theoretic Approach to Theory and Control. [Elsevier: New York, 1976.]

Arnold, L., Stochastic Differential Equations: Theory and Applications, (Wiley: New York, 1974].

Arora, S.R., and Lele, P.T., A Note on Optimal Maintenance Policy and Sale Date of a Machine, Management Science 17 (1970) 170-173.

Arrow, K.J., Applications of Control Theory to Economic Growth, in Lectures in Applied Mathematics, Vol. 12. (American Mathematical Society: Providence, R.I., 1968).

Arrow, K.J., and Chang, S., Optimal Pricing, Use, and Exploration of Uncertain Natural Resource Stocks, Technical Report No. 31, Department of Economics, Harvard University (1978).

Arrow, K.J., and Kurz, M., Public Investment, The Rate of Return, and Optimal Fiscal Policy, (The Johns Hopkins Press: Baltimore, Maryland, 1970).

Arthur, W.B., and McNicoll, G., Optimal Time Paths with Age Dependence: A Theory of Population Policy, Review of Economic Studies, 44 (1977), 111-123.

Bell, D.J., and Jacobson, D.H., Singular Optimal Control (Academic Press, New York, 1975).

Bellman, R., Dynamic Programming (Princeton University Press: Princeton, N.J. 1957).

Bellman, R., and Kalaba, R., Quasilinearization and Boundary Value Problems, (American Elsevier: New York, 1965).

Bellman, R.E., and Kalaba, R.E., Dynamic Programming and Modern Control Theory (Academic Press: New York, 1965).

Bensoussan, A., Hurst, Jr., E.G., and Näslund, B., Management Applications of Modern Control Theory (American Elsevier: New York, 1974).

Bensoussan, A., and Lions, J.L., Nouvelles Methodes en Contrôle Impulsionnel, Applied Mathematics and Optimization 1 (1975) 289-312.

Bensoussan, A., Nissen, G., and Tapiero, C., Optimum Inventory and Product Quality Control with Deterministic and Stochastic Deterioration - An Application of Distributed Parameter Control Systems, IEEE Transactions on Automatic Control 20 (1975) 407-412.

Berkovitz, L.D., Variational Methods in Problems of Control and Programming, Journal of Mathematical Analysis and Applications, 3 (1961) 145-169.

Berkovitz, L.D., and Dreyfus, S.E., The Equivalence of Some Necessary Conditions for Optimal Control in Problems with Bounded State Variables, Journal of Mathematical Analysis and Applications 10 (1965) 275-283.

Black, F. and Scholes, M., The Pricing of Options and Corporate Liabilities, Journal of Political Economy 81 (1973) 637-659.

Blaquiere, A., Necessary and Sufficiency Conditions for Optimal Strategies in Impulsive Control, Presented at the Third Kingston Conference on Differential Games and Control Theory, Kingston, R.I., June 5-8, 1978.

Boltyanskii, V.G., Mathematical Methods of Optimal Control (Holt, Rinehard & Winston: New York, 1971).

Bookbinder, J.H., and Sethi, S.P., The Dynamic Transportation Problem: A Survey Naval Research Logistics Quarterly 27 (1980) 65-88.

Breakwell, J.V., Stochastic Optimization Problems in Space Guidance, in H.F. Karreman (ed.) Stochastic Optimization and Control (Wiley: New York, 1968 91-100.

Brito, D.L., and Oakland, W.H., Some Properties of The Optimal Income Tax, International Economic Review 18 (1977) 407-423.

Bryant, G.F., and Mayne, D.Q., The Maximum Principle, International Journal of Control 20 (1974) 1021-1054.

Bryson, Jr., A.E., and Ho, Y.C., Applied Optimal Control (Blaisdell: Waltham, Massachusetts, 1969).

Buchanan, L.F., and Norton, F.E., Optimal Control Applications in Economic Systems, in C.T.Leondes (ed.) Advances in Control Systems, Vol. 8 (Academic Press: New York, 1971) 141-187.

Burdet, C.A., and Sethi, S.P., On the Maximum Principle for a Class of Discrete Dynamical System with Lags, Journal of Optimization Theory & Applications 19 (July 1976) 445-454.

Burmeister, E. and Dobell, A.R., Mathematical Theories of Economic Growth (MacMillan: London, 1970).

Butkowskiy, A.G., Distributed Control Systems, (American Elsevier: New York, 1969).

Canon, M.D., Cullum, C.D., and Polak, E., Theory of Optimal Control and Mathematical Programming (McGraw-Hill: New York, 1970).

Case, J.H., Economics and the Competitive Process, (New York University Press: New York, 1979).

Charnes, A., and Kortanek, K., A Note on the Discrete Maximum Principle and Distribution Problems, Journal of Mathematics and Physics 45 (1966) 121-126.

Chen, S.F., and Leitmann, G., Labour-Management Bargaining Modelled as a Dynamic Game, Optimal Control Applications and Methods, 1 (1980) 11-25.

Chow, G.C., Analysis and Control of Dynamic Economic Systems (Wiley: New York, 1975).

Clark, C.W., Mathematical Bioeconomics (Wiley: New York, 1976).

Clarke, F.H., Necessary Conditions for a General Control Problem, in David Russel (ed.), Calculus of Variations and Control Theory (Academic Press: New York, 1976).

Coddington, E.A., and Levinson, N.L., Theory of Ordinary Differential Equations (McGraw-Hill: New York, 1955).

Connors, M.M., and Teichroew, D., Optimal Control of Dynamic Operations Research Models (International Textbook: Scranton, Pa., 1967).

Dantzig, G., Linear Control Processes and Mathematical Programming, SIAM J. Control 4 (1966) 56-60.

Dasgupta, P., and Heal, G., The Optimal Depletion of Exhaustible Resources, The Review of Economic Studies, Symposium Issue (1974).

Davis, B.E., Investment and Rate of Return for the Regulated Firm, The Bell Journal of Economics and Management Science (1970) 245-270.

Davis, B.E., and Elzinga, D.J., The Solution of an Optimal Control Problem in Financial Modeling, Operations Research 19 (1972) 1419-1433.

Deal, K.R., Optimizing Advertising Expenditures in a Dynamic Duopoly, Operations Research 27 (1979) pp. 682-692.

Deal, K., Sethi, S.P. and Thompson, G.L., A Bilinear-Quadratic Differential Game in Advertising in P.T. Liu and J.G. Sutinen (eds.) Control Theory in Mathematical Economics (Marcel Dekker, Inc.: New York, 1979).

Derzko, N. and Sethi, S.P., Optimal Exploration and Consumption of a Natural Resource - Stochastic Case, Working Paper, University of Toronto (1980).

Derzko, N.A., Sethi, S.P., and Thompson, G.L., Necessary and Sufficient Condition for Optimal Control of Quasilinear Partial Differential Systems, Working Paper, University of Toronto (1979).

Derzko, N., Sethi, S.P. and Thompson, G.L., Distributed Parameter Systems Approac to the Optimal Cattle Ranching Problem, Optimal Control Applications and Methods 1 (1980) 3-10.

Dhrymes, P.J., On Optimal Advertising Capital and Research Expenditures Under Dynamic Conditions, Economica 39 (1962) 275-279.

Dorfman, R., Economic Interpretation of Optimal Control Theory, American Economi Review 49 (Dec. 1969) 817-831.

Drews, W., Hartberger, R., and Segers, R., On Continuous Mathematical Programmin in R. W. Cottle and J. Krarup (eds.), Optimization Methods for Resource Allocation (The English Universities Press: London, 1974).

Elton, E. and Gruber, M., Finance as a Dynamic Process (Prentice-Hall: Englewood Cliffs, N.J., 1975).

Fan, L.T., and Wang, C.S., An Application of the Discrete Maximum Principle to a Transportation Problem, Journal of Mathematics and Physics 43 (1964) 255-260.

Feldman, A.A., Optimal Control Systems (Academic Press: New York, 1965).

Fleming, W.H., and Rishel, R.W., Deterministic and Stochastic Optimal Control (Springer-Verlag: New York, 1975).

Fletcher R., and Reeves, C.M., Function Minimization by Conjugate Gradients, Computer Journal 7 (1964) 149-154.

Friedman, A., Differential Games (Wiley: New York, 1971).

Gelfand, I.M., and Fomin, S.V., Calculus of Variations (Prentice-Hall: Englewood Cliffs, N.J., 1963).

Gihman, I.I., and Skorohod, A.V., Stochastic Differential Equations, (Springer-Verlag: New York, 1971).

Glad, S.T., A Combination of Penalty Function and Multiplier Methods for Solving Optimal Control Problems, Journal of Optimization Theory and Applications 28 (1979) 303-329.

Goh, B.S., Leitmann, G., and Vincent, T.L., Optimal Control of a Prey-Predator System, Mathematical Biosciences 19 (1974) 263-286.

Gordon, H.S., Economic Theory of a Common-Property Resource: The Fishery, Journal of Political Economy 62 (1954) 124-142.

Gordon, M.J., The Investment, Financing and Valuation of the Corporation (Richard D. Irwin, Inc.: Homewood, Ill., 1962).

Gould, J.P., Diffusion Processes and Optimal Advertising Policy, in Microeconomic Foundation of Employment and Inflation Theory, edited by E.S. Phelps et. al. (Norton: New York, 1970) pp. 338-368.

Hadley, G., and Kemp, M.C., Variational Methods in Economics (North-Holland: Amsterdam, 1971).

Halkin, H., A Maximum Principle of the Poltryagin Type for Systems Described by Nonlinear Difference Equations, SIAM J. Control 4 (1966) 90-111.

Halkin, H., On the Necessary Condition for Optimal Control of Nonlinear Systems, in G. Leitmann (ed.), Topics in Optimization (Academic Press: New York, 1967).

Hartberger, R.J., A Proof of the Pontryagin Maximum Principle for Initial-Value Problem, Journal of Optimization Theory and Applications, 11 (1973) 139-145.

Harris, H., Optimal Planning under Transaction Costs: The Demand for Money and Other Assets, Journal of Economic Theory 12 (1976) 298-314.

Hestenes, M.R., Calculus of Variations and Optimal Control Theory (Wiley: New York, 1966).

Ho, Y.C., Differential Games, Dynamic Optimization and Generalized Control Theory, Journal of Optimization Theory and Applications 6 (1970) 179-209.

Holt, C., Modigliani, F., Muth, J., and Simon, H., Planning Production, Inventories and Work Force (Prentice-Hall: Englewood Cliffs, N.J., 1960).

Holtzman, J.M., On the Maximum Principle for Nonlinear Discrete-Time Systems, IEEE Transaction Automatic Control 11 (1966) 273-274.

Hotelling, H., The Economics of Exhaustible Resources, Journal of Political Economy, 39 (1931) 137-175.

Hwang, C.L., Fan, L.T., and Erickson, L.E., Optimum Production Planning by the Maximum Principle, Management Science 13 (1967) 750-755.

Ijiri, Y., and Thompson, G.L., Applications of Mathematical Control Theory to Accounting and Budgeting (The Continuous Wheat Trading Model), The Accounting Review 45 (1970) 246-258.

Ijiri, Y., and Thompson, G.L., Mathematical Control Theory Solution of an Interactive Accounting Flows Model, Naval Research Logistics Quarterly 19 (1972) 411-422.

Intrilligator, M., Mathematical Optimization and Economic Theory (Prentice-Hall: Englewood Cliffs, N.J., 1971).

Isaacs, R., Differential Games (Wiley: New York, 1965).

Jacobson, D.H., Lele, M.M., and Speyer, J. L., New Necessary Conditions of Optimality for Control Problems with State-Variable Inequality Constraints, Journal of Mathematical Analysis and Applications 35 (1971) 255-284.

Jacquemin, A.P., Optimal Control and Advertising Policy, Metro-economica 25 (1973) 200-209.

Johnson, C.D., and Gibson, J.E., Singular Solutions in Problems of Optimal Control, IEEE Transactions on Automatic Control AC-8 (1963) 4-15.

Joseph, P.D., and Tou, J.T., On Linear Control Theory, Trans. AIEE, Part II 80 (1961) 193-196.

Kalman, R.E., A New Approach to Linear Filtering and Prediction Problems, Trans. ASME, 83D (1961) 35-45.

Kalman, R.E., amd Bucy, R., New Results in Linear Filtering and Prediction, Trans. ASME, 83D (1961) p. 95.

Kamien, M.I., and Schwartz, N.L., Optimal Maintenance and Sale Age for a Machine Subject to Failure, Management Science 17 (1971) 427-449.

Kamien, M.I., and Schwartz, N.L., Limit Pricing and Uncertain Entry, Econometrica 39 (1971a) 441-454.

Kamien, M.I., and Schwartz, N.L., Dynamic Optimization, Graduate School of Management, Northwestern University (1979), (Multilithed notes.)

Keeler, E., Spence, M., and Zeckhauser, R., The Optimal Control of Pollution, Journal of Economic Theory 4 (1971) 19-34.

Keller, H.B., Numerical Methods for Two-Point Boundary Value Problems, (Blaisdell Waltham, Mass., 1968).

Kilkki, P., and Vaisanen, U., Determination of Optimal Policy for Forest Stands by Means of Dynamic Programming, Acta Forestalia Fennica 102 (1969) 100-112.

Kirk, D.E., Optimal Control Theory: An Introduction (Prentice-Hall: Englewood Cliffs, N.J., 1970).

Kleindorfer, P.R., Stochastic Control Models in Management Science: Theory and Computation, TIMS Studies in the Management Sciences 9 (1978) 69-88.

Kleindorfer, P.R., Kriebel, C.H., Thompson, G.L., and Kleindorfer, G.B., Discrete Optimal Control of Production Plans, Management Science 22 (1975) 261-273.

Kleindorfer, P.R., and Lieber, Z., Algorithms and Planning Horizon Results for Production Planning Problems with Separable Costs, Operations Research 27 (1979) 874-887.

Krouse, C.G., Optimal Financing and Capital Structure Programs for the Firm, Journal of Finance 27 (1962) 1057-1071.

Krouse, C.G., and Lee, W.Y., Optimal Equity Financing of the Corporation, Journal of Financial and Quantitative Analysis 8 (1973) 539-563.

Kushner, H.J., Probability Methods for Approximations in Stochastic Control and for Elliptic Equations (Academic Press: New York, 1977).

Lansdowne, Z.F., The Theory and Applications of Generalized Linear Control Processes, Technical Report No. 10, Department of Operations Research, Stanford University (1970).

Lasdon, L.S., Mitter, S.K., and Warren, A.D., The Conjugate Gradient Method for Optimal Control Problems, IEEE Transactions on Automatic Control AC-12 (1967) 132-138.

Lee, E.B., and Markus, L., Foundations of Optimal Control Theory (Wiley: New York, (1968).

Legey, L., Ripper, M., and Varaiya, P., Effects of Congestion on the Shape of the City, Journal of Economic Theory, 6 (1973) 162-179.

Lesourne, J., Croissance Optimale des Entreprises (Dunod: Paris, 1973).

Lieber, Z., An Extension to Modigliani and Hohn's Planning Horizon Results, Management Science 20 (1973) 319-330.

Linter, J., The Cost of Capital and Optimal Financing of Corporate Growth, Journal of Finance, 23 (1963) 292-310.

Lions, J.L., Optimal Control of Systems Governed by Partial Differential Equations, (Springer-Verlag: New York, 1971).

Lucas, Jr., R.E., Optimal Management of a Research and Development Project, Management Science 17 (1971) 679-697.

Luenberger, D.G., Mathematical Programming and Control Theory: Trends of Interplay, in A.M. Geoffrion (ed.), Perspectives on Optimization (Addison-Wesley: Reading, Mass., 1972).

Lundin, R.A., and Morton, T.E., Planning Horizons for the Dynamic Lot Size Model: Protective Procedures and Computational Results, Operations Research, 23 (1975) 711-734.

McIntyre, J., and Paiewonsky, B., On Optimal Control with Bounded State Variables in C. T. Leondes (ed.) Advances in Control Systems, Vol. 5 (Academic Press New York, 1967) 389-419.

Mangasarian, O.L., Sufficient Conditions for the Optimal Control of Nonlinear Systems, SIAM J. Control 4 (1966) 139-152.

Mangasarian, O.L., Nonlinear Programming (McGraw-Hill: New York, 1969).

Mangasarian, O.L., and Fromovitz, S., A Maximum Principle in Mathematical Programming, in A.V. Balakrishman and L.W. Neustadt (eds.), Mathematical Theory of Control (Academic Press: New York, 1967).

Manh-Hung, N., Essays on the Optimal Dynamic Exploitation of Natural Resources and the Social Rate of Discount, Ph.D. Dissertation, University of Toronto (1974).

Masse, P., Optimal Investment Decisions (Prentice-Hall: Englewood Cliffs, N.J., 1962).

Menra, R.K., An Optimal Control Approach to National Settlement System Planning, RM-75-58, International Institute of Applied Systems Analysis, Laxenburg, Austria (Nov. 1975).

Mehra, R.K., and Davis, R.E., Generalized Gradient Method for Optimal Control Problems with Inequality Constraints and Singular Arcs, IEEE Transactions on Automatic Control, AC-17 (1972) 69-79.

Merton, R.C., Optimum Consumption and Portfolio Rules in a Continuous-Time Model, Journal of Economic Theory, 3 (1971) 373-413.

Merton, R.C., An Intertemporal Capital Asset Pricing Model, Econometrica 5 (1973) 867-888.

Miele, A., Extremization of Linear Integral Equations by Green's Theorem, in G. Leitman (ed.), Optimization Techniques (Academic Press: New York, 1962).

Miller, M.H., and Modigliani, F., Dividend Policy, Growth, and Valuation of Shares, The Journal of Business 34 (1961) 411-433.

Modigliani, F., and Hohn, F., Production Planning Over Time, Econometrica 23 (1955) 46-66.

Moiseev, N.N., Numerical Methods in Optimal Control Theory (Nauka: Moscow, 1971).

Morton, T.E., Universal Planning Horizons for Generalized Convex Production Scheduling, Operations Research, 26 (1978) 1046-1057.

Näslund, Bertil, Simultaneous Determination of Optimal Repair Policy and Service Life, The Swedish Journal of Economics 68 (1966) 63-73.

Näslund, Bertil, Optimal Rotation and Thinning, Forest Science 15 (1969) 446-451.

Nelson, R.T., Labor Assignment as a Dynamic Control Problem, Operations Research 14 (1960) 369-376.

Nepomiastchy, P., Application of Optimal Control Theory and Penalty Function Techniques to Solution of a Particular Scheduling Problem, Zhurnal Vychiflitel'noi Matematiki i Matematichefkoi Fiziki (in Russian), Vol. 10, No. 4, (1970).

Nerlove, M., and Arrow, K.J., Optimal Advertising Policy Under Dynamic Conditions, Economica 39 (1962) 129-142.

Norstrom, C.J., The Continuous Wheat Trading Model Reconsidered. An Application of Mathematical Control Theory with a State Constraint, W.P. #58-77-78, GSIA, Carnegie-Mellon University, Pittsburgh, Pa., (1978).

Ozga, S., Imperfect Markets Through Lack of Knowledge, Quarterly Journal of Economics, 74 (1960) 29-52.

Palda, K.S., The Measurement of Cumulative Advertising Effects (Prentice-Hall: Englewood Cliffs, N.J., 1964).

Pekelman, D., Simultaneous Price-Production Decision, Operations Research 22 (1974) 788-794.

Pekelman, D., Production Smoothing with Fluctuating Price, Management Science 21 (1975) 576-590.

Pekelman, D., A Survey of Adaptive Control Schemes, TIMS Studies in the Management Sciences 9 (1978) 89-120.

Pekelman, D., On Optimal Utilization of Production Processes, Operations Research 27 (1979) 260-278.

Pekelman, D., and Sethi, S.P., Advertising Budgeting, Wearout and Copy Replacement, J. Opl. Res. Soc. 29 (1978) 651-659.

Peterson, D.W., The Economic Significance of Auxiliary Functions in Optimal Control, International Economic Review, 14 (1973) 234-52.

Pierskalla, W.P., and Voelker, J.A., A Survey of Maintenance Models: The Control and Surveillance of Deteriorating Systems, Naval Research Logistics Quarterly 23 (1976) 353-388.

Pindyck, R.S., (ed.), Advances in the Economics of Energy and Resources, Vol. II, (J.A.I. Press: Greenwich, Conn., 1978).

Pindyck, R.S., The Optimal Exploration and Production of Nonrenewable Resources, Journal of Political Economy, 86 (1978a) 841-862.

Pindyck, R.S., Uncertainty and Exhaustible Resource Markets, Journal of Political Economy, forthcoming (1980).

Polak, E., Computational Methods in Optimization (Academic Press: New York, 1971)

Polak, E., A Historical Survey of Computational Methods in Optimal Control, SIAM Review 15 (1973) 553-584.

Pontryagin, L.S., Boltyanskii, V.G., Gamkrelidze, R.V., and Mischenko, E.F., The Mathematical Theory of Optimal Processes (Wiley: New York, 1962).

Ramsey, F., A Mathematical Theory of Savings, Economic Journal 38 (1928) 543-559.

Rapp, B., Models for Optimal Investment and Maintenance Decisions (Halsted Press: Stockholm, 1974).

Raviv, A., The Design of an Optimal Insurance Policy, The American Economic Review 69 (1979) 84-96.

Richard, S.F., A Generalized Capital Asset Pricing Model, TIMS Studies in the Management Sciences 11 (1979) 215-232.

Roberts, S.M. and Shipman, J.S., Two-Point Boundary Value Problems: Shooting Methods (American Elsevier: New York, 1972).

Rockafellar, R.T., State Constraints in Convex Control Problems of Bolza, SIAM Journal of Control, 19 (1972) 691-715.

Russak, B., Relations Among the Multipliers for Problems with Bounded State Constraints, SIAM J. Control and Optimization 14 (1976) 1151-1155.

Russell, D.L., Penalty Functions and Bounded Phase Coordinates, SIAM J. Control 2 (1965) 409-422.

Sage, A.P., Optimum Systems Control (Prentice-Hall: Englewood Cliffs, N.J. 1968)

Schaefer, M.B., Some Considerations of Population Dynamics and Economics in Relation to the Management of Marine Fisheries, Journal of the Fisheries Research Board of Canada 14 (1957) 669-681.

Sarma, V.V.S., and Alam, M., Optimal Maintenance Policies for Machines Subject to Deterioration and INtermittent Breakdowns, IEEE Transactions on System, Man, and Cybernetics SMC-5 (1975) 396-398.

Seierstad, A., and Sydsaeler, K., Sufficient Conditions in Optimal Control Theory International Economic Review, 18 (1977) 367-391.

Sethi, S.P., Optimal Control of the Vidale-Wolfe Advertising Model, Operations Research 21 (1973) 998-1013.

Sethi, S.P., Simultaneous Optimization of Preventive Maintenance and Replacement Policy for Machines: A Modern Control Theory Approach, AIIE Transactions 5 (1973a) 156-163.

Sethi, S.P., An Application of Optimal Control Theory in Forest Management, Journal of Management Science and Applied Cybernetics (New Delhi) 2 (1973b) 9-16.

Sethi, S.P., Sufficient Conditions for the Optimal Control of a Class of Systems with Continuous Lags, Journal of Optimization Theory and Applications 13 (1974) 545-552.

Sethi, S.P., Some Explanatory Remarks on the Optimal Control of the Vidale-Wolfe Advertising Model, Operations Research 22 (1974a) 1119-1120.

Sethi, S.P., Quantitative Guidelines for Communicable Disease Control Program: A Complete Synthesis, Biometrics, 30 (1974b) 681-691.

Sethi, S.P., Optimal Control of a Logarithmic Advertising Model, Operations Research Quarterly 26 (1975) 317-319.

Sethi, S.P., Dynamic Optimal Control Models in Advertising: A Survey, SIAM Review 19 (1977) 685-725.

Sethi, S.P., Nearest Feasible Paths in Optimal Control Problems: Theory, Examples, and Counterexamples, Journal of Optimization Theory and Applications 23 (1977a).

Sethi, S.P., Optimal Advertising for the Newlove-Arrow Model under a Budget Constraint, Operational Research Quarterly, 28 (1977b) 683-693.

Sethi, S.P., A Linear Bang-Bang Model of Firm Behavior and Water Quality, IEEE Transactions on Automatic Control AC-22 (1977c) 706-714.

Sethi, S.P., Optimal Equity Financing Model of Krouse and Lee: Corrections and Extensions, Journal of Financial and Quantitative Analysis, 13 (1978) 487-505.

Sethi, S.P., A Survey of Management Science Applications of the Deterministic Maximum Principle, TIMS Studies in the Management Sciences 9 (1978a) 33-67.

Sethi, S.P., Optimal Depletion of Exhaustible Resources, Applied Mathematical Modelling, 3 (1979) 367-378.

Sethi, S.P., Optimal Filtering Policy for Dynamic Continuous Thieves, Management Science 25 (1979a) 535-542.

Sethi, S.P., Optimal Advertising Policy with the Contagion Model, Journal of Optimization Theory and Applications 29 (1979b) 615-627.

Sethi, S.P., and Chand, S., Planning Horizon Procedures in Machine Replacement Models, Management Science, 25 (1979) 140-151.

Sethi, S.P., Drews, W.P., and Segers, R.G., A Unified Framework for Linear Control Problems with State Variable Inequality Constraints, Journal of Optimization Theory and Applications, forthcoming (1981).

Sethi, S.P., and Lee, S.C., Optimal Advertising for the Nerlove-Arrow Model under a Replenishable Budget, Working Paper, University of Toronto, (1979).

Sethi, S.P., and McGuire, T.W., Optimal Skill Mix: An Application of the Maximum Principle for Systems with Retarded Controls, Journal of Optimization Theory and Applications 23 (1977) 245-275.

Sethi, S.P., and Morton, T.E., A Mixed Optimization Technique for the Generalized Machine Replacement Problem, Naval Research Logistics Quarterly 19 (1972) 471-481.

Sethi, S.P., and Thompson, G.L., Applications of Mathematical Control Theory to Finance, Journal of Financial and Quantitative Analysis 5 (1970) 381-394.

Sethi, S.P., and Thompson, G.L., Christmas Toy Manufacturer's Problem: An Application of the Stochastic Maximum Principle, Opsearch 14 (1977) 161-17

Sethi, S.P., and Thompson, G.L., Simple Models in Stochastic Production Planning, in A. Bensoussan, P. Kleindorfer, and C. Tapiero, Applied Stochastic Control in Econometrics and Management Science (North-Holland: Amsterdam, 1980).

Shell, K., (ed.)., Essays on the Theory of Optimal Economic Growth, (The MIT Press: Cambridge, Mass., 1967).

Shell, K., Applications of Pontryagin's Maximum Principle to Economics, in H. W. Kuhn and G. P. Szego (eds.) Mathematical Systems Theory and Economics (Springer-Verlag: Berlin, 1969).

Simon, H.A., Dynamic Programming under Uncertainty with a Quadratic Criterion Function, Econometrica, 24 (1956) 74-81.

Solow, R.M., Intergenerational Equity and Exhaustible Resources, Review of Economic Studies, Symposium Issue (1974).

Sorenson, H., Kalman Filtering, in C.T. Leondes (ed.) Advance in Control Systems, Vol. 3,(Academic Press: New York, 1966) 219-292.

Southwick. L., and Zionts, S., An Optimal-Control-Theory Approach to the Education-Investment Decision, Operations Research 22 (1974) 1156-1174.

Sprzeuzkouski, A.Y., A Problem in Optimal Stock Management, Journal of Optimization Theory and Applications 1 (1967) 232-241.

Srinivasan, V., Decomposition of a Multi-period Media Scheduling Model in Terms of Single Period Equivalents, Management Science 23 (1976) 349-360.

Starr, A.W., and Ho, Y.C., Nonzero-sum Differential Games, Journal of Optimization Theory and Applications, 3 (1969) 184-206.

Stoeppler, S., Dynamische Produktionstheorie, (West Deutscher-Verlag: Opladen, West Germany) 1975

Tapiero, C.S., Optimal Simultaneous Replacement and Maintenance of a Machine with Process Discontinuities, Revue Francaise d'informatique et Recherche Opérationnelle 2 (1971) 79-86.

Tapiero, C.S., and Soliman, M.A., Multi-commodities Transportation Schedules Over Time, Networks 2 (1972) 311-327.

Tapiero, C.S., Optimal Maintenance and Replacement of a Sequence of Machines and Technical Obsolescence, Opsearch 19 (1973) 1-13.

Tapiero, C.S., Managerial Planning: An Optimum and Stochastic Control Approach (Gordon Breach: New York, 1977).

Taylor, J.G., Lanchester-type Models of Warfare and Optimal Control, Naval Research Logistics Quarterly 21 (1974) 79-106.

Terborgh, G., Dynamic Equipment Policy (McGraw-Hill: New York, 1949).

Thompson, G.L., Optimal Maintenance Policy and Sale Date of a Machine, Management Science 14 (1968) 543-550.

Thompson, G.L., Continuous Expanding and Contracting Economies, Working Paper No. 71-79-80, GSIA, Carnegie-Mellon University (1980).

Thompson, G.L., and Sethi, S.P., Turnpike Horizons for Production Planning, Management Science 26 (1980) 229-241.

Treadway, A.B., Adjustment Costs and Variable Inputs in the Theory of the Competitive Firm, Journal of Economic Theory 2 (1970) 329-347.

Tsurumi, H., and Tsurumi, Y., Simultaneous Determination of Market Share and Advertising Expenditure under Dynamic Conditions: The Case of a Firm Within the Japanese Pharmaceutical Industry, The Economic Studies Quarterly 22 (1971) 1-23.

Tu, P.N.V., Optimal Educational Investment Program in an Economic Planning Model, Canadian Journal of Economics 2 (1969) 52-64.

Turner, R.E., and Neuman, C.P., Dynamic Advertising Strategy: A Managerial Approach, Journal of Business Administration 7 (1976) 1-21.

Vanthienen, L., Simultaneous Price-Reduction Decision Making with Production Adjustment Costs, Proceedings XX International Meeting of TIMS held in Tel Aviv, Israel on June 24-29, 1973, Vol. 1 (Jerusalem Academic Press: Jerusalem, 1975) 249-254.

Varaiya, P.P., N-Person Nonzero-Sum Differential Games with Linear Dynamics, SIAM J. Control 8 (1970) 441-449.

Vickson, J.J., Schedule Control for Randomly Drifting Production Sequences, Department of Management Sciences, Univ. of Waterloo, 1979.

Vidale, M.L., and Wolfe, H.B., An Operations Research Study of Sales Response to Advertising, Operations Research 5 (1957) 370-381.

Vousden, N., International Trade and Exhaustible Resources: A Theoretical Model, International Economic Review, 15 (1974) 149-167.

Wagner, H.M. and Whitin, T.M., Dynamic Version of the Economic Lot Size Model, Management Science 5 (1958) 89-96.

Weinstein, M.C. and Zeckhauser, R.J., The Optimum Consumption of Depletable Natural Resources, The Quarterly Journal of Economics, 89 (1975) 371-392.

Westphal, L.C., Toward the Synthesis of Solutions of Dynamic Games, in C. T. Leondes (ed) Advances in Control Systems Vol. 11, (Academic Press: New York, 1974) 389-489.

Wickwire, K., Mathematical Models for the Control of Pests and Infectious Diseases: A Survey, Theoretical Population Biology 11 (April 1977) 182-232

Wiener, N., The Interpolation and Smoothing of Stationary Time Series (MIT Press: Cambridge, Mass., 1949).

Wright, C., Some Political Aspects of Pollution Control, Journal of Environmental Economics and Management, 1 (1974), 173-186.

Young, L.C., Calculus of Variations and Optimal Control Theory (W. B. Saunders Co Philadelphia, Pa., 1969).

Zimin, I.N., and Ivanilov, Y.P., Solution of Network Planning Problems by Reducin them to Optimal Control Problems, Zhurnal Vychiflitel'noi Matematiki i Matematichefkoi Fiziki (in Russian), Vol. 11, No. 3 (1971).

ANSWERS TO SELECTED EXERCISES

Completely worked solutions to all exercises in this book are contained in the teachers manual, which is made available to instructors by the publisher.

Chapter 1

1.1. (a) Feasible. $J = -333,333$.

1.2. $J = 36$.

1.3. (a) $C = \$157,861$/year. (b) $J = 103.41$ utilities

(c) \$15,000/year

1.4. (b) $W(20) = 985,648$ $\qquad J = 104.34$

1.12. $\text{Imp}(G_1,G_2;t) = G_1 - G_2$.

Chapter 2

2.2. The optimal control is

$$u^*(t) = \begin{cases} 2 & \text{if } 0 \le t \le 2 - \ln 2.5 \\ \text{undefined} & \text{if } t = 2 - \ln 2.5 \\ 0 & \text{if } t > 2 - \ln 2.5 \end{cases} .$$

2.8. Optimal control $u^* = \text{bang}(0,1;\lambda_1-\lambda_2)$

where $\lambda(t) = (8e^{-2(t-18)}, 4e^{-2(t-18)})$.

2.10. (a) $u^*(t) = \text{bang}[0,1;(g_1K_1 + g_2K_2)(\lambda_1-\lambda_2)]$

(c) $\hat{t} - T - \ln[(g_2b_1 - g_1b_2)/(g_2-g_1)b_2]$

2.12. (a) $x(100) = 30 - 20e^{-10} \approx 30.$

(b) $u^* = 3$ for $t \in [0,100]$.

(c) $u^*(t) = \begin{cases} 3 & \text{for } t \in [0,100-10 \ln 2] \\ 0 & \text{otherwise.} \end{cases}$

2.14. (a) $C^*(t) = \rho W_0 e^{(r-\rho)t}/(1-e^{-\rho T})$

(b) $\dot{C}^*(t) = K(r-\rho)$.

2.17. $\dot{\lambda} = x + 3\lambda x^2$, $\lambda(1) = 0$, and $\dot{x} = -x^3 + \lambda$, $x(0) = 1$.

2.18. $\dot{x} = f(x) + b(x)u$, $x(0) = x_0$, $x(T) = 0$

$\dot{u} = [b(x)^2 g'(x) - \{b(x)f'(x) - b'(x)f(x)\} 2c^2u]/[2c^2b(x)]$.

Chapter 3

3.1. $x - u_1 \geq 0$, $u_1 - u_2 \geq 0$, $u_1 \geq 0$, $1 + u_2 \geq 0$.

3.2. $X = [-1,5]$.

3.4. $u^*(t) = -1$; $\mu_1 = -\lambda = \frac{1}{2} - t$, $\mu_2 = \eta = 0$.

3.6. One solution appears in Figure 3.1. Another solution is $u(t) = \frac{1}{2}$ for $t \in [0,2]$. There are many others.

3.8. $L = F(x,u) + \lambda f(x,u,t) + \mu g(x,u,t)$

$$\dot{\lambda} = (\dot{\alpha}/\alpha)\lambda - \frac{\partial L}{\partial x}, \quad \mu \geq 0, \quad \mu g = 0 .$$

3.12. $\lambda(t) = t - 1$.

3.13. (a) $\lambda(t) = 10[1 - e^{.1(t-100)}]$

$$\mu = \begin{cases} 0 & \text{if } K = 300 \\ -10[1-e^{.1(K/3-100)}] & \text{if } K < 300 \end{cases}$$

$u^*(t) = \text{bang}[0,3;\ \lambda+\mu]$

The problem is infeasible for $K > 300$.

(b) $t^{**} = \text{Min}[0, 100 - \frac{K}{3}]$

$$u^*(t) = \begin{cases} 0 & \text{for } t \leq t^{**} \\ 3 & \text{for } t > t^{**} \end{cases}$$

3.19. 11.87 minutes.

3.20. $u^* = -1$, $T^* = 5$.

3.21. $u^* = -2$, $T^* = \frac{5}{2}$.

3.30. (a) $u^* = 0$

(c)
$$u^* = \begin{cases} 1 & 0 \leq t \leq 1 - T \\ 0 & 1 - T < t \leq T. \end{cases}$$

(e) $J = -(\frac{1}{8} + \frac{1}{8K})$

(f) $J = -\frac{1}{8}$.

3.31. (a) $\{\bar{I}, \bar{P}, \bar{\lambda}\} = \{I_1 - \rho(S - P_1), S, 2(S - P_1)\}$

(b) $I = I_1$.

Chapter 4

4.1. (a)
$$u^*(t) = \begin{cases} 5 & t \leq 1 + 6 \ln .99 \approx 0.94 \\ 0 & t > .094 \end{cases}$$

(b) $\lambda_2(t)/\lambda_1(t) = e^{(3t^2-4t+1)/12}$

$$u^* = \begin{cases} -5 & 0 \leq t \leq .28 \\ 0 & .28 < t \leq .4 \\ 5 & .4 < t \leq .93 \\ 0 & .93 < t \leq 1.0 \end{cases}$$

4.5. $u^* = v^* = 0$ for all t.

4.7. $u^* = 0, \quad v^* = \frac{4}{5}$ for $t \in [0,49]$

$u^* = 0, \quad v^* = 0$ for $t \in [49,60]$

$J^* = 34{,}420.$

4.10. (b) $f(t^*) = t^* - 10 \ln(1-.3e^{.1t^*})$.

(c) $t^* = 1.969327, \quad J(t^*) = 19.0374.$

Chapter 5

5.5. $Q(t) = t^4 - 160t^3 + 1740t^2 - 7360t + 9639.$

5.7. $v^* = \text{Sat}[-V_2, V_1; (\lambda_2 - \lambda_1 p)/2\beta\lambda_1].$

5.9. $J^* = 10.56653.$

5.10. $v^*(t) \approx 3e^{-3t}, \quad y^*(t) \approx 1 - e^{-3t}.$

5.12.

$$u^*(t) = \begin{cases} 0, & 0 \le t \le 7/3 \\ 2, & 7/3 < t < 3 \\ -1, & 3 \le t < 13/3 \\ 0, & 13/3 \le t \le 6 \end{cases} .$$

5.13.

$$\mu_1 = \begin{cases} -\frac{5}{2}t + \frac{5}{2} & \text{in } [0,1] \\ 0 & \text{in } (1,3] \end{cases} .$$

$$\mu_2 = \begin{cases} 0 & \text{in } [0,1.8) \\ -\frac{1}{2}t + \frac{3}{2} & \text{in } [1.8,3] \end{cases} .$$

$$\eta = \begin{cases} 0 & \text{in } [0,1) \text{ and } (1.8,3] \\ -\frac{5}{2}t + \frac{5}{2} & \text{in } [1,1.8). \end{cases}$$

5.14. (a) $v^*(t) = \begin{cases} -1 & \text{for } t \in [0,1.8) \\ 1 & \text{for } t \in (1.8,3] \end{cases}$.

(b) $v^*(t) = 1$ for all $t \in [0,10]$.

5.15. $u^* = \begin{cases} -1 & \text{for } t \in [0,1/2] \\ 0 & \text{for } t \in (1/2,t_1] \\ +1 & \text{for } t \in (t_1, t_1 + 1/2] \\ 0 & \text{for } t \in (t_1 + 1/2, 3] \end{cases}$.

5.17. $u^*(t) = \begin{cases} 0 \text{ for } 0 \le t \le t_1 \\ \dfrac{h(t-t_1)}{c} \quad \text{for } t_1 < t \le T \end{cases}$.

where $t_1 = T - \dfrac{\sqrt{2Bc}}{h}$.

Chapter 6

6.1. $p^* = 102.5 + 0.2G.$

6.7. $\dfrac{\bar{u}}{pS} = \dfrac{\delta\beta}{\eta(\rho+\delta)}$.

6.10. $\dot{G} + \delta G = \text{Bang}[0,\infty;\ \lambda-1]$

$-\dot{\lambda} + (\rho+\delta)\lambda = \pi'(G)$.

6.12. The equations corresponding to (6.28) and (6.29) can be obtained by replacing ρ by $\rho + \dot{r}/r$. The form of (6.30) remains unchanged.

6.17. (b) $t_1 = \frac{1}{}\ln\dfrac{x_0}{x^S}$, $t_2 = \dfrac{1}{rQ+\delta}\ln\dfrac{\bar{x} - x^S}{\bar{x} - x_T}$.

6.18. $T \geq \frac{1}{rQ+\delta} \ln \frac{rQ(1-x_0)-\delta x_0}{rQ(1-x^S)-\delta x^S} + \frac{1}{\delta} \ln \frac{x^S}{x_T}$

6.19. The reachable set is $[x_0 e^{-\delta T}, (x_0 - \bar{x})e^{-(\delta+rQ)T} + \bar{x}]$ where $\bar{x} = rQ/(\delta+rQ)$.

6.25. $\mathrm{Imp}(A,B;t) = -\frac{1}{r} \ln \left(\frac{1-A}{1-B}\right)$

6.26. (b) $J = 0.6325$

Chapter 7

7.1. (a) $y = 1$, $z = 3$ (b) $y = 2$, $z = 10$.

7.2. (a) (1,3) is a relative maximum
(b) (2,10) is a relative minimum.

7.3. $x = 50$; $x = 80$.

7.6. (a) $x = 4$ is a local maximum
(b) $x = 8$ is a local minimum and $x = 20$ is a local and global maximum.

7.7. (a) (0,0) is the nearest point.
(b) (1/2, 1/2) is the nearest point.

7.8. $\left(\frac{1}{\sqrt{5}}, \frac{2}{\sqrt{5}}\right)$ is the closest point.

7.9. (a) $(2\sqrt{2},0)$ (b) (0,2) (c) (0,2).

7.10. $\lambda_i^T = \frac{\partial F}{\partial x_i^T}$ for $i = 1,2,\ldots,n$; $\lambda_{n+1}^T = 1$.

7.14.

$$u^{k*} = \begin{cases} +1 & \text{if } \lambda^{k+1}b > 1 \\ -1 & \text{if } \lambda^{k+1}b < -1 \\ 0 & \text{if } |\lambda^{k+1}b| < 1 \end{cases}$$

where $\lambda^k = (I + A)^{T-k} \lambda^T$.

Chapter 8

8.1. $t^s = 5.25, \quad T = 11.$

8.3. $T = t^s = 2.47.$

8.4. $t^s = 0, \quad T = 30.$

8.5. $u^*(t) = \text{Sat}[0,1;u^0(t)]$ where

$$u^o(t) = [2 - e^{.05(t-34.8)}]^2/(1+t)$$

$t_1 \approx 3; \quad t_2 - T = 34.8.$

Chapter 9

9.3. $\bar{x} = .734$

9.4. (a) $$\bar{x} = \frac{X}{4}\left[\left(1 - \frac{\rho}{r} + \frac{c}{Xp}\right) + \sqrt{\left(1 - \frac{\rho}{r} + \frac{c}{Xp}\right)^2 + \frac{8c\rho}{prX}}\right.$$

(b) For $\rho = 0$, $\bar{x} = 220{,}000$. For $\rho = .1$, $\bar{x} = 86{,}000$.
For $\rho = \infty$, $\bar{x} = 40{,}000$.

9.5. $[g'(x) - \rho][p-c(x)] - c'(x)g(x) = 0.$

9.7. $[g'(x) - \rho][p(t)-c(x)] - c'(x)g(x) + \dot{p} = 0.$

Chapter 10

10.1. $\lambda(t) = \lambda_0 e^{(\rho-\beta)t}$; where

$$\lambda_0 = \frac{[K_0 e^{\beta T} + \frac{\bar{C}}{\beta}(1 - e^{\beta T}) - K_T]\ (2\rho-\beta)}{e^{\beta T} - e^{2(\beta-\rho)T}}$$

$$K(t) = K_0\ e\beta^t + \frac{\bar{C}}{\beta}(1 - e^{\beta t}) - \frac{\lambda_0}{\beta-2\rho}(e^{2(\beta-\rho)t} - e^{\beta t}) .$$

Chapter 11

11.1. $I(t) = a_1e^{-t} + a_2e^{t} + \frac{5\pi}{1+\pi^2}\cos \pi t$

$P(t) = -a_1e^{-t} + a_2e^{t} + \frac{5}{1+\pi^2}\sin \pi t$

$a_1 = \frac{b_1}{1+e^{-2}}, \quad a_2 = \frac{b_1e^{-2}}{1+e^{-2}} \quad .$

11.2. J = -13.09084

11.3. (a) J = -14.00367

(b) J = -14.07014

11.4. J = -5.030427

11.5. J = -688.0697

11.6. J = -954.3504

Chapter 12

12.7. $0 \leq T + \frac{1}{D}\ln(1 - \frac{Q}{P}) = \hat{t}$.

12.10. $t_1 = T/2$.

NAME INDEX

SUBJECT INDEX